AS CHEMISTRY for AQA

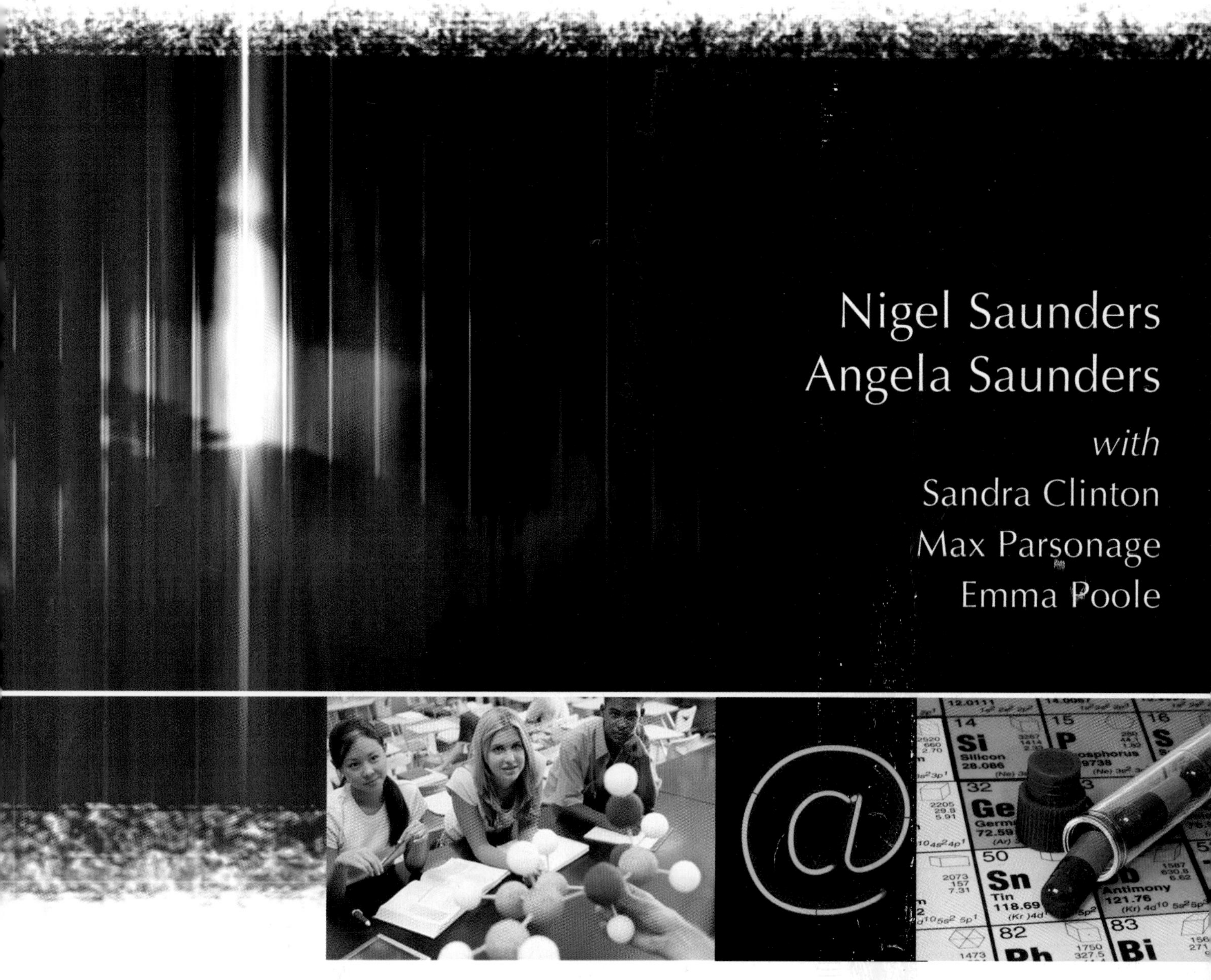

Nigel Saunders
Angela Saunders
with
Sandra Clinton
Max Parsonage
Emma Poole

OXFORD
UNIVERSITY PRESS

OXFORD
UNIVERSITY PRESS

Great Clarendon Street, Oxford OX2 6DP

Oxford University Press is a department of the University of Oxford.
It furthers the University's objective of excellence in research, scholarship, and education by publishing worldwide in

Oxford New York

Auckland Cape Town Dar es Salaam Hong Kong Karachi Kuala Lumpur Madrid Melbourne Mexico City Nairobi New Delhi Shanghai Taipei Toronto

With offices in

Argentina Austria Brazil Chile Czech Republic France Greece Guatemala Hungary Italy Japan Poland Portugal Singapore South Korea Switzerland Thailand Turkey Ukraine Vietnam

First published 2007

Reprinted 2008

British Library Cataloguing in Publication Data

Data available

ISBN-13: 9780-19-915273-5

10 9 8 7 6 5 4 3 2

Printed in Spain by Cayfosa

Paper used in the production of this book is a natural, recyclable product made from wood grown in sustainable forests. The manufacturing process conforms to the environmental regulations to the country of origin.

Acknowledgments

The Press wishes to acknowledge the contribution of Dr Mike Clugston in the original preparation of some the illustrations used herein.

We are grateful for permission to reproduce the following photographs:

p14 Mehau Kulyk/Science Photo Library; p18 (top) Popperfoto/Alamy; p27 (top) NASA/ Science Photo Library; p18 (lower) Science Photo Library; p35 Keenan and Wood: *General College Chemistry*, 5th edn, Harper and Rowe New York 1976; p40 Charles D. Winters/ Science Photo Library; p43 Martyn F. Chillmaid/Science Photo Library; p44 Andrew Lambert/LGPL; p55 (top) David Taylor/SPL; p55 (lower) Jerry Mason/SPL; p56 Andrew Lambert LGPL; p58 (top) sciencephotos/Alamy; p58 (lower) Grant Heilman/ Alamy; p59 Lawrence Migdale/Science Photo Library; p60 (top) David Hoffman/Photo Library/Alamy; p60 (lower) Crown Copyright/Health & Safety Laboratory/Science Photo Library; p61 CC Studio/Science Photo Library; p62 (top) Robert Brook/Science Photo Library; p62 (lower) GIPhotoStock Z/Alamy; p64 NASA/Science Photo Library; p65 (middle) Martin Sookias; p65 (lower) Auto Express/Quadrant Picture Library; p66 BSIP, Alexandre/Science Photo Library; p68 Craig Lovell/AGSTOCKUSA/Science Photo Library; p70 (left) Andrew Lambert Photography/Science Photo Library; p70 (right) University of Wisconsin – Madison; p71 (left) Martyn F. Chillmaid/Science Photo Library p71 (right) Martyn F. Chillmaid/Science Photo Library; p72 Yoav Levy/Phototake Inc./Alamy; p75 (top) David Levenson/Alamy, David Levenson/Alamy; p75 (lower) Charles D Winters/ Timeframe Photography; p77 Sur/Dreamstime.com; p80 Martin Sookias; p87 Roger Ressmeyer/ Corbis; p88, Charles D. Winters/Science Photo Library; p90, Claude Nuridsany & Marie Perennou/Science Photo Library; p96 Adam Hart Davis/SPL; p99 Volker Steger/ Science Photo Library; p108 Andrew Lambert, LGPL; p111 (lower) Klaus Guldbrandsen/ Science Photo Library; p112 (top) Phillipe Hays/Still Pictures; p112, (lower)Roger Harris/ Science Photo Library; p115, P. Baeza,Publiphoto Diffusion/Science Photo Library; p116 (top) Amy Nichole Harris/Shutterstock; p116 (lower) Tina Manley/Environment/Alamy; p119 Simon Fraser/ Science Photo Library; p120 Russell Knightley/Science Photo Library; p122 Science Photo Library/NASA; p137 Hertfordshire Constabulary; p141 (left) Johnson Matthey; p141 (right) Thomas Eisner, Daniel Aneshansley, Cornell University; p142 Charles D. Winters/Science Photo Library; p144 (left) sciencephotos/Alamy; p144 (right) Louis Lanzano/Associated Press; p145 (left) Yoshikazu Tsuno/AFP/Getty Images; p145 (right) Yoshikazu Tsuno/AFP/Getty Images; p146 Simon Fraser/Science Photo Library; p149 Geschattsbereich Linde Engineering; p150 Steve Yeater/Associated Press; p151 Ian R. MacDonald, Texas A&M University Corpus Christi; p155 (top) Andrew Lambert LGPL; p155 (lower) Thermitt Welding GB (Ltd.); p156 Andrew Lambert Photography/ Science Photo Library; p157 (top and lower) Andrew Lambert Photography/Science Photo Library; p158 James Scherer/Houghton Mifflin Company; p163 NASA; p164 Charles D Winters/Science Photo Library; p166 Vestergaard Frandsen; p167 Raul Touzon/ Gettyimages; p169 James Scherer/Houghton Mifflin Company; p170 (top) Alex Bartel/ Science Photo Library; p170 (lower) Andrew Lambert Photography/Science Photo Library; p172 Andrew Lambert Photography/Science Photo Library; p173 Carlos Dominguez/Science Photo Library/Photo Library; p174 sean sprague/Alamy; p175 (top) Frank Herholdt/Tony Stone Images; p174 (lower) Jenny Matthews/Alamy; p176 Andrew Lambert Photography/Science Photo Library; p178 (top and lower left) Bochsler Photographics and Imaging, Burlington, Canada; p178 (right) Andrew Lambert Photography/ Science Photo Library; p179 (top) Phil Degginger/Alamy; p179 (lower) Andrew Lambert Photography/ Science Photo Library p180 David R. Frazier/Photolibrary, Inc./Alamy; p181 (top) geogphotos/Alamy p181 (lower) Andrew Lambert/Leslie Garland Picture Library/Alamy; p182 (top) Construction Photography/ Corbis; p182 (lower) CNRI/Science Photo Library; p183 Andrew Lambert Photography/Science Photo Library; p186 (top) Scenics & Science/Alamy; p186 (lower) Pixel Youth movement/Alamy; p187 Paul Glendell/Alamy; p188 GSF; p190 Hannu/Big Stock Photo; p191 Martin Bond/Science Photo Library; p193 Jodi Jacobson/Still Pictures; p194 Dartboards.com; p195 David J. Green/Alamy; p196 (right) Howard Davies/Corbis; p196 (left) Pamela Moore/iStockphoto; p198 Stockbyte/Alamy; p200 (left) John Greim/Science Photo Library; p200 (right) John Cole/Science Photo Library; p202 James King-Holmes/Science Photo Library; p203 (top) NASA/Science Photo Library; p203 (left) geogphotos/Alamy; p203 (right) Mason Jerry/ Science Photo Library; p206 Peter Gould; p209 Andrew Lambert LGPL; p213 Andrew Lambert Photography; p214 Shangara Singh/Alamy; p219 (top) Michael Marten/Science Photo Library; p219 (lower) Gareth Wyn-Jones/Photofusion Picture Library/Alamy; p220 (top) Lil Lang/Alamy; p220 (lower) Michael Marten/Science Photo Library; p222 Anthony Blake/Photo Library; p223 (top) Mitch Kezar/Phototake Inc/Photo Library; p223 (lower) Claude Thibault/Alamy; p224 Rolf/Stockfolio/Alamy; p225 (top) sciencephotos/ Alamy; p225 (lower) Dave Reede/AGSTOCKUSA/Science Photo Library; p231 (lower) Wild Pictures/Alamy; p236 Paul Rapson/Science Photo Library; p238 NASA; p244/255 Theodore W. Gray.

Cover images: (background) Photodisc; (front left) Jupiter Images/Comstock Premium/ Alamy; (front middle) Jasmin Awad/iStockphoto; (front right) webking/iStockphoto; (back left) cb34inc/iStockphoto; (back middle) Andrew Green/iStockphoto; (back right) Emrah Turudu/iStockphoto.

In a few cases we have been unable to trace the copyright holder prior to publication. If notified the publishers will be pleased to amend the acknowledgements in any future addition.

Introduction

Chemistry is all around you. In a world without chemistry, there would be no advanced fuels, plastics, or medicines. There would be no flat screen colour displays for mobile phones. Millions of people might go hungry for lack of fertilizer for crops, and our water would not be safe to drink. Even something as simple as cooking a meal involves chemistry. But in the twenty-first century, chemists can design and control substances with astonishing precision. Chemists have a vital role to play in the modern world, and this is an exciting time to study chemistry.

Your AS Chemistry studies will give you the opportunity to develop an understanding of many important chemical ideas. You will discover how early discoveries became the shoulders on which modern chemistry stands, and you will analyze the present and look forward to the future. Chemistry is a refreshingly practical subject, and your laboratory skills will improve during the course.

This book covers all the subject content you need for the AQA AS Level Chemistry Specification. It aims to build on your GCSE studies, showing you how the knowledge and understanding of different aspects of chemistry can be woven together. Above all, *AS Chemistry for AQA* aims to help you enjoy and take part fully in this exciting and challenging subject.

Nigel and Angela Saunders,
Harrogate, North Yorkshire 2007

Contents

* practical based material

Welcome: how to use *AS Chemistry for AQA*

AS Chemistry for AQA has been written specifically for students taking the new specifications in GCE Advanced Subsidiary Level Chemistry for AQA, the Assessment and Qualifications Alliance. AQA is the largest of the three English examination boards.

The main part of the book covers all the subject content you will need to know for the assessments of Units 1 and 2. For easy access, the topics are covered in a series of double-page spreads that follow the order given in the specifications. Each spread is self-contained and could be studied in a different order than the one in the book. But if you do this, you should check any prior knowledge needed to understand the concepts covered in the spread. Use the objectives listed in the left-hand margin to help you decide this.

It is assumed that students taking AS Chemistry will have a good background in chemistry at GCSE level. It is also assumed that students have some knowledge of basic biology and physics. Where calculations are needed, the mathematics involved are explained. A chapter called *Basic Chemistry* is provided towards the back of the book. This covers key topics from GCSE chemistry and is intended to help you make the move to studying chemistry at a higher level as easy as possible.

Each main double-page spread contains a boxed list of prior knowledge and:

- a list of objectives which relate directly to the AQA Specification
- text covering the subject content, with key terms shown in **bold** print and defined in the *Glossary* at the back of the book
- *Check your understanding* answers to calculations are printed in the book; answers to all questions will bc found in the e-book and the teacher's resources.

Flexible use of the spreads means that you can easily follow their own route through the course, but it is assumed that all of Unit 1 has been studied before starting Unit 2. This is because the Specification state that a knowledge of the Chemistry in Unit 1 is assumed in Unit 2.

How Science Works is an integral part of the new GCE AS Level Chemistry specifications, and is an integral part of this book. Many of the spreads contain a *Science@Work* section that deals with one or more of the related criteria. These are discussed more fully in the spread called *How Science Works* on page 10.

AS Chemistry for AQA also includes seven double-page spreads covering investigative and practical skills. They cover the tasks indicated in the specifications, and are intended to help you prepare for the investigative skills of Unit 3. These are discussed more fully in the section called *Preparing for assessment* on page 12.

AQA AS Chemistry

AQA is the largest of the three English examination boards. It sets and marks examinations, including GCSEs and A Levels. AQA has a very useful website at **www.aqa.org.uk**. You can download examination papers and mark schemes from there to help you with your studies.

The chemistry specification

AQA's *AS and A Level Chemistry Specifications* build on its earlier chemistry course and aims to encourage you to:

- study chemistry in a modern context
- become enthusiastic about chemistry
- show that you can bring different ideas together
- develop your practical skills and data analysis skills
- appreciate how science works and its importance in the wider world

You will learn about the chemistry behind contemporary issues such as how burning fossil fuels may cause global warming, biofuels, and recycling metals and plastics. You also discover how CFCs damage the Earth's ozone layer, and how chemists were able to develop less harmful alternatives.

The AS Chemistry course is divided into three Units. Units 1 and 2 cover chemical knowledge and understanding, and Unit 3 covers practical skills. But you can also expect to be assessed on your performance in class practicals that support the chemical ideas in Units 1 and 2.

Unit 1: Foundation chemistry

Unit 1 should be studied first. It builds on your GCSE studies and covers the key chemical ideas you need to understand chemistry. These include:

- atomic structure and isotopes
- the use of a machine called the mass spectrometer
- chemical calculations
- different types of bonding, such as ionic, covalent, and metallic bonding
- patterns in the periodic table
- organic chemistry (the chemistry of carbon-based compounds such as alkanes)

The Specification has been written so that you can complete this unit during the autumn term of Year 12. This is so you can enter for the Unit 1 examination in January. But you do not have to enter you then. You can enter you for the Unit 1 examination in June instead.

Unit 2: Chemistry in action

Unit 2 should be studied after Unit 1. It extends the ideas met in that unit, and shows how they can be applied to new situations. Unit 2 includes topics such as:

- collision theory and rates of reaction
- oxidation and reduction
- the extraction of metals
- the properties of chlorine and the other halogens
- alkenes (hydrocarbons with a carbon-carbon double bond)
- haloalkanes, including CFCs and their effects on the ozone layer
- alcohols and biofuels

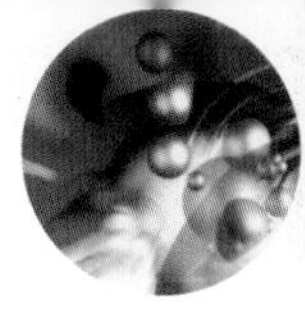

The Unit 2 examination is only available in June. If you take the Unit 1 examination in January, this is the only Chemistry examination you should have to do in the summer. Otherwise you take both examinations (Units 1 and 2) in one sitting in June.

Unit 3: Investigative and practical skills

Unit 3 assesses your practical skills and data analysis skills. It is designed so that your teacher can assess you in ordinary laboratory practicals. If you studied AQA's GCSE science courses, you will be familiar with how this works. There are two routes: *Route T* with a Practical Skills Assessment **PSA**, and an Investigative Skills Assignment **ISA**; *Route X* with a Practical Skills Verification **PSV** and an Externally Marked Practical Assignment **EMPA**. Your institution will choose which route to follow.

The PSA/PSV

Your teachers assess your practical skills during several ordinary laboratory practicals. They must do this at least twice in each of three areas of chemistry. These are inorganic chemistry, physical chemistry, and organic chemistry. If there is time, you could do lots of these practicals and your teachers could choose the best marks.

The ISA/EMPA

When your teachers are confident that you can use standard laboratory equipment skilfully, and know how to analyze and evaluate your results, they give you an ISA/EMPA to do. This is practical/written test carried out under controlled conditions. This ensures that the answers are your own work, even if you collected the results as a class.

Use of ICT

The use of ICT is encouraged in developing practical skills.

The examinations

The designers of the course know that AS Chemistry can be quite different from GCSE science for many students. They know that students often make a slow start to their AS studies as a result. So Unit 1 is deliberately a bit shorter than Unit 2, and it carries fewer marks. The table summarizes how Units 1 and 2 are assessed.

Unit	Form of assessment	Length of paper	Type of questions	% of Total AS marks
1	Written paper	1¼ hours	4 to 6 short questions, and one or two longer structured questions	33.33
2	Written paper	1¾ hours	6 to 8 short questions, and two longer structured questions	46.67

The total duration of both papers is three hours. They contribute 80% of your AS marks. The rest comes from Unit 3 practical: Route T or Route X.

The Specification set out exactly the knowledge and understanding you need to be successful in your chemistry. In this book, each double page spread contains learning objectives based on the Specification. Look carefully at what you need to be able to do. It might be something relatively straightforward, like recalling a fact or a definition. But often it is something that needs you to really think. You might need to understand something, and be able to describe or explain it. Or you may be given information to interpret. Whatever it is, work hard, and above all enjoy it – chemistry really can be fun!

How Science Works, and you

It can be easy when you are learning new ideas in chemistry to forget that these ideas were discovered and developed by people. *How Science Works* seeks to explain how scientists carry out their investigations. It examines how their beliefs can influence their thinking and approach to their research, and shows how scientists contribute to decision-making in society. You can expect examination questions to assess your understanding of *How Science Works,* so you must be prepared for this. The Specification identifies twelve key aspects of *How Science Works* and *AS Chemistry for AQA* contains numerous examples of these, particularly in the *Science@work* sections.

Specification summary

A Theories, models, and ideas

Scientists use theories and models to explain their observations. These form the basis for scientific investigations. Progress is made when there is valid evidence to support a new theory. (eg Ionization plots: spread 1.10)

B Questions, problems, scientific arguments and ideas

Scientists use their knowledge and understanding in their work. This includes when they make their observations, identify a scientific problem, and when they question scientific explanations. Progress is made when scientists contribute to new ideas and theories. (eg Ozone layer: spread 12.02)

C Using appropriate methods

Scientists develop explanations or hypotheses of their observations. They can make predictions from their hypotheses, and these can be tested using carefully planned experiments. Data can be quickly collected, recorded, and analyzed with the help of ICT. You learn about this aspect of How Science Works through your practical work in chemistry. (eg plus your Practical Skills Assessment–PSA, and your Investigative Skills Assessment–ISA)

D Carrying out experiments and investigations

Scientists use a wide range of practical skills, including carrying out experiments and examining the data through graphs and statistical analysis. They need to choose the right equipment and record their results carefully. You learn about this aspect of How Science Works through your practical work in chemistry. (plus your Practical Skills Assessment–PSA, and your Investigative Skills Assessment–ISA)

E Analyzing data

Scientists analyze observations and results. They look for patterns and correlations. They make informed decisions about what to do with anomalous results that fall outside the expected range. Data that matches scientific predictions increases the confidence that scientists have about their models. (eg Enthalpy of combustion/mean bond energies: spread 5.05)

F Evaluating methods and data

Scientists question the validity of new evidence and the conclusions drawn from it. Different research teams may come up with different results, even when using similar methods. In trying to resolve these differences, scientists may improve their methods or develop new hypotheses that can be tested. (eg Hydrolysis of haloalkanes/bond enthalpy: spread 12.05)

G The nature of scientific knowledge

Scientific knowledge and understanding rarely stands still. Scientific explanations are based on experimental evidence, supported by the scientific community. But new evidence may be found that needs a better explanation than the existing one, and scientific knowledge changes as a result. (eg Global warming: spread 4.09)

H Communicating information and ideas

Scientists share their findings, allowing other scientists to evaluate their work. It is important that precise scientific language is used to avoid confusion. You will need to develop your ability to use the appropriate words and phrases in your explanations and answers. (eg IUPAC rules: spread 4.01/14.04)

I Benefits and risks of science

Developments in science, medicine, and technology have improved the quality of life for most people. But there may also be risks involved with the methods that scientists develop and the way in which their research might be used. Scientists consider these benefits and risks in their work. (eg Chlorine in water treatment: spread 9.06)

J Ethical and environmental issues

Scientific research is funded through public funding or by private companies, and scientists must consider ethical and environmental issues in their research. Individual scientists have their own moral or religious beliefs, and they contribute to making decisions about what research should be allowed. (eg Recycling scrap mental: spread 11.06)

K The role of the scientific community

Scientific research is published in scientific journals. Before this happens, it is examined by other scientists in the same field to check its validity. This is called *peer review*. The scientific community is then able to study new research. This is important because it may be possible for the research and its conclusions to be influenced by the organization that funded the research. (eg Acid rain: spread 4.08)

L Society uses science to inform its decisions

Politicians and other decision-makers may use scientific findings to inform their decisions. Scientists may take part in making decisions, especially if the scientific evidence is incomplete. The final decisions are often influenced by the existing beliefs of the decision makers, by special interest groups, public opinion, and the media. (eg CO and NO/catalytic converters: spread 4.07)

Chemistry resources on the Internet

There are many Internet resources that might support your chemistry studies. The US has a larger population than the UK, so a large proportion of sites are biased towards American courses. These sites can still be useful to you if use them with care.

Many UK sites support the AQA course. Other sites' support might be helpful, but other courses are *not* the same. This might mean that sites may cover topics you do not need. They might also miss out topics that you do need. Sometimes the chemistry vocabulary used is slightly different, too. Finally, there are some very good general chemistry websites that take care to explain things clearly to you.

Here is a list of 13 useful chemistry websites. It is not exhaustive and you may find others that you prefer. Remember that websites do move or disappear. You can use search engines to find other sites. Try key words and phrases such as 'chemistry revision', 'chemistry worksheets', and 'chemistry help'. Modify your search by adding 'UK', 'as level', or 'a level' to make it more specific to the UK. Remember that you can also search for particular topics.

Supporting AS/A2 Level Chemistry		
Site	**Content**	**Web address**
Creative Chemistry	worksheets, practical guides, revision notes, and interactive molecular models	www.creative-chemistry.org.uk/alevel
Knockhardy Publishing Science Notes	notes and PowerPoint® presentations	www.knockhardy.org.uk/sci.htm
Chemguide	generally helpful pages	www.chemguide.co.uk
Chemistry in Perspective	web-based chemistry text book	www.chembook.co.uk
Rod's Pages	topic information, laboratory tips, and examination tips	www.rod.beavon.clara.net
S-Cool	topic guides, questions, and revision summaries	www.s-cool.co.uk
General Chemistry Online	general chemistry resources	http://antoine.frostburg.edu/chem/senese/101
Greener Industry	for sustainable industry	www.greener-industry.org
RSC Video Clips	video clips of chemistry experiments and reactions	www.chemsoc.org/networks/learnnet/videoclips.htm
WebElements™	chemistry information relating to the periodic table	www.webelements.com

Online revision quizzes and questions

These three sites have interactive or downloadable quizzes and questions to check your knowledge and understanding:

Site	Content	Web address
LearnNet	quizzes/questions	www.chemsoc.org/networks/learnnet/questions.htm
Revision Resources for A Level Chemistry	revision	www.mp-docker.demon.co.uk
Timberlake's Chemistry	revision	www.karentimberlake.com/quizzes.htm

Preparing for assessment

Units 1 and 2

Units 1 and 2 are assessed by two examination papers, marked by external examiners rather than your teachers. The table summarizes the duration of these papers, and their contribution to your final marks.

Unit	Duration (hours)	Total marks on paper	UMS marks	Percent of AS marks
1 *(Foundation chemistry)*	1¼	70	100	33.33
2 *(Chemistry in action)*	1¾	100	140	46.67

Units 1 and 2 contribute 80% of your total AS marks (and 40% of your total A Level marks).

The double-page spreads in *AS Chemistry for AQA* cover all the content listed in the Specifications. One of your main tasks during your AS Level course will be to learn and understand all the subject content. Unlike GCSE, there is no Foundation Tier or Higher Tier. You must be prepared to apply your knowledge and understanding to answer questions on any topic from the Specifications.

Unit 3

Unit 3 tests your investigative and practical skills. It is marked by your teachers, rather by external examiners. Their marking is checked by an external Moderator to make sure the marks have been given fairly and accurately. The table summarizes how Unit 3 is assessed.

Focus of assessment	Method of assessment
PSA/PSV	Your teacher grades your practical ability in certain tasks: two from each of three areas of chemistry: inorganic, physical, and organic chemistry.
ISA/EMPA	You carry out practical work, collect and process data. You then complete a written test under controlled conditions.

AS Chemistry for AQA contains spreads with guidelines for carrying out practical work. These are based on the tasks and suggested contexts in the AS Chemistry Specification. The double-page spreads involved are:

Unit	Spread	Task(s)
1	2.06	Experiments to find M_r
1	2.17	Carrying out a titration
2	5.06	Measuring an enthalpy change
2	6.03	Investigating rates of reaction
2	10.05	Testing for anions
2	14.03	Investigating the combustion of alcohols
2	14.07	Distillation and organic tests

Use the *Objectives*

Use the objectives for each spread to help you organize your learning. They relate directly to the subject content in the AQA specifications. They tell you what you should be able to do when you have finished studying each topic. Before an assessment in January or June, check that you can recall and show an understanding of the scientific knowledge in the Unit or Units you will be sitting. You will also be expected to be able to apply your knowledge and understanding in unfamiliar contexts, including the ideas behind How Science Works.

Use the *Check your understanding* questions

These questions are designed to help you check that you understand and can recall key ideas covered in each spread. Immediately after you have finished a spread, read the questions and write down your answers. Check through the text to make sure that your answers are correct. If you cannot find the answer, or if you do not understand the question, make a note of it and discuss it with your lecturer or teacher. You may also check the answers in the support material available separately from Oxford University Press.

Use the *Glossary*

One of the challenges of studying Chemistry is getting to grips with lots of specialist words and phrases. It can be a bit daunting when first confronted with a lot of new words. It is a good idea to begin to build up your chemical vocabulary right from the start of your studies. You will find it much easier to explain chemical ideas if you have a good grasp of chemical vocabulary. For each topic, you might like to put together your own list of key words and phrases, and use the glossary or a chemistry dictionary to define them. Your examiners will be looking to see that you can use specialist vocabulary when appropriate.

Make concept maps

Concept maps (sometimes called mind maps or patterned notes) are a useful way of organizing your ideas for revision. It is often best to keep them simple:

1. **Choose** a theme, and then identify related concepts as key words or phrases. Write simple definitions for each one. The Glossary will help here. Then start looking for connections between the concepts. Some connections will be obvious, but you might need to look at the relevant spread again to see all the connections.
2. **Link concepts** with lines or arrows. Where appropriate, add simple joining phrases between them to make the connections clear. Some phrases you could use are: *such as*, *necessary for*, *leads to*, *produces*, *and*, *or*, *during*.
3. **Draw** a box around each concept so you can tell it from the connecting phrases.

You might need to use a concept more than once, but concept maps are usually clearer and more useful if concepts are used only once. You might find it helpful to use different colours to group similar concepts together.

Use past examination papers

The AQA website provides links to specimen question papers, past papers and mark schemes. These are free and are great way to check your revision progress. Download and print out the papers. Have a go at answering the questions, then use the mark scheme to check your answers. It might take several hours the first time you try a question paper. But stick at it: you will get faster and more accurate with practice. You can also download the Examiners' Reports for past examinations. These tell you how many marks you need for each AS grade.

Discuss your work

Talk to other students about the chemistry you are studying, if you can. Form a small group that meets to discuss chemistry topics. There are thousands of other students studying your course. These are the people you are competing against, not your friends and class mates. Encourage and help each other to develop a better understanding of chemistry. Get together, order a pizza, and make your revision fun!

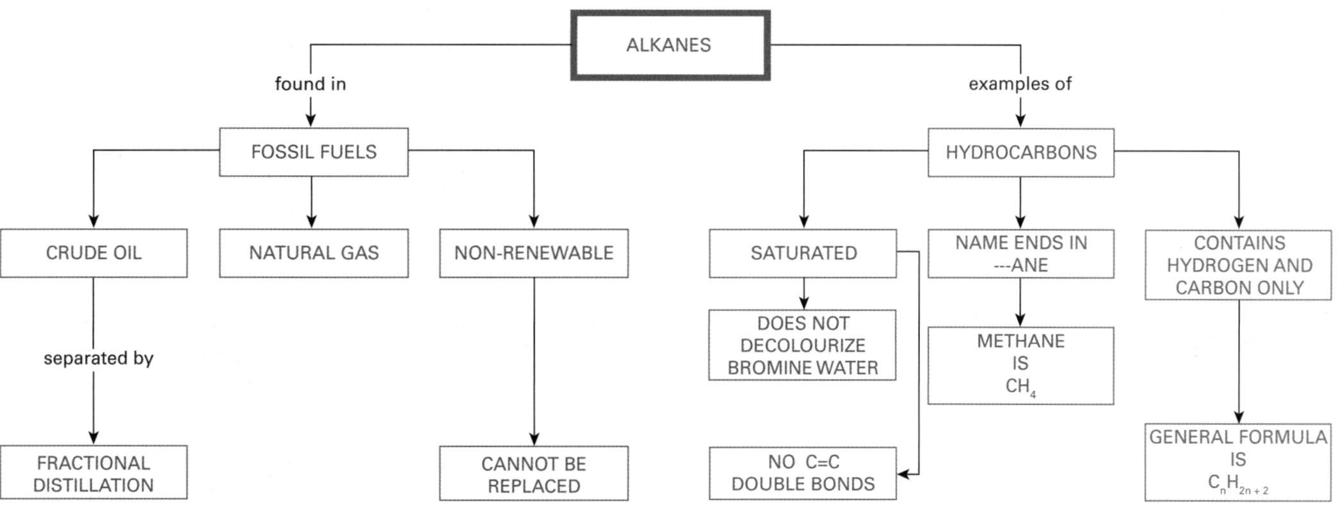

An example of a concept map for alkanes.

Unit 1: *Foundation Chemistry* explores the fundamental principles that are the basis of chemistry. Familiar ideas are taken from your GCSE studies and extended. You find out more about atomic structure and bonding, and discover how to work out the shapes of simple molecules and ions. There is further work on chemical calculations, including calculations involving gases. The properties of elements depend upon their structure and bonding, and you find out about how these properties change across a period of the periodic table. Starting with the simple alkanes used as fuels and studied at GCSE, you learn how to name organic compounds.

AQA Approved Specification (July 2007)

Foundation Chemistry

A computer-generated model of sub-atomic particles.

1.01 Atoms

OBJECTIVES

already from GCSE, you know

- atoms have a small central nucleus, made up of protons and neutrons and surrounded by electrons
- different sub-atomic particles have different properties
- atoms are given mass numbers and atomic numbers

and after this spread you should be able to

- understand the importance of these particles in the structure of the atom
- recall the meaning of mass number (*A*) and atomic (proton) number (*Z*)
- explain the existence of isotopes

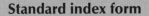

Standard index form

Standard index form is also called standard form or scientific notation. It is a useful way to show very small or very large numbers. The numbers are written like this: $A \times 10^n$.

A is a number between 1 and 10 (but not 10 itself), and *n* is a whole number that shows you how many places to move the decimal point. So you write 246.8 as 2.468×10^2 and you write 0.02468 as 2.468×10^{-2}.

Atoms are incredibly small. If people were the size of carbon atoms, the world's population would fit into a box less than a thousandth of a millimetre across. You will discover that this makes it difficult to study the structure of atoms. But first you need to check that you understand the idea of the nuclear atom.

Sub-atomic particles

Atoms are made from three **sub-atomic particles**. These are called **protons**, **neutrons**, and **electrons**. The table shows the mass and electric charge of each sub-atomic particle.

particle	mass (kg)	charge (C)
proton	1.672×10^{-27}	$+1.602 \times 10^{-19}$
neutron	1.674×10^{-27}	0
electron	9.109×10^{-31}	-1.602×10^{-19}

Notice how very small these numbers are.

It is difficult to understand the properties of sub-atomic particles if you look at their actual masses and charges. Instead, it is often more useful to look at their masses and charges compared to the proton. These numbers are called the **relative mass** and the **relative charge**. The table shows the relative mass and relative charge of each sub-atomic particle.

particle	relative mass	relative charge
proton	1	+1
neutron	1	0
electron	1/1836	−1

In GCSE science the relative mass of the electron is sometimes given as zero. Its actual value is 1/1836 or 5.4×10^{-4}.

You can see that the proton and neutron have the same relative mass. They have much more mass than the electron. The proton is positively charged. The electron has the same size of charge but this is negative instead. The neutron is neutral (it has no charge).

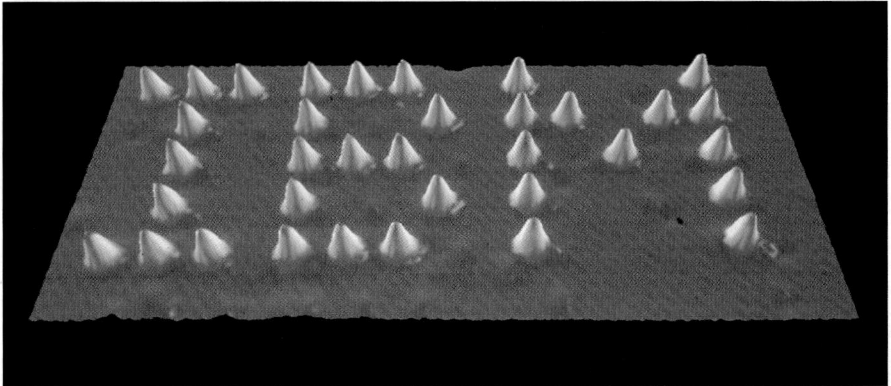

The scanning tunnelling electron microscope analyses tiny electric currents flowing between a surface and a very fine probe to build up images like this. Here, the tip of the microscope has been used to place xenon atoms on a nickel surface.

The nuclear atom

The word 'nuclear' means anything to do with the **nucleus**. The nuclear atom has a nucleus at its centre, with electrons arranged in energy levels around it. The nucleus is made of protons and neutrons. Most of the mass of the atom is found in the nucleus. This is because the protons and neutrons in the nucleus have much more mass than the electrons.

Any atom has equal numbers of protons and electrons. This means that atoms are neutral overall. For example, fluorine atoms each have nine protons and nine electrons. The nine positive charges carried by the protons in the nucleus are balanced by the nine negative charges carried by the electrons. Because neutrons are neutral, there can be any number of neutrons in the nucleus without making any difference to the overall charge of the atom.

Atomic number and mass number

The **atomic number** of an atom is the number of protons in its nucleus. It is also called the proton number. The atomic number is shown by the symbol ***Z***. Every **element** has its own unique atomic number. All the atoms of a particular element have the same atomic number.

The **mass number** of an atom is the number of sub-atomic particles in its nucleus. The mass number is shown by the symbol ***A***. It is the number of neutrons added to the number of protons.

The full symbol for an atom is written like this: $^{A}_{Z}\mathbf{X}$. The atomic number is shown as a subscript and the mass number as a superscript.

You can work out the number of neutrons in an atom by subtracting the atomic number from the mass number. So for $^{19}_{9}F$, the number of neutrons is $19 - 9 = 10$. The number of protons (and the number of electrons) in the atom is 9.

Isotopes

Isotopes are atoms that have the same number of protons but different numbers of neutrons in their nuclei. For example, $^{35}_{17}Cl$ and $^{37}_{17}Cl$ are both isotopes of chlorine. Their atoms each contain 17 protons. But every $^{35}_{17}Cl$ atom has $35 - 17 = 18$ neutrons, and every $^{37}_{17}Cl$ atom has $37 - 17 = 20$ neutrons. The different isotopes of an element are chemically identical to each other because they contain the same number of protons and electrons.

Check your understanding

1. Work out the numbers of protons, neutrons, and electrons in these atoms:
   a $^{4}_{2}He$
   b $^{27}_{13}Al$
   c $^{56}_{26}Fe$
2. A certain atom X has one less proton and two more neutrons than $^{39}_{19}K$. What are its atomic number and mass number?
3. Explain, in terms of sub-atomic particles, why $^{12}_{6}C$ and $^{14}_{6}C$ are isotopes of carbon.

1.02 Discovering the atom

OBJECTIVES

already from GCSE, you know

- all substances are made of atoms
- the structure of the atom

and after this spread you should

- be able to recall some historical ideas about atoms
- understand how the electron was discovered
- understand more how science (works) develops over time

The philosophers of ancient Greece thought a lot about the world around them. They had no way to carry out experiments to find out what everything is made from, so they came up with several different ideas. There were two main ones. Matter was either continuous or made up of very tiny bits. If matter were continuous, you could imagine dividing it up into smaller and smaller pieces without end. But if it were made up of very tiny bits, there would be a limit to how far you could divide it up. The Greek word *atomos* means indivisible or 'cannot be divided'. This is where the modern word **atom** comes from.

Arguing through the centuries

Leucippus (of Miletus, c 490 BCE) and his pupil Democritus (of Abdera, c 470–380 BCE) developed a detailed theory of atoms in the fifth century BCE.

- All matter is made of atoms that are too small to see and cannot be cut into anything smaller.
- There is just empty space between atoms.
- Atoms are completely solid with no empty space inside them.
- Atoms are the same all the way through.
- Atoms have different sizes, shapes, and masses.

Aristotle's ideas held up the progress of science for centuries.

But a very influential Greek philosopher called Aristotle (of Stagira, c 384–322 BCE) disagreed with these ideas. He believed that everything was made from four 'elements': earth, air, fire, and water. Different substances were made through different combinations of these 'elements' and changes to them. Aristotle's ideas were highly thought of, and the idea of atoms fell into the background.

The rise of the atom

One of the first cracks in Aristotle's ideas was uncovered in 1600s. John Mayow discovered that the volume of air in a jar decreased a little if he burned a candle in it. This meant that air must contain at least two different gases, so it could not be a true chemical element.

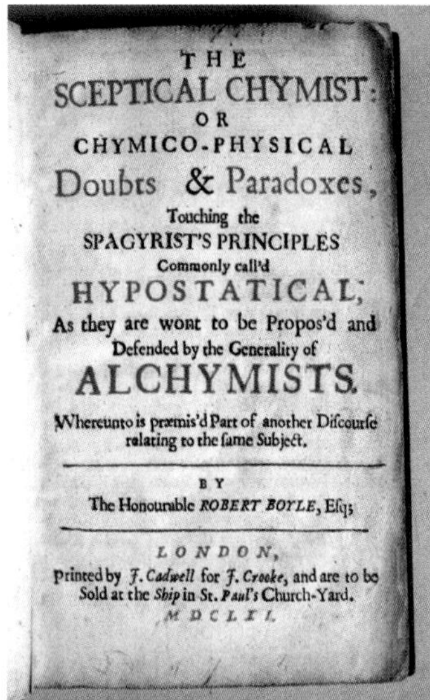

THE
SCEPTICAL CHYMIST:
OR
CHYMICO-PHYSICAL
Doubts & Paradoxes,
Touching the
SPAGYRIST'S PRINCIPLES
Commonly call'd
HYPOSTATICAL;
As they are wont to be Propos'd and
Defended by the Generality of
ALCHYMISTS.
Whereunto is præmis'd Part of another Diſcourſe
relating to the ſame Subject.

BY
The Honourable ROBERT BOYLE, Esq;

LONDON,
Printed by J. Cadwell for J. Crooke, and are to be
Sold at the Ship in St. Paul's Church-Yard.
MDCLXI.

You can read Robert Boyle's The Sceptical Chymist *on the Internet.*

Robert Boyle (1627–1691) published his book *The Sceptical Chymist* in 1661. He wrote about tiny particles that were identical and could not be divided. They made up all matter and joined together in different ways to make compounds. Boyle encouraged scientists to carry out experiments.

An English scientist called John Dalton (1766–1844) carried out experiments on gases during the nineteenth century. He improved on Boyle's ideas and called the tiny particles atoms. By 1808 Dalton had developed some important ideas through his experiments.

- Atoms in an element are all the same.
- A compound is made from atoms of two or more elements joined together in fixed proportions.
- Atoms do not change in a chemical reaction.
- Atoms in the **reactants** re-arrange during chemical reactions to form the **products**.

Dalton's work gradually became accepted during the nineteenth century. It was generally believed that atoms were indivisible. But that was set to change.

Indivisible?

Invisible rays are produced when electricity is passed through gases under low pressure. The rays come from the negative electrode, called the cathode. In 1895, Jean Perrin (1870–1942) discovered that these *cathode rays* are negatively charged. Two years later, Joseph John (J.J.) Thomson (1856–1940) deflected cathode rays in precise ways using electric and magnetic fields. He discovered that the rays had a tiny mass, about 2000 times less the mass of a hydrogen atom. This showed that cathode rays must be made of particles from the atoms of the cathode. Thomson called the particles 'corpuscles', but you know them as electrons. From then on, atoms were no longer believed to be indivisible.

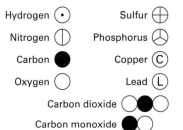

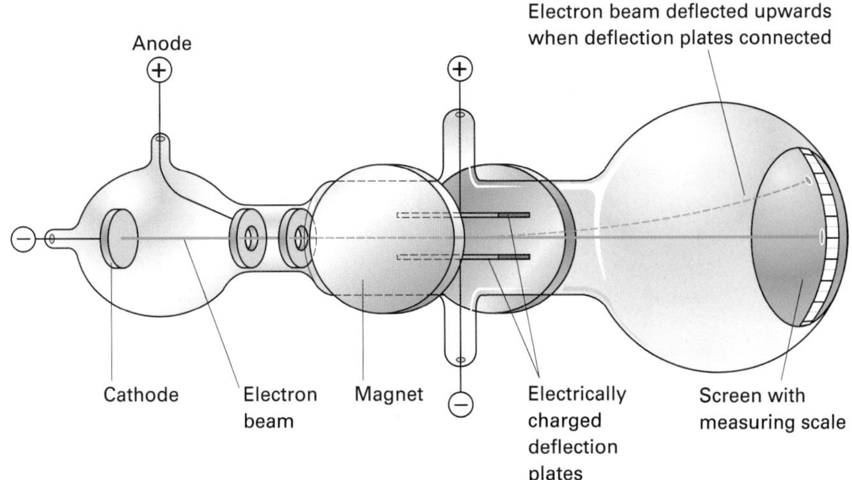

Thomson's discharge tube apparatus enabled him to measure the mass to charge ratio of electrons. The fluorescent screen glows when electrons hit it.

Hydrogen ⊙ Sulfur ⊕
Nitrogen Phosphorus
Carbon ● Copper Ⓒ
Oxygen ○ Lead Ⓛ
Carbon dioxide ○●○
Carbon monoxide ●○

John Dalton invented a way to represent elements and compounds using symbols. The modern system was begun by Jöns Jacob Berzelius in 1811.

The plum pudding model

Thomson knew that electrons carried a negative charge. He also knew that atoms are neutral, so he understood that there must be positive charges in atoms to provide the balance. In 1906, Thomson suggested a structure for atoms that became known as the **plum pudding model**. In Thomson's model, an atom is a ball of positive charge with electrons moving around inside. The electrons are the 'plums' in the 'pudding'. Before long, scientists were carrying out experiments to see if this model was correct.

Models in science

Scientists use ideas called *models* to explain their observations. Models can be tested by doing experiments. Scientific progress happens when a model is either supported by experimental results or disproved by them.

Check your understanding

1. In what ways were the ideas of Leucippus and Democritus similar to modern ideas about the atom? In what ways were they different?
2. In what ways were Dalton's ideas similar to modern ideas about the atom? In what ways were they different?
3. Explain why J. J. Thomson was able to conclude that atoms must contain even smaller particles but the Greek philosophers could not.

1.03 Investigating atoms

OBJECTIVES

already from GCSE, you know

- all substances are made of atoms
- the structure of the atom

and after this spread you should be able to

- understand how modern models of the atom were developed
- recall how the proton and neutron were discovered
- recall how atomic number and proton number were linked

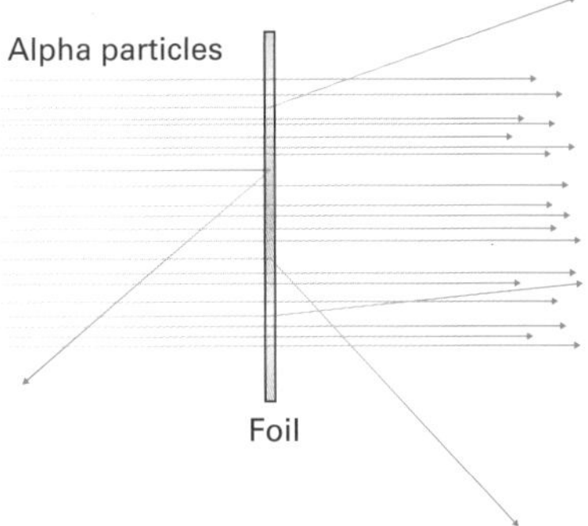

This summarizes Geiger and Marsden's results.

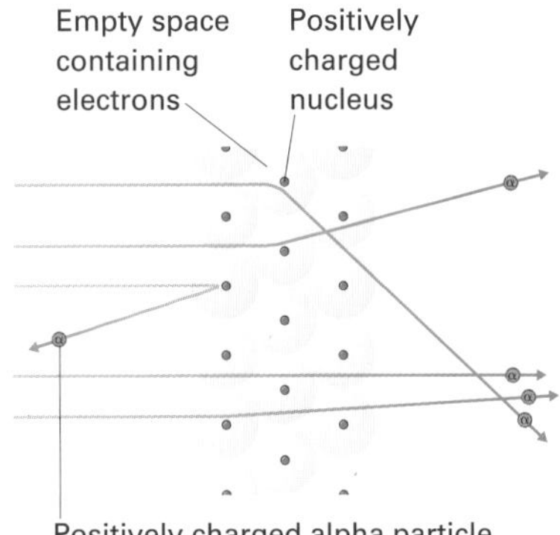

This is Rutherford's interpretation of Geiger and Marsden's results. The nucleus is not to scale. It would be much smaller compared to the rest of the atom.

The Geiger–Marsden experiment

Thomson's plum pudding model of the atom was put to the test in 1909 by two young scientists, Hans Geiger (1882–1945) and Ernest Marsden. They carried out an experiment suggested by Ernest Rutherford (1871–1937), who was the professor of physics at the university where they worked. In their experiment, a narrow beam of **alpha particles** was produced by **radioactive** radium. The particles were aimed at very thin gold foil. Alpha particles are now known to contain two protons and two neutrons. They are positively charged and will be repelled by any positive charges in the gold atoms. So the scientists expected them to be slightly deflected off course as they went through the foil.

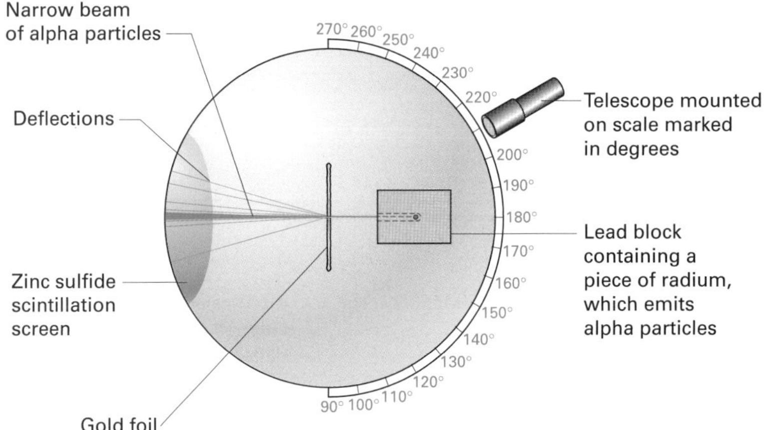

The Geiger–Marsden experiment. Alpha particles make tiny flashes of light when they hit the zinc sulfide screen. The small telescope let the scientists look for these flashes. Each angle of deflection was worked out by measuring the position of a flash on the screen.

Geiger and Marsden made two observations that could not be explained by the plum pudding model.

- Most of the alpha particles were deflected very little, suggesting that atoms are mostly empty space.
- Some alpha particles were deflected a lot, some by more than 90°.

Rutherford said later that the second observation was almost as incredible 'as if you fired a fifteen-inch [artillery] shell at a piece of tissue paper and it came back and hit you'.

Rutherford's model

Rutherford analysed the results from the gold foil experiment. He published a scientific paper in 1911 setting out a new model for the structure of the atom. In this model, most of the mass of the atom was concentrated into a tiny nucleus at the centre of the atom. The electrons went around the nucleus, rather like planets around the sun. This explained the results. The electrons did not affect the alpha particles, and most alpha particles passed through the empty space in the gold atoms. Some of them went close enough to the nucleus for it to deflect them.

Protons and neutrons

Rutherford continued experimenting. He bombarded nitrogen atoms with alpha particles. Some of the atoms gave off a positively charged particle that was identical to a hydrogen nucleus. Rutherford published his results in 1919. Shortly afterwards he called the particle the proton. This comes from the Greek word *protos*, which means 'the first'.

Rutherford realized that protons do not have enough mass to be the only particle in the nucleus. He suggested that there should be another particle in the nucleus, with the same mass but no charge. This particle was discovered in 1932 by James Chadwick (1891–1974). Once again, alpha particles were used to bombard atoms. In Chadwick's experiment, beryllium atoms gave out particles with the properties predicted by Rutherford 11 years earlier. These particles were neutrons.

Atomic number

Henry Moseley (1887–1915) linked atomic number to the number of protons in the nucleus. Atoms give off X-rays when they are bombarded by high-speed electrons. Moseley discovered in 1913 that he got a straight line graph if he plotted the atomic number against the square root of the X-ray frequencies. Before Moseley's experiment, atomic numbers were worked out by looking at the element's position on the periodic table. Moseley showed that the atomic number is the same as the number of protons in the nucleus.

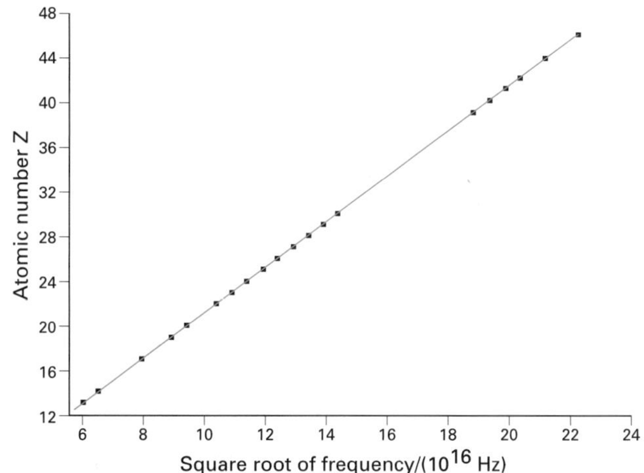

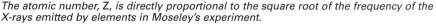

The atomic number, Z, is directly proportional to the square root of the frequency of the X-rays emitted by elements in Moseley's experiment.

Improving Rutherford's model

Niels Bohr (1885–1962) improved Rutherford's model in 1913. He suggested that electrons can only have certain amounts of energy and only move around the nucleus at certain distances. Bohr's model really only worked for hydrogen, the simplest atom. In 1925 Erwin Schrödinger (1887–1961) developed a way to describe electrons using ideas from a branch of physics called quantum mechanics. This worked for other atoms, not just hydrogen. It led to the modern atomic orbital model of the atom that we use today.

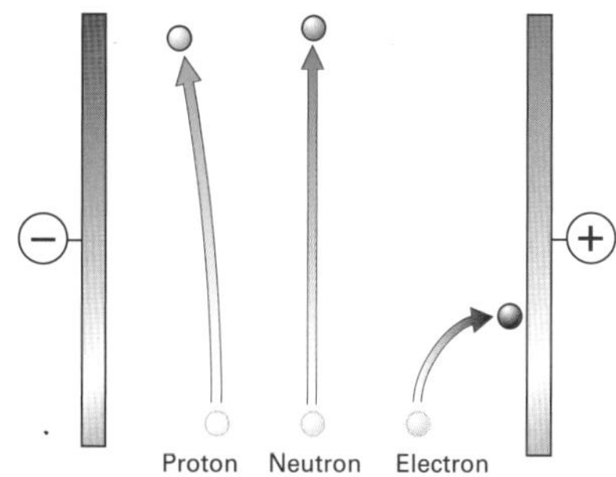

Charged particles are deflected in an electric field. Neutrons are neutral so they are not deflected. Protons and electrons are deflected in opposite directions because they have opposite charges. Protons are deflected less than electrons because they have more mass.

Check your understanding

1. Explain why Geiger and Marsden's results were unexpected.
2. Describe how alpha particles have helped scientists to explore the structure of the atom.
3. Explain the importance of using models of the atom to make predictions that can be tested by experiments.
4. Suggest why atomic number is also called proton number.

1.04 The mass spectrometer

OBJECTIVES

already from GCSE, you know

- ions are atoms or groups of atoms that have lost or gained electrons
- positively charged ions are attracted to negative charges

and after this spread you should be able to

- describe how the mass spectrometer works

The **mass spectrometer** is a machine that can analyse samples of elements. It will give you accurate information about which isotopes are in a sample, and their relative amounts. It is very sensitive and will detect even very tiny amounts of individual isotopes. This is so useful that space probes carry mass spectrometers to analyse samples of the soil and atmospheric gases on other planets.

The sample is injected into one end of the mass spectrometer. The sample must be in the gaseous state, so liquid or solid samples are heated to vaporize them (turn them into a gas). The sample then passes through four main stages. These are called **ionization**, **acceleration**, **deflection**, and **detection**. You need to be able to explain what happens at each stage.

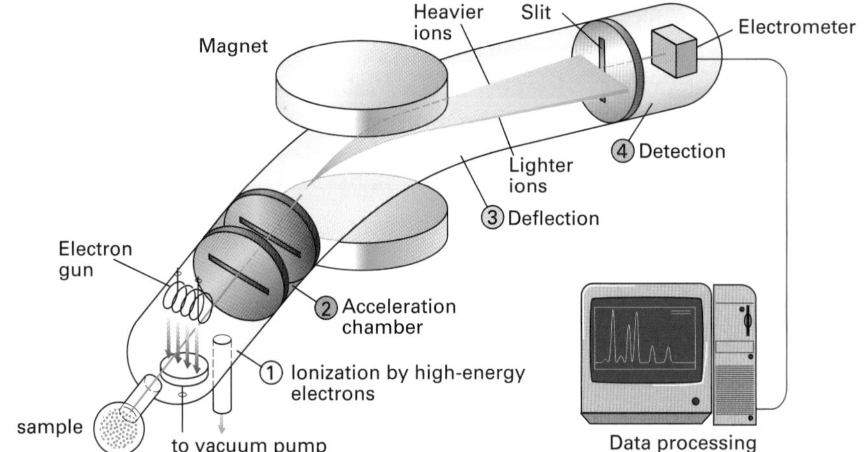

The main parts of a mass spectrometer.

Ionization

The sample has to be **ionized**. This is so that particles in the sample can be accelerated and deflected in the middle two stages. It also means that the particles can be detected in the last stage.

High-energy electrons stream off a hot metal wire in an **electron gun**. These electrons hit atoms in the sample and knock out electrons from them. The conditions are adjusted so that only one electron is removed from each atom. This means that they become ions with single positive charges.

This is a general equation to show the process:

$$X(g) + e^- \rightarrow X^+(g) + 2e^-$$

(e^- is the symbol for an electron)

This is the equation for ionizing a sample of magnesium:

$$Mg(g) + e^- \rightarrow Mg^+(g) + 2e^-$$

The symbol (g) shows that the atoms and ions are in the gas **state**.

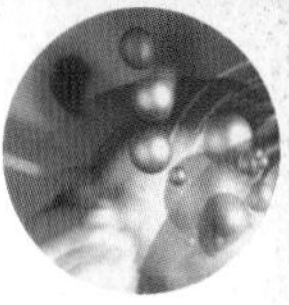

Acceleration

The ions have to be accelerated so that they move through the mass spectrometer. This is done using an electric field. The ions are positively charged, so they are attracted to negatively charged plates and pass through narrow slits in them.

This stage does two jobs.

- It makes the ions in the sample move very quickly.
- It focuses the ions into a narrow beam.

The mass spectrometer is connected to a vacuum pump. The pump removes air from inside it. This stops molecules in the air getting in the way of the speeding ions.

Deflection

The ions will keep moving in a straight line unless a force is applied to them. A magnetic field produces a force that pushes (deflects) the ions sideways. This force depends only on the size of the charge on the ion. So if all the ions have one charge each, and are travelling at the same speed, the amount of deflection depends on the mass of the ions. Lighter ions will be deflected more than heavier ions. Only the ions that have the right combination of mass and charge will be deflected round the bend in the mass spectrometer, and on to the final stage.

Detection

The particles in the sample are ionized so that they can be accelerated and deflected. But it also means that they can produce a tiny current in the detector.

The detector contains a negatively charged plate. A tiny electric current is produced each time an ion hits it. The more ions, the bigger the current. The detector is connected through an amplifier to a computer. The current is recorded as a peak in a **mass spectrum**. Different ions in the sample are brought to the detector by altering the strength of the deflecting magnetic field.

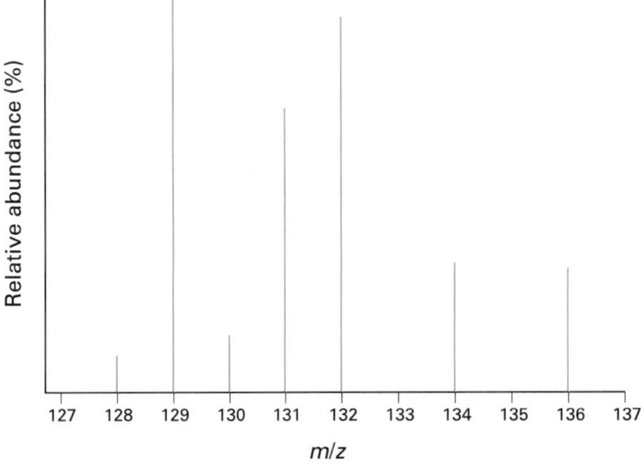

The mass spectrum of a sample of xenon gas.

Check your understanding

1. Give three reasons why the sample must be ionized in the mass spectrometer.
2. Describe what happens in each of the four main stages in the mass spectrometer.
3. How are different ions in a sample brought to the detector?

1.05 Detecting isotopes

OBJECTIVES

already from AS Level, you

- can explain the existence of isotopes
- know how the mass spectrometer works

and after this spread you should be able to

- use information from mass spectra to work out a relative isotopic mass

The standard atom

You will remember that the actual masses of sub-atomic particles are very small indeed. So they are compared to the mass of a proton to give their relative masses. You can do a similar thing for atoms. Instead of using the actual mass of an atom, you can use its mass compared to a standard atom.

The standard atom chosen is $^{12}_{6}C$, often just written as ^{12}C. Its relative mass is given as 12 exactly. All the other atoms are compared to this one. So an atom of ^{4}He is a third of the mass of an atom of ^{12}C, and an atom of ^{24}Mg is twice the mass of an atom of ^{12}C. For individual isotopes the relative mass is called the relative isotopic mass.

Relative isotopic mass

The relative isotopic mass of an atom is its mass compared to one-twelfth the mass of a ^{12}C atom. You can also write this definition as an equation:

$$\text{relative isotopic mass} = \frac{\text{mass of one atom of the isotope}}{\text{mass of one atom of } ^{12}\text{C}} \times 12$$

The *m/z* ratio

The amount that an ion is deflected in the mass spectrometer depends on its **mass to charge ratio**. You write this as ***m/z***. It is the relative mass of the ion, *m*, divided by its charge, *z*. An ion with a small *m/z* ratio will be deflected more than an ion with a large *m/z* ratio. The table shows you three examples.

ion	relative mass, *m*	charge, *z*	*m/z* ratio
$^{16}O^+$	16	1	16 ÷ 1 = 16
$^{32}S^+$	32	1	32 ÷ 1 = 32
$^{32}S^{2+}$	32	2	32 ÷ 2 = 16

You can work out the m/z *ratio of an ion if you know its relative mass and its charge.*

The $^{16}O^+$ ion has a smaller *m/z* ratio than the $^{32}S^+$ ion. It will be deflected more. Notice that the $^{16}O^+$ and $^{32}S^{2+}$ ions have the same *m/z* ratio. They will be deflected by the same amount, so the mass spectrometer will be unable to tell them apart. This shows why it is important that the ions all have single positive charges.

Relative masses of atoms and ions

Remember that the mass of an electron is 1836 times less than the mass of a proton or neutron. This means that you can ignore the mass of any electrons lost or gained by an atom to form an ion. You can assume that the relative mass of an ion is the same as the relative mass of the atom that formed it.

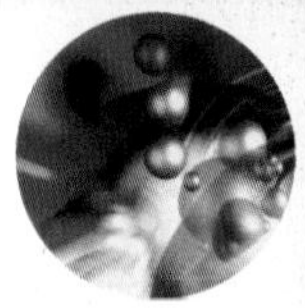

Using a mass spectrum

A mass spectrum shows the *m*/*z* ratio of each ion on the horizontal axis. The *m*/*z* ratio is the same as the relative isotopic mass if the charge *z* on the ion is +1. This is useful because it means that the mass spectrometer can show you which isotopes are in a particular sample.

A mass spectrum shows the **relative abundance** of each ion on the vertical axis. This is the proportion of each ion in the sample. It is usually written as a percentage. So if a certain ion made up only half of the sample, its relative abundance would be 50%. This is useful because it means that the mass spectrometer can show which isotopes in a sample are most common.

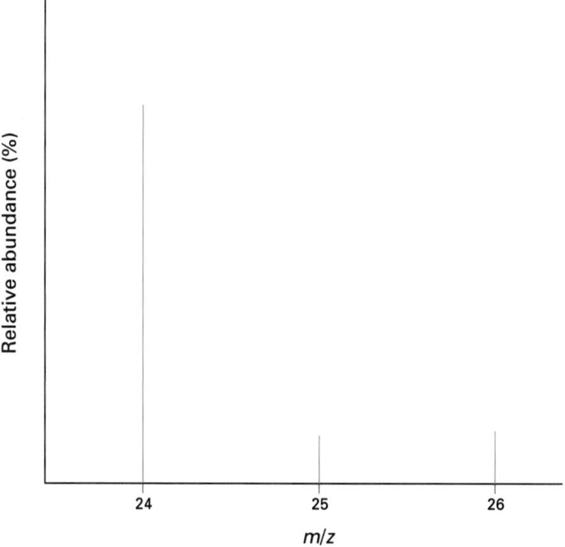

If you use the information from both axes on a mass spectrum, you can work out the **relative atomic mass** of the element in your sample. You will find out about this in the next section.

Check your understanding

1. Calculate the *m*/*z* ratios of the following ions:
   - **a** $^{7}Li^{+}$
   - **b** $^{14}N^{+}$
   - **c** $^{28}Si^{+}$
   - **d** $^{28}Si^{2+}$
2. Two of the ions in question 1 cannot be separated from each other by the mass spectrometer. Which ions are they, and why can they not be separated?
3. Which ion, $^{16}O^{+}$ or $^{18}O^{+}$, will be deflected more in the mass spectrometer, assuming they are travelling at the same speed and at the same time? Explain your answer.
4. What is the relative isotopic mass of an ion with a single positive charge and an *m*/*z* ratio of 24?

1.06 More from the mass spectrometer

OBJECTIVES

already from AS Level, you know

- how to work out the relative isotopic mass using data from a mass spectrum data

and after this spread you should be able to

- use mass spectrum data to work out relative atomic mass and relative molecular mass
- recall that mass spectrometers in planetary space probes can be used to identify elements

Relative molecular mass as an equation

You can also write the definition for relative molecular mass as an equation:

$$\text{relative molecular mass} = \frac{\text{mean mass of a molecule}}{\text{mass of one atom of } {}^{12}\text{C}} \times 12$$

Relative atomic mass

A natural sample of an element usually contains more than one of its isotopes. You take into account the different isotopes in an element by using its relative atomic mass. This is given the symbol A_r. An element's **relative atomic mass** is the mean mass of an atom of the element compared to one-twelfth the mass of a ^{12}C atom.

Relative atomic mass as an equation

You can also write the definition for relative atomic mass as an equation.

$$\text{relative atomic mass} = \frac{\text{mean mass of an atom of the element}}{\text{mass of one atom of } {}^{12}\text{C}} \times 12$$

Take care when you write down the symbol A_r. Do not confuse it with Ar, which is the chemical symbol for argon.

Working out relative atomic mass

You can work out the relative atomic mass of an element using data from a mass spectrum. You need the m/z ratio of each ion and its relative abundance. This is how to work out A_r.

$$A_r = \frac{\text{sum of } (m/z \times \text{relative abundance})}{\text{sum of relative abundances}}$$

m/z	24	25	26
relative abundance (%)	79.0	10.0	11.0

Results from the mass spectrum of a sample of magnesium.

This is how you work out the A_r of magnesium using the results in the table.

$$A_r \text{ of Mg} = \frac{(24 \times 79.0) + (25 \times 10.0) + (26 \times 11.0)}{(79.0 + 10.0 + 11.0)}$$

$$A_r \text{ of Mg} = \frac{(1896) + (250) + (286)}{(100.0)} = \frac{2432}{100.0} = 24.32$$

You are usually expected to write A_r values to one decimal place. So in this example you would write 24.3 for your final answer. Notice that the A_r is close to 24. This is because ^{24}Mg is much more abundant than the other two isotopes in the sample.

Relative molecular mass

Relative molecular mass deals with compounds, and elements consisting of more than one atom. It is given the symbol M_r. Relative molecular mass is the mean mass of a molecule compared to one-twelfth the mass of a ^{12}C atom.

The mass spectrometer can also provide information about molecules. The high-energy electrons in the ionization stage can remove electrons from molecules. When this happens, the ion formed is called the molecular ion. If there is a single positive charge on the **molecular ion**, its m/z ratio is the same as its relative molecular mass.

The peak caused by the molecular ion is the peak on the far right of the mass spectrum. It is the one with the highest *m*/*z* ratio. So the M_r of ethanol is 46. The tallest peak is called the **base peak**.

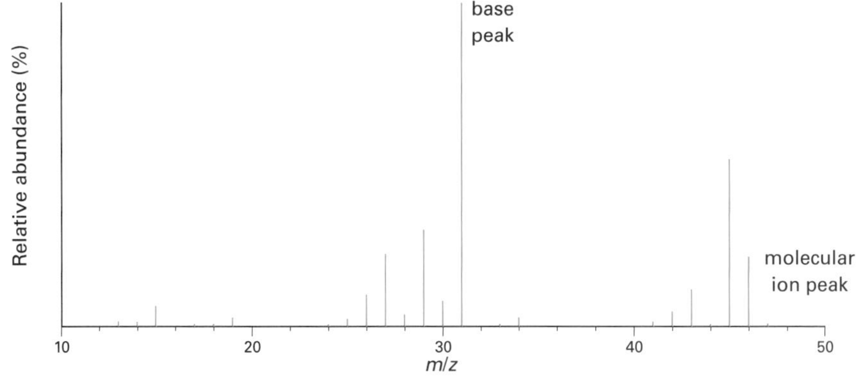

This is the mass spectrum for ethanol, C_2H_5OH, the alcohol found in alcoholic drinks.

Mass spectrometers in space

Mass spectrometers are not just used here on Earth. Miniature versions are often fitted into space probes. They are used to analyse soil and atmospheric gases on other planets.

Cassini-Huygens was a spacecraft launched in 1997. It consisted of an orbiter called Cassini and a probe called Huygens. Its mission was to investigate Saturn and Titan, one of the giant planet's moons. The spacecraft took almost seven years to reach Saturn. It carried a mass spectrometer to investigate the composition of the atoms and ions in Titan's upper atmosphere. A miniature mass spectrometer was also carried by the Huygens probe that actually landed on Titan. It was able to identify gases such as neon and carbon monoxide. It also analysed the relative isotopic abundances of elements such as carbon, nitrogen, hydrogen, oxygen, and argon.

Check your understanding

1. Calculate the relative atomic mass of lead, Pb, using these mass spectrum data. Give your answer to one decimal place.

m/*z*	204	206	207	208
relative abundance (%)	1.4	24.1	22.1	52.4

2. What is the M_r of the compound represented by the mass spectrum shown here?

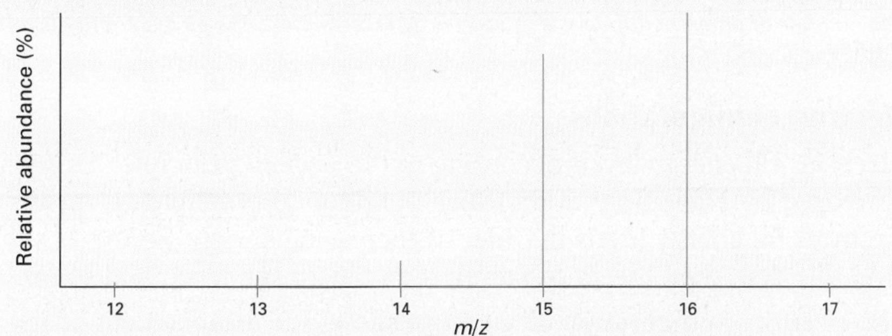

3. What information can automated mass spectrometers on space probes give us that would be difficult to obtain by making observations from Earth?

Isotopes and the molecular ion peak

Sometimes you see two peaks that could be the molecular ion peak, a tall peak with a short peak one unit to the right of it. The short peak is caused by heavy isotopes in the molecule. For example, natural carbon consists of 98.93% ^{12}C and 1.07% ^{13}C. Molecules in which an atom of ^{12}C is replaced by an atom of ^{13}C will produce a second peak. This will be short and one unit to the right of the main molecular ion peak. In such cases, use the tall peak to work out the M_r of the substance.

The Cassini-Huygens spacecraft being given its final checks. It reached Saturn on 1st July 2004.

This is the mass spectrometer carried by the Cassini orbiter.

1.07 Arranging electrons

OBJECTIVES

already from GCSE, you know

- electrons occupy particular energy levels
- the electrons in an atom occupy the lowest available energy levels
- how to represent the electron arrangements of the first twenty elements

and after this spread you should be able to

- represent the electron configurations of atoms up to $Z = 18$ in terms of levels and sub-levels

Energy levels

In the modern atomic orbital model of the atom, electrons can only occupy certain **energy levels**. In general, the energy of an energy level increases the further away from the nucleus it is. For this course you need to know about the first four energy levels. These are numbered from 1 to 4, with 1 having the lowest energy.

Sub-levels

Each energy level contains one or more **sub-levels**. These are given letters instead of numbers.You need to know about s, p, and d sub-levels. The different sub-levels can hold different maximum numbers of electrons:

- **s sub-levels** can hold 2 electrons
- **p sub-levels** can hold 6 electrons
- **d sub-levels** can hold 10 electrons

The energy of the sub-levels in an energy level increases as you go from s to p, and then to d.

How many electrons?

The table shows the maximum number of electrons that can occupy each energy level. The higher the energy level, the more electrons it can hold.

energy level	type of sub-level	maximum number of electrons in sub-level	maximum number of electrons in level
1	s	2	2
2	s	2	8
	p	6	
3	s	2	18
	p	6	
	d	10	
4	s	2	32
	p	6	
	d	10	
	f	14	

The number of different sub-levels in an energy level is the same as the energy level's number.

Electron configurations

Electrons fill energy levels according to the ***Aufbau* principle**. This was developed by Niels Bohr and means 'building-up principle'. It means that electrons fill energy levels in order of increasing energy. So the 1s sub-level fills first, then the 2s sub-level, then the 2p sub-level, and so on.

The electron configuration of an atom shows the arrangement of its electrons in the different sub-levels. You show the number of electrons in each sub-level using a superscript number. This is written at the top right of the letter.

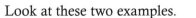

Look at these two examples.

- The helium atom, $Z = 2$, has two electrons.
  You write its electron configuration as $1s^2$.
  This shows that there are two electrons in the 1s sub-level.
- The aluminium atom, $Z = 13$, has 13 electrons.
  You write its electron configuration as $1s^2\ 2s^2\ 2p^6\ 3s^2\ 3p^1$.
  This shows that there are two electrons in the 1s sub-level, two in the 2s sub-level, six in the 2p sub-level, two in the 3s sub-level, and one electron in the 3p sub-level.

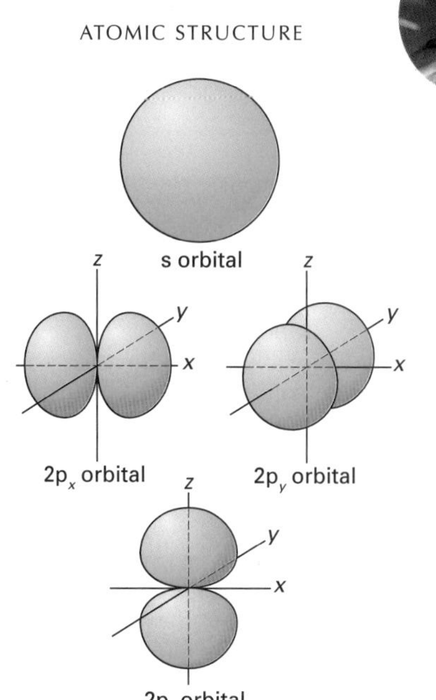

Different orbitals have different shapes.

Orbitals

Electrons are moving all the time. You cannot know exactly where an electron will be at any moment. But you can work out where electrons are most likely to be found. These regions are called **orbitals**.

Pauli exclusion principle: Each orbital can hold a maximum of two electrons. Electrons can exist in two states, called spin up and spin down. When two electrons occupy one orbital, they must have opposite spins.

The different types of sub-level contain different numbers of orbitals:

- s sub-levels have one orbital
- p sub-levels have three orbitals
- d sub-levels have five orbitals

You show the electrons in an orbital using arrows. One arrow points up, and the other (if needed) points down.

Hund's rule: Electrons occupy orbitals in a particular sub-level following Hund's rule. They occupy the orbitals as single unpaired electrons, all with the same spin. They only begin to pair up when there are no more empty orbitals in that sub-level. An empty orbital is called a **vacant orbital**.

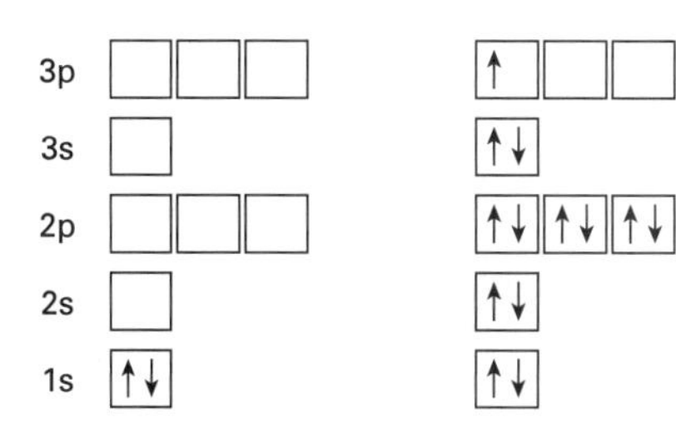

These are spin diagrams for helium and aluminium.

Element	Electron configuration	Spin diagrams 1s	2s	2p	3s	3p
H	$1s^1$	[↑]	[ ]	[ ][ ][ ]	[ ]	[ ][ ][ ]
He	$1s^2$	[↑↓]	[ ]	[ ][ ][ ]	[ ]	[ ][ ][ ]
Li	$1s^2\ 2s^1$	[↑↓]	[↑]	[ ][ ][ ]	[ ]	[ ][ ][ ]
Be	$1s^2\ 2s^2$	[↑↓]	[↑↓]	[ ][ ][ ]	[ ]	[ ][ ][ ]
B	$1s^2\ 2s^2\ 2p^1$	[↑↓]	[↑↓]	[↑][ ][ ]	[ ]	[ ][ ][ ]
C	$1s^2\ 2s^2\ 2p^2$	[↑↓]	[↑↓]	[↑][↑][ ]	[ ]	[ ][ ][ ]
N	$1s^2\ 2s^2\ 2p^3$	[↑↓]	[↑↓]	[↑][↑][↑]	[ ]	[ ][ ][ ]
O	$1s^2\ 2s^2\ 2p^4$	[↑↓]	[↑↓]	[↑↓][↑][↑]	[ ]	[ ][ ][ ]
F	$1s^2\ 2s^2\ 2p^5$	[↑↓]	[↑↓]	[↑↓][↑↓][↑]	[ ]	[ ][ ][ ]
Ne	$1s^2\ 2s^2\ 2p^6$	[↑↓]	[↑↓]	[↑↓][↑↓][↑↓]	[ ]	[ ][ ][ ]
Na	$1s^2\ 2s^2\ 2p^6\ 3s^1$	[↑↓]	[↑↓]	[↑↓][↑↓][↑↓]	[↑]	[ ][ ][ ]
Mg	$1s^2\ 2s^2\ 2p^6\ 3s^2$	[↑↓]	[↑↓]	[↑↓][↑↓][↑↓]	[↑↓]	[ ][ ][ ]
Al	$1s^2\ 2s^2\ 2p^6\ 3s^2\ 3p^1$	[↑↓]	[↑↓]	[↑↓][↑↓][↑↓]	[↑↓]	[↑][ ][ ]
Si	$1s^2\ 2s^2\ 2p^6\ 3s^2\ 3p^2$	[↑↓]	[↑↓]	[↑↓][↑↓][↑↓]	[↑↓]	[↑][↑][ ]
P	$1s^2\ 2s^2\ 2p^6\ 3s^2\ 3p^3$	[↑↓]	[↑↓]	[↑↓][↑↓][↑↓]	[↑↓]	[↑][↑][↑]
S	$1s^2\ 2s^2\ 2p^6\ 3s^2\ 3p^4$	[↑↓]	[↑↓]	[↑↓][↑↓][↑↓]	[↑↓]	[↑↓][↑][↑]
Cl	$1s^2\ 2s^2\ 2p^6\ 3s^2\ 3p^5$	[↑↓]	[↑↓]	[↑↓][↑↓][↑↓]	[↑↓]	[↑↓][↑↓][↑]
Ar	$1s^2\ 2s^2\ 2p^6\ 3s^2\ 3p^6$	[↑↓]	[↑↓]	[↑↓][↑↓][↑↓]	[↑↓]	[↑↓][↑↓][↑↓]

increasing energy →

Electron configurations and spin diagrams for the first eighteen elements.

Check your understanding

1. Write down the electronic configurations for the elements with these atomic numbers:
   a $Z = 4$ **b** $Z = 8$
   c $Z = 12$ **d** $Z = 17$
2. Draw the spin diagrams for each of the atoms in question 1.
3. Explain these ideas: the Aufbau principle, the Pauli exclusion principle, Hund's rule.

1.08 Tackling blocks

OBJECTIVES

already from GCSE, you know

- that the elements can be arranged into the periodic table
- which elements are in Group 0

already from AS Level, you know

- how to represent the electron configurations of atoms up to $Z = 18$ in terms of levels and sub-levels
- about orbitals, the Pauli exclusion principle, and Hund's rule

and after this spread you should be able to

- represent the electron configurations of atoms from $Z = 19$ to $Z = 36$ in terms of levels and sub-levels
- classify an element as belonging to the s, p, or d block in the periodic table

A shortened electron configuration

You can use a shortened version of the electron configurations when there are many electrons involved. You find the first Group 0 element with a smaller atomic number, and write its chemical symbol inside square brackets. You then write the remaining electrons next to this. So argon is the Group 0 element to use for potassium and calcium, as its atomic number is 18. Its full electronic configuration is

$1s^2\ 2s^2\ 2p^6\ 3s^2\ 3p^6$

(written here as [Ar]), so the shortenedelectron configurations for potassium and calcium are

- potassium: [Ar] $4s^1$
- calcium: [Ar] $4s^2$

Overlapping energy levels

The electronic structure of argon, $Z = 18$, is $1s^2\ 2s^2\ 2p^6\ 3s^2\ 3p^6$. You would probably expect the electronic structure of potassium, $Z = 19$, to be $1s^2\ 2s^2\ 2p^6\ 3s^2\ 3p^6\ 3d^1$. It is not.

Energy level 3 is lower in energy overall than energy level 4. But there is some overlap between the 3d and 4s sub-levels. The 4s sub-level is lower in energy than the 3d sub-level. This means that for elements after argon, the 4s sub-level fills first:

- potassium, $Z = 19$, has the electron configuration $1s^2\ 2s^2\ 2p^6\ 3s^2\ 3p^6\ 4s^1$
- calcium, $Z = 20$, has the electron configuration $1s^2\ 2s^2\ 2p^6\ 3s^2\ 3p^6\ 4s^2$

Once the 4s sub-level is filled, electrons can occupy the 3d sub-level.

Scandium to zinc

From scandium to vanadium, the 3d sub-level is gradually filled.

Z	element	symbol	shortened electron configuration
21	scandium	Sc	[Ar] $3d^1\ 4s^2$
22	titanium	Ti	[Ar] $3d^2\ 4s^2$
23	vanadium	V	[Ar] $3d^3\ 4s^2$

Shortened electron configurations for scandium, titanium, and vanadium.

You would probably expect the shortened electron configuration for the next element to be [Ar] $3d^4\ 4s^2$. Instead, it is [Ar] $3d^5\ 4s^1$ for chromium, Cr. This is because one of the 4s electrons is promoted to the 3d sub-level. This way both the 3d and 4s sub-levels are half full, which is a more stable arrangement.

The 4s sub-level is filled again for the next element, manganese. From iron to nickel, the 3d sub-level is gradually filled.

Z	element	Symbol	shortened electron configuration
25	manganese	Mn	[Ar] $3d^5\ 4s^2$
26	iron	Fe	[Ar] $3d^6\ 4s^2$
27	cobalt	Co	[Ar] $3d^7\ 4s^2$
28	nickel	Ni	[Ar] $3d^8\ 4s^2$

Shortened electron configurations for manganese to nickel.

You would probably expect the shortened electron configuration for the next element to be [Ar] $3d^9\ 4s^2$. Instead, it is [Ar] $3d^{10}\ 4s^1$ for copper, Cu. This is because one of the 4s electrons is promoted to the 3d sub-level, just as with chromium. This way the 3d sub-level is full, which is a more stable arrangement.

The 4s sub-level is filled again for the next element, zinc. From gallium to krypton, the 4p sub-level is gradually filled.

Z	element	symbol	shortened electron configuration
30	zinc	Zn	[Ar] $3d^{10}$ $4s^2$
31	gallium	Ga	[Ar] $3d^{10}$ $4s^2$ $4p^1$
32	germanium	Ge	[Ar] $3d^{10}$ $4s^2$ $4p^2$
33	arsenic	As	[Ar] $3d^{10}$ $4s^2$ $4p^3$
34	selenium	Se	[Ar] $3d^{10}$ $4s^2$ $4p^4$
35	bromine	Br	[Ar] $3d^{10}$ $4s^2$ $4p^5$
36	krypton	Kr	[Ar] $3d^{10}$ $4s^2$ $4p^6$

Shortened electron configuration for zinc to krypton.

Blocks in the periodic table

The periodic table contains all the elements in order of increasing atomic number. You will remember that the columns are called groups and the rows are called periods. The electron configuration of the elements give you another way to view the periodic table. It can be divided up into blocks, according to which sub-levels are being filled:

- The **s block** is on the left, where the s sub-levels are filling. It contains hydrogen, helium, and the elements of groups 1 and 2.
- The **p block** is on the right, where the p sub-levels are filling. It contains the elements of groups 3 to 7, and the remainder of group 0.
- The **d block** is between the s block and the p block, where the d sub-levels are filling.

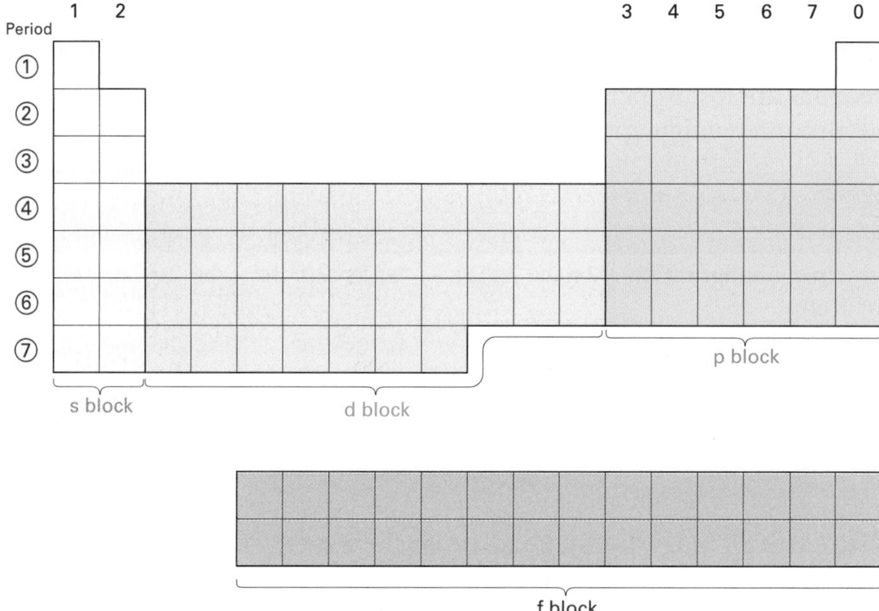

An outline of the periodic table showing the s, p, and d blocks.

Group numbers

There are several ways to number the groups in the periodic table. You are most likely to have used the numbering seen here, where the groups are numbered from 1 to 7, with the noble gases being group 0. The noble gases are also sometimes numbered as group 8. Two numbering systems exist in which roman numbers and capital letters are used. For example, group 7 is VIIA in one system, and VIIB in the other.

To avoid confusion, IUPAC (see page 103) now recommend a system in which the groups are numbered from 1 to 18. In the periodic table seen here groups 1 and 2 have the same numbers, but groups 3 to 0 become groups 13 to 18. So group 7 becomes group 17. The short groups in the d block are numbered from 3 to 12.

Check your understanding

1. **a** Write down the full electron configurations for chromium and copper.
   b Explain what is unusual about the electron configurations of these two elements.
2. Write down the full electron configurations for the elements with these atomic numbers:
   a $Z = 25$ **b** $Z = 35$
3. Which blocks in the periodic table do these elements belong to?
   a lithium
   b nitrogen
   c gold

1.09 Electron configurations of ions

OBJECTIVES

already from GCSE, you know

- atoms that lose electrons become positively charged ions
- atoms that gain electrons become negatively charged ions

already from AS Level, you know

- how to represent the electron configurations of atoms from $Z = 1$ to $Z = 36$ in terms of levels and sub-levels

and after this spread you should be able to

- represent the electron configurations of ions from $Z = 1$ to $Z = 36$ in terms of levels and sub-levels

Ions

Atoms may lose or gain electrons during chemical reactions. When this happens, they become charged particles called **ions**. In general, hydrogen and metals tend to lose electrons to form positively charged ions. Non-metals tend to gain electrons to form negatively charged ions.

Hydrogen ions

Hydrogen has the simplest atoms. Each hydrogen atom has one electron in the 1s sub-level. Remember that each electron carries a single negative charge. When the electron in a hydrogen atom is lost, the remaining particle carries a single positive charge. The particle is a hydrogen ion. It has the symbol H^+. The superscript + sign shows you that the ion carries a single positive charge.

Science @ Work

Metal or non-metal?

Hydrogen atoms are unusual. They can behave both as a metal atom and as a non-metal atom. The hydrogen atom can lose an electron, just as atoms of metals do. It forms a positively charged hydrogen ion, H^+. But the hydrogen atom can also gain an electron, just as atoms of non-metals do. It forms a negatively charged hydride ion, H^-. You will discover later that hydrogen atoms can share electrons, forming covalent bonds. This is also a feature of non-metals.

Metal ions in the s and p blocks

Metal atoms can lose electrons to form positive ions. In all cases electrons are lost from the highest occupied energy level. The table shows the electron configurations of the atoms Na, Mg, Al, and their ions.

group	1	2	3
element	sodium	magnesium	aluminium
electron configuration of atom	$1s^2\ 2s^2\ 2p^6\ 3s^1$	$1s^2\ 2s^2\ 2p^6\ 3s^2$	$1s^2\ 2s^2\ 2p^6\ 3s^2\ 3p^1$
electron configuration of ion	$1s^2\ 2s^2\ 2p^6$	$1s^2\ 2s^2\ 2p^6$	$1s^2\ 2s^2\ 2p^6$
symbol of ion	Na^+	Mg^{2+}	Al^{3+}

A useful rule of thumb for metals in groups 1, 2, and 3 is that the charge on the ion is the same as the group number.

Notice that all three atoms lose electrons from energy level 3. This is because that is their highest energy level occupied by electrons. You should also notice that the three ions have the same electron configuration. This is the same electron configuration as the group 0 element neon. You would say that these ions are **isoelectronic** with neon. But remember that they contain different nuclei, so they have not become neon atoms.

Metal ions in the d block

The 3d sub-level is higher in energy than the 4s sub-level. So you might expect metal atoms from the d block to form ions by losing electrons from the 3d sub-level. But remember that energy level 3 is lower in

energy overall than energy level 4. This means that when d block atoms form ions, electrons are lost first from the 4s sub-level, and then from the 3d sub-level. The table shows four common examples.

element	chromium	iron	iron	copper
electron configuration of atom	[Ar] $3d^5\ 4s^1$	[Ar] $3d^6\ 4s^2$	[Ar] $3d^6\ 4s^2$	[Ar] $3d^{10}\ 4s^1$
electron configuration of ion	[Ar] $3d^3$	[Ar] $3d^6$	[Ar] $3d^5$	[Ar] $3d^9$
symbol of ion	Cr^{3+}	Fe^{2+}	Fe^{3+}	Cu^{2+}

Shortened electron configurations for four common ions from the d block.

Elements from the d block often form more than one ion. For example, iron can form two different ions, Fe^{2+} and Fe^{3+}. They are not isoelectronic with each other. Remember to always take away electrons from the 4s sub-level first when working out electron configurations of d block elements.

Non-metal ions

Non-metal atoms can gain electrons to form negative ions. In all cases it is the highest occupied energy level that gains electrons. The table shows the electron configurations of the atoms P, S, Cl, and their ions.

group	5	6	7
element	phosphorus	sulfur	chlorine
electron configuration of atom	$1s^2\ 2s^2\ 2p^6\ 3s^2\ 3p^3$	$1s^2\ 2s^2\ 2p^6\ 3s^2\ 3p^4$	$1s^2\ 2s^2\ 2p^6\ 3s^2\ 3p^5$
electron configuration of ion	$1s^2\ 2s^2\ 2p^6\ 3s^2\ 3p^6$	$1s^2\ 2s^2\ 2p^6\ 3s^2\ 3p^6$	$1s^2\ 2s^2\ 2p^6\ 3s^2\ 3p^6$
symbol of ion	P^{3-}	S^{2-}	Cl^-

A useful rule of thumb for non-metals in groups 5, 6, and 7 is that the charge on the ion is equal to the group number minus 8.

Notice that all three atoms gain electrons in energy level 3. This is because that is their highest energy level occupied by electrons. You should also notice that the three ions are isoelectronic with the group 0 element argon. They contain different nuclei, so they have not become argon atoms.

Check your understanding

1. What are ions and how do they form?
2. Write down the full electron configurations for these ions:
   a Li^+
   b Ca^{2+}
   c O^{2-}
   d Cl^-
3. a Which ions from question 2 are isoelectronic?
   b Which element from Group 0 are they isoelectronic with?
4. Write down the full electron configurations for these ions:
   a Ni^{2+}
   b Ti^{2+}
   c Ti^{4+}

1.10 Evidence for energy levels

OBJECTIVES

already from AS Level, you know

- how to represent the electron configurations of atoms and ions from $Z = 1$ to $Z = 36$ in terms of levels and sub-levels

and after this spread you should

- know the meaning of the term ionization energy
- understand how ionization energies give evidence for energy levels
- know that early models of atomic structure predicted that atoms and ions with noble gas electron configurations should be stable

The mole

The **mole** is the unit for **amount of substance**. Its symbol is **mol**. In chemistry, the word *amount* implies the number of atoms, ions, or molecules there are. Since these particles are very small and have little mass, there are very many of them in the objects around you. One mole is the number of atoms in exactly 12 g of ^{12}C. This is 6.022×10^{23} – a very large number indeed!

Ionization energy

Ionization is the process of removing electrons from atoms to form ions. This needs energy. The first ionization energy is the energy needed to remove one electron from a gaseous atom. A general equation for first ionization is:

$$M(g) \rightarrow M^+(g) + e^-$$

where M stands for the chemical symbol of the element.

This is the equation that represents the first ionization of aluminium:

$$Al(g) \rightarrow Al^+(g) + e^-$$

There are two things you should notice about this equation:

- You write the symbol (g) for a gas, even though aluminium is a solid at room temperature.
- You write the ion as Al^+ even though aluminium forms Al^{3+} ions in reactions.

Electrons are negatively charged, so they are attracted to the positively charged nucleus. Overcoming that force of attraction needs energy. The energy needed to ionize one atom is very small, so the energy needed to ionize a very large number of atoms is given instead. Ionization energies have units of kilojoule per mole, written as kJ mol^{-1}. For example, the first ionization energy of aluminium is +577.5 kJ mol^{-1}. The + sign shows you that energy is supplied. It is an **endothermic** process (one that needs energy).

Evidence from groups

The first ionization energy decreases as you go down a group. For example in group 2, the first ionization energy decreases from 900 kJ mol^{-1} for beryllium at the top of the group, to 503 kJ mol^{-1} for barium near the bottom of the group.

In each case, an electron from the highest occupied s sub-level is being removed. As you go down group 2, each successive element has more occupied energy levels. The distance between the outer electrons and the nucleus increases. Electrons in higher energy levels are less strongly attracted to the nucleus than electrons closer to it. So, even though the nuclear charge increases down the group, less energy is needed to remove an outer electron.

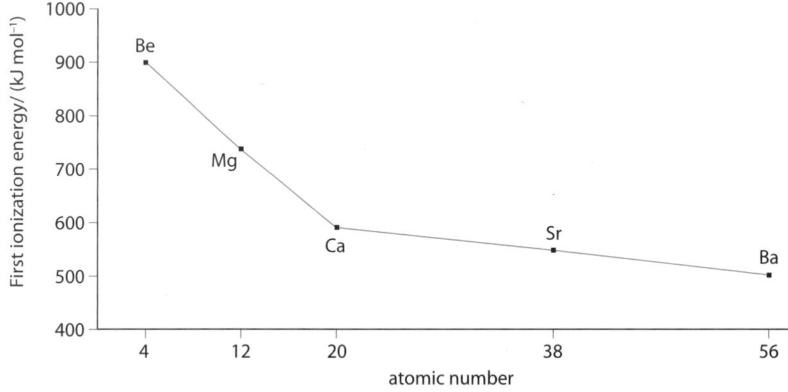

The first ionization energy decreases as you go down group 2.

Evidence from periods

The first ionization energy increases as you go across each period from left to right, then abruptly decreases at the start of the next period. The group 0 elements at the ends have the highest first ionization energies. This is because their highest occupied s and p energy sub-levels are full. They have a stable electron configuration, so a relatively large amount of energy is needed to remove one of their electrons.

The elements of group 1 at the start of each new period have much lower first ionization energies. A single unpaired electron from the highest occupied s orbital is removed when these atoms are ionized. Relatively little energy is needed to remove this electron. It is in a higher energy level, and the electrons in the lower occupied energy levels partly shield it from the attraction of the nucleus.

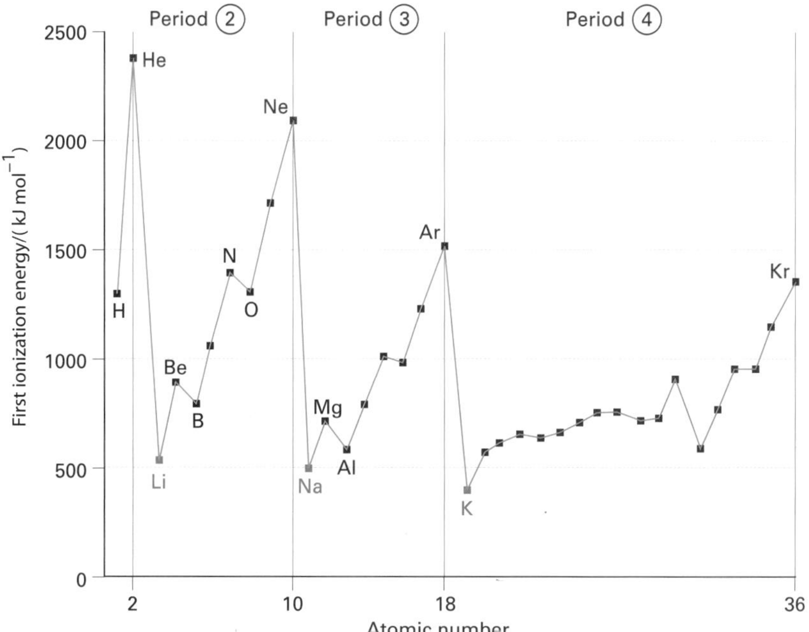

This graph shows the first ionization energies for elements Z = 1 to 36 (hydrogen to krypton).

Check your understanding

1. a What does *first ionization energy* mean?
   b Write the equation for the first ionization of magnesium, Mg.
2. Describe two pieces of evidence for the existence of energy levels, as shown by first ionization energies.
3. Researchers at the University of Helsinki made the first argon compound in 2000. They shone ultraviolet light onto a mixture of frozen argon and hydrogen fluoride. Explain why it is surprising that argon hydrofluoride, HArF, was made as a result.

Noble gases can react, too

Group 0 contains helium, neon, argon, krypton, xenon, and radon. These are very unreactive and are called the noble gases. As early as 1904, a German chemist called Richard Abegg (1869–1910) realised that this must be because they have a very stable electron configuration. The highest occupied s and p energy sub-levels of noble gas atoms are full. Atoms from other groups lose, gain, or share electrons to achieve a noble gas electron configuration. So it was a big surprise when Neil Bartlett made a xenon compound in 1962.

Bartlett mixed xenon with a very reactive compound called platinum hexafluoride. To everyone's amazement, a reaction happened and he made $XePtF_6$. Compounds of all the noble gases except helium and neon have now been made.

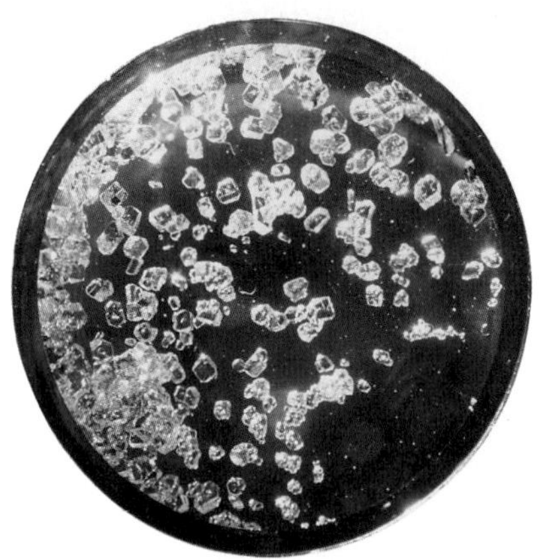

These are crystals of xenon tetrafluoride, XeF_4.

1.11 Evidence for energy sub-levels

OBJECTIVES

already from AS Level, you

- know the meaning of the term ionization energy
- understand how ionization energies give evidence for energy levels

and after this spread you should

- understand how ionization energies give evidence for energy sub-levels
- understand how successive ionization energies give more evidence for energy levels

First ionization energies and sub-levels

Evidence for the existence of energy sub-levels comes from looking at the first ionization energies of period 3 elements in more detail. Notice that there is a general increase in first ionization energy as you go across the period. But there are also two abrupt drops, between magnesium and aluminium, and between phosphorus and sulfur. You will learn about the general trend first.

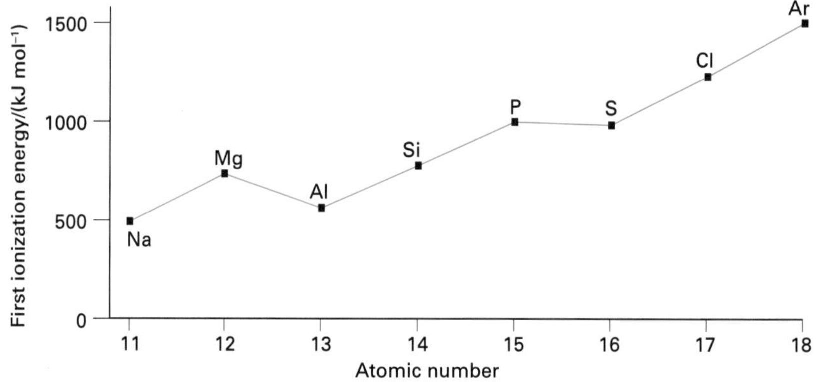

First ionization energy plotted against atomic number for the elements in period 3.

As you go across period 3, the number of protons in the nucleus steadily increases. This means that the charge of the nucleus increases too. The force of attraction between the nucleus and the outer electron increases, and increasing amounts of energy are needed to remove the outer electron. Note that even though the number of electrons also steadily increases, they enter the same energy level. This means that they do not significantly increase the shielding of the outer electron from the nucleus.

You need to study the electron configurations of magnesium and aluminium to explain the drop in first ionization energy between them. They are:

- magnesium: $1s^2\ 2s^2\ 2p^6\ 3s^2$
- aluminium: $1s^2\ 2s^2\ 2p^6\ 3s^2\ 3p^1$

The outer electron in aluminium is in a p sub-level. This is higher in energy than the outer electron in magnesium, which is in an s sub-level. As a result, less energy is needed to remove it.

What about the drop in first ionization energy between phosphorus and sulfur? Their electron configurations are:

- phosphorus: $1s^2\ 2s^2\ 2p^6\ 3s^2\ 3p^3$
- sulfur: $1s^2\ 2s^2\ 2p^6\ 3s^2\ 3p^4$

	phosphorus	sulfur
3p	↑ | ↑ | ↑	↑↓ | ↑ | ↑
3s	↑↓	↑↓
2p	↑↓ | ↑↓ | ↑↓	↑↓ | ↑↓ | ↑↓
2s	↑↓	↑↓
1s	↑↓	↑↓

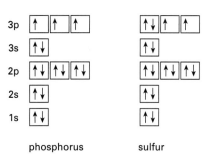

Spin diagrams for phosphorus and sulfur.

The outer electron is in the 3p sub-level in both atoms. You need to study their spin diagrams to understand why there is a drop in first ionization energy.

The 3p electrons in phosphorus are all unpaired. In sulfur, two of the 3p electrons are paired. There is some repulsion between paired electrons in the same sub-level. This partly counteracts the force of their attraction to the nucleus. So less energy is needed to remove one of the paired electrons from a sulfur atom than is needed to remove one of the paired electrons from a sulfur atom.

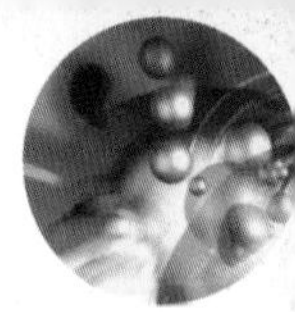

More evidence for energy levels

More evidence for energy levels comes from the energies needed to remove each electron from an atom in turn. The table gives some examples for the sodium atom.

ionization	equation	Energy (kJ mol^{-1})
first	$Na(g) \rightarrow Na^{+}(g) + e^{-}$	496
second	$Na^{+}(g) \rightarrow Na^{2+}(g) + e^{-}$	4563
third	$Na^{2+}(g) \rightarrow Na^{3+}(g) + e^{-}$	6913

Each individual ionization involves removing one electron from an atom or ion.

There is a general increase in the energy needed to remove each successive electron from sodium. This is because the negatively charged electron is being removed from ions with an increasing positive charge. But there are very big increases between the first and second ionization energies, and between the ninth and tenth ionization energies. This is evidence for the existence of energy levels. The big increases happen when the electron is being removed from a fully occupied energy level that is lower in energy and closer to the nucleus than the previous one.

You can tell which group an element belongs to by looking at its successive ionization energies. Sodium is in group 1: the first big increase in ionization energy happens between the first and second ionization energies. Magnesium is in group 2: the first big increase in ionization energy happens between the second and third ionization energies.

Logarithms

The ionization energy values on the graph are converted into their logarithms. You can do this using the "log" button on your calculator. It is done so some very large numbers fit on comfortably.

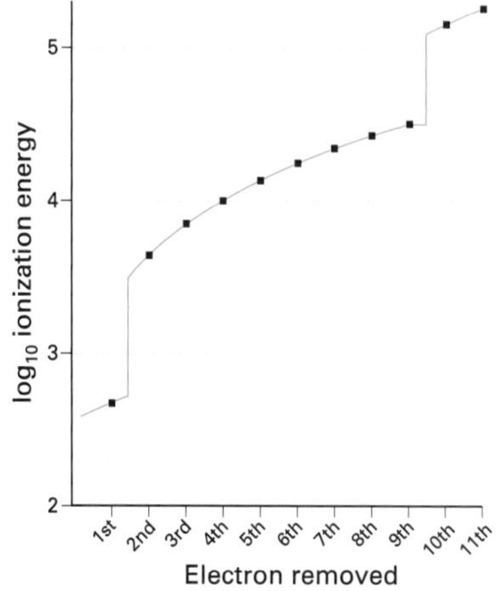

Successive ionization energies for the sodium atom, $1s^2\ 2s^2\ 2p^6\ 3s^1$.

Check your understanding

1. Why does the first ionization energy increase as you go across period 3?
2. Explain why:
   a The first ionization energy of aluminium is lower than the first ionization energy of magnesium.
   b The first ionization energy of sulfur is lower than the first ionization energy of phosphorus.
3. The table shows some ionization energies for an element.

ionization	first	second	third	fourth	fifth	sixth
ionization energy (kJ mol^{-1})	1086	2353	4621	6223	37832	47278

   a Write the equation for the fourth ionization, where M stands for the symbol of the element.
   b Which group in the periodic table does element M belong to? Explain how you know.

2.01 Masses of atoms and molecules

OBJECTIVES

already from GCSE, you know

- how to use the periodic table to find relative atomic masses
- how to work out relative formula masses

already from AS Level, you know

- the definitions of relative atomic mass and relative molecular mass

and after this spread you should

- be able to work out relative formula masses using the periodic table
- understand the difference between relative molecular mass and relative formula mass

RAM and RMM

If you read older text books you may come across other ways of writing A_r and M_r. RAM means *relative atomic mass* and RMM means *relative molecular mass.*

Checkpoint

It is important that you know the definitions of relative atomic mass A_r and relative molecular mass M_r.

Relative atomic mass, A_r

An element's relative atomic mass is the mean mass of an atom of the element compared to one-twelfth the mass of a ^{12}C atom:

$$\text{relative atomic mass} = \frac{\text{mean mass of one atom of the element}}{\text{mass of one atom of } ^{12}\text{C}} \times 12$$

Relative molecular mass, M_r

Relative molecular mass is the mean mass of a molecule compared to one-twelfth the mass of a ^{12}C atom.

$$\text{relative atomic mass} = \frac{\text{mean mass of a molecule}}{\text{mass of one atom of } ^{12}\text{C}} \times 12$$

Elements and A_r

The atoms in an element all have the same atomic number. The elements are arranged in order of increasing atomic number in the periodic table. The relative atomic mass of an element is usually shown in the top of its box in the periodic table, with the atomic number at the bottom. Some tables show this information the other way round. If you are not sure which number to use, the relative atomic mass will always be the larger number of the two.

The periodic tables used at GCSE show relative atomic masses as whole numbers (chlorine is a common exception at 35.5). The periodic table you use at this level shows relative atomic masses to one decimal place. So the A_r of H is 1.0 not 1, and the A_r of Fe is 55.8 not 56. Make sure you use the correct periodic table in your studies.

Group / Period	1	2											3	4	5	6	7	0
①	H 1.0																	He 4.0
②	Li 6.9	Be 9.0											B 10.8	C 12.0	N 14.0	O 16.0	F 19.0	Ne 20.2
③	Na 23.0	Mg 24.3											Al 27.0	Si 28.1	P 31.0	S 32.1	Cl 35.5	Ar 39.9
④	K 39.1	Ca 40.1	Sc 45.0	Ti 47.9	V 50.9	Cr 52.0	Mn 54.9	Fe 55.8	Co 58.9	Ni 58.7	Cu 63.5	Zn 65.4	Ga 69.7	Ge 72.6	As 74.9	Se 79.0	Br 79.9	Kr 83.8
⑤	Rb 85.5	Sr 87.6	Y 88.9	Zr 91.2	Nb 92.9	Mo 95.9	Tc 98.9	Ru 101.1	Rh 102.9	Pd 106.4	Ag 107.9	Cd 112.4	In 114.8	Sn 118.7	Sb 121.8	Te 127.6	I 126.9	Xe 131.3
⑥	Cs 132.9	Ba 137.3	La 138.9	Hf 178.5	Ta 180.9	W 183.9	Re 186.2	Os 190.2	Ir 192.2	Pt 195.1	Au 197.0	Hg 200.6	Tl 204.4	Pb 207.2	Bi 209.0	Po 210.0	At 210.0	Rn 222.0
⑦	Fr 223.1	Ra 226.0	Ac 227.0	Rf	Db	Sg	Bh	Hs	Mt									
					Ce 140.1	Pr 140.9	Nd 144.2	Pm 144.9	Sm 150.4	Eu 152.0	Gd 157.3	Tb 158.9	Dy 162.5	Ho 164.9	Er 167.3	Tm 168.9	Yb 173.0	Lu 175.0
					Th 232.0	Pa 231.0	U 238.0	Np 237.0	Pu 239.1	Am 243.1	Cm 247.1	Bk 247.1	Cf 252.1	Es (252)	Fm (257)	Md (258)	No (259)	Lr (260)

The periodic table shows the relative atomic masses of the elements. There is a larger version at the back of the book. You will be given a periodic table in the examination.

Elements and M_r

You can assume when writing formulae and equations that most elements exist as single atoms. This includes all the metals and the noble gases of Group 0. You need to take care with the other gaseous elements and the elements in Group 7. They exist as **diatomic molecules** with two atoms joined together by chemical bonds. This means for example that the formula of hydrogen gas is H_2 not H, and the formula of chlorine gas is Cl_2 not Cl.

You calculate the relative molecular mass of a substance by adding together the relative atomic masses of all the atoms in its formula. For example:

The formula of chlorine gas is Cl_2.

The A_r of chlorine is 35.5 (from the Periodic Table).

The M_r of chlorine gas = 35.5 + 35.5 = 71.0

Note that you give the answer to one decimal place.

Compounds and M_r

You calculate the relative molecular mass of compounds by adding together the relative atomic masses of all the atoms in the formula. For example:

The formula of carbon dioxide is CO_2

The A_r of carbon is 12.0 and the A_r of oxygen is 16.0

The M_r of carbon dioxide = 12.0 + 16.0 + 16.0
= 44.0

Remember that the formulae of compounds may have brackets in them. For example, the formula of magnesium hydroxide is $Mg(OH)_2$. Where this happens, multiply the A_r of each atom inside the brackets by the number outside. For example:

The formula of magnesium hydroxide is $Mg(OH)_2$

The A_r of magnesium is 24.3, the A_r of oxygen is 16.0 and the A_r of hydrogen is 1.0

The M_r of magnesium hydroxide = 24.3 + (2 × 16.0) + (2 × 1.0)
= 24.3 + 32.0 + 2.0
= 58.3

Check your understanding

1. Calculate the relative molecular masses of these substances.
   - a oxygen, O_2
   - b bromine, Br_2
   - c hydrogen chloride, HCl
   - d water, H_2O
   - e methane, CH_4
   - f tetrafluoroethene, C_2F_4
2. Calculate the relative formula masses of these compounds.
   - a sodium hydroxide, NaOH
   - b nitric acid, HNO_3
   - c sulfuric acid, H_2SO_4
   - d calcium hydroxide, $Ca(OH)_2$
   - e ammonium carbonate, $(NH_4)_2CO_3$
   - f aluminium nitrate, $Al(NO_3)_3$

Relative formula mass

Some compounds contain ions and not molecules. The phrase relative formula mass covers all substances other than simple atoms. For this course you should use the relative molecular mass for molecular substances and relative formula mass for ionic compounds. The symbol for both is M_r.

Take care

For many elements, the number given as the relative atomic mass is the same as the mass number. They are not the same thing. The mass number is the number of protons and neutrons in the nucleus. The relative atomic mass takes into account the presence of different isotopes. For example, the A_r of bromine is 79.9 and it would be tempting to think that its mass number was 80. But bromine contains two isotopes (^{79}Br and ^{81}Br) in almost equal amounts.

2.02 Amount of substance

OBJECTIVES

already from AS Level, you know

- how to work out relative formula masses using the periodic table

and after this spread you should

- understand the concept of the mole
- be able to work out the mass of a substance from its relative mass and amount

Chemists need to know the amount of each substance in a chemical reaction. This lets them calculate just how much of each reactant they need, and how much useful product is likely to be made. In everyday life, the amount of something is usually its mass in grams if it is solid, or its volume in cubic centimetres if it is a liquid or gas. Other measures might be used for bigger quantities, such as kilograms, tonnes, cubic metres, and so on. Chemists need to know how many atoms, molecules, and ions they have in a reaction. This is what a chemist means by the amount of a substance. This might seem difficult to find out, as atoms, molecules, and ions are incredibly small and have very little mass. This is where the concept of the **mole** helps.

The mole and the Avogadro constant

The mole is the unit for **amount of substance**. It is given the symbol mol.

The number of atoms in 12 g of ^{12}C is called the **Avogadro constant**, which is given the symbol L or N_A. The Avogadro constant is 6.022×10^{23}. This is approximately six hundred thousand billion billion, a number that does not make much sense in everyday life. If you could stack the Avogadro number of £1 coins one on top of the other, the pile would reach from the Earth to the centre of the galaxy and back again – four times. The mole and Avogadro's constant only make sense when applied to incredibly tiny particles with very little mass.

What's in a mole?

One mole is the amount of a substance that contains the same number of particles as there are atoms in exactly 12 g of ^{12}C. These particles can be atoms, molecules, ions, or electrons.

Science @ Work

Amedeo Avogadro (1776–1856)

Amedeo Avogadro was an Italian scientist who proposed an idea about gases in 1811. Avogadro's principle states that equal volumes of gases, at the same temperature and pressure, contain the same number of particles. There was a lot of confusion between atoms and molecules at the time, and Avogadro's idea was not accepted straightaway by other scientists. It was only accepted after the results of many experiments supported it. But sadly Avogadro was dead by then.

Masses and moles

The mass of one mole of a substance is its A_r or M_r in grams. So one mole of ^{12}C atoms has a mass of 12.0 g, and one mole of oxygen molecules O_2 has a mass of 32.0 g. These two equations show how mass, relative mass, and amount of substance are related:

For atoms: mass $= A_r \times n$

For molecules: mass $= M_r \times n$

The symbol n represents the amount of substance in mol.

All these dishes contain one mole of an element or compound

Worked example of calculating mass from A_r

What is the mass of 3 mol of helium atoms, $A_r = 4.0$?

mass $= A_r \times n$

mass $= 4.0 \times 3$
$= 12.0$ g

Worked example of calculating mass from M_r

What is the mass of 1.5 mol of nitrogen molecules, $M_r = 28.0$?

mass = $M_r \times n$

mass = 28.0×1.5
= 42.0 g

You can also find the mass of ions if you know the relative formula mass and the amount of substance. Remember that the electron has 1836 times less mass than the proton, so you can ignore its mass when you work out the M_r of an ion. For example:

The formula of the sulfate ion is SO_4^{2-}

The A_r of sulfur is 32.1 and the A_r of oxygen is 16.0

The M_r of the sulfate ion = 32.1 + (4 × 16.0)
= 32.1 + 64.0
= 96.1

Worked example of calculating the mass of ions from M_r

What is the mass of 2.0 mol of sulfate ions, $M_r = 96.1$?

mass = $M_r \times$ *mol*

mass = 96.1×2.0
= 192.2 g

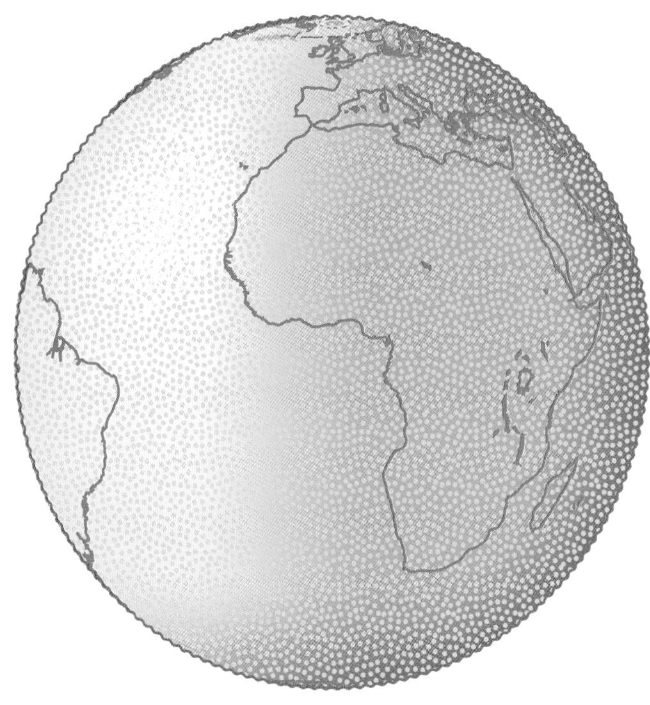

To help you with a sense of scale: if you could fill the Earth's volume with grapefruit you would need a number of them approaching Avogadro's number.

Check your understanding

1. What is the mass of:
   - **a** 1 mol of magnesium atoms, $A_r = 24.3$
   - **b** 2.5 mol of neon atoms, $A_r = 20.2$
   - **c** 0.5 mol of water molecules, $M_r = 18.0$
   - **d** 2.0 mol of calcium carbonate, $CaCO_3$
   - **e** 10.0 mol of ammonium ions, NH_4^+
2. Why do we use the mole for atoms, molecules, and ions but not for everyday objects such as tables and chairs?

2.03 More moles

OBJECTIVES

already from AS Level, you

- understand the concept of the mole
- can work out the mass of a substance from its relative mass and amount

and after this spread you should

- be able to work out the amount and relative mass of a substance using the concept of the mole
- understand that there were other standard atoms before ^{12}C

Remember:

For atoms: mass $= A_r \times n$

For molecules: mass $= M_r \times n$

These equations can be re-arranged in two different ways. This lets you find one unknown quantity if you know the other two.

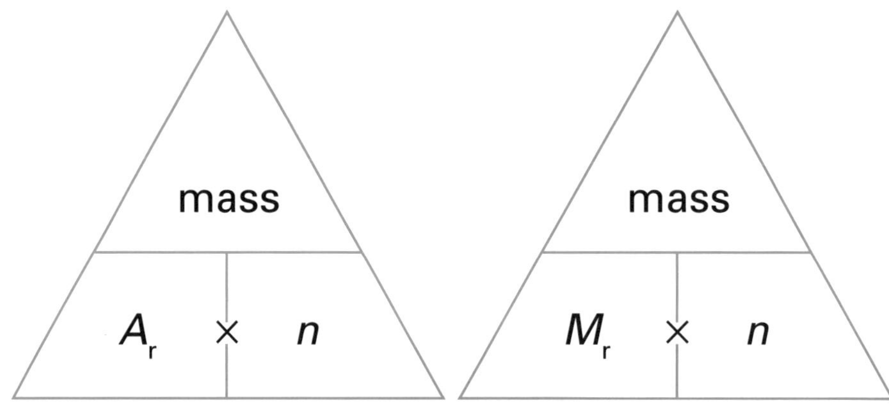

"Magic triangles" can help you to see how equations can be rearranged.

Number of moles

You can work out the number of moles of a substance if you know its mass and its relative mass.

For atoms: $n =$ mass $\div A_r$

For molecules: $n =$ mass $\div M_r$

Worked example of calculating moles from A_r

How many moles of atoms are there in 27.0 g of beryllium, $A_r = 9.0$?

$n =$ mass $\div A_r$

$n = 27.0 \div 9.0$
$= 3.00$ mol

Worked example of calculating moles from M_r

How many moles of molecules are there in 11.0 g of carbon dioxide, $M_r = 44.0$?

$n =$ mass $\div M_r$

$n = 11.0 \div 44.0$
$= 0.250$ mol

Calculations like these are particularly useful when designing experiments or analysing results.

Finding A_r or M_r

You can work out the relative mass of a substance if you know its mass and the number of moles it contains.

For atoms: $A_r =$ mass $\div n$

For molecules: $M_r =$ mass $\div n$

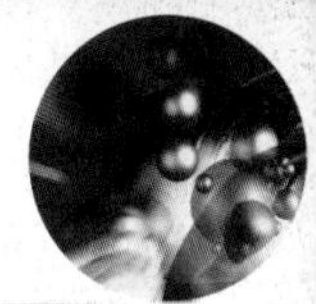

Worked example of calculating M_r

What is the relative formula mass of compound Y, if 0.5 mol has a mass of 20.0 g?

M_r = mass ÷ n

M_r = 20.0 ÷ 0.5
= 40.0

Worked example of calculating A_r

What is the relative atomic mass of element X, if 2.0 mol has a mass of 90.0 g?

A_r = mass ÷ n

A_r = 90.0 ÷ 2.0
= 45.0

Towards a standard atom

John Dalton and others weighed elements and compounds very accurately. They were able to show that the different atoms in a compound were always combined in fixed proportions. The simplest element is hydrogen. It was given a weight of 1, which let the scientists work out the atomic weight of other substances. Other standard atoms were in use, too.

Jöns Jacob Berzelius used oxygen as his standard atom. He argued that oxygen was a better standard than hydrogen because it formed more compounds. This allowed scientists to make more atomic weight measurements, which improved their accuracy. By 1901, oxygen was adopted as the standard atom worldwide.

Arguments continued but things really became difficult in 1929, when oxygen isotopes were discovered. Physicists used a scale where ^{16}O had an atomic weight of 16. Chemists used a scale where oxygen was given an atomic weight of 16, even though it contained small proportions of ^{16}O and ^{17}O. The two groups needed a common standard.

The ^{12}C atom was adopted as the standard atom in 1961. There were several reasons for this. It was already a standard in mass spectrometry and changing to it would cause only small changes to published atomic weights. The terms atomic weight and *atomic mass were replaced by relative atomic mass* in 1979.

Molar mass

The equation M_r = mass ÷ n is not strictly accurate, although it works well enough for most purposes. The problem is that M_r is a relative quantity, so it has no units, whereas mass is measured in grams and amount of substance in moles. This means that M_r really ought to have units of grams per mole, g mol^{-1}. Some examination boards and text books mention "molar mass". This has the same value as M_r but has units of g mol^{-1}. The equivalent equation is molar mass = mass ÷ n.

M_r or molar mass:

Whichever way you do these calculations you get the same numerical answer.

Check your understanding

1. How many moles of atoms are there in:
   - **a** 12.0 g of carbon
   - **b** 11.5 g of sodium
2. How many moles of molecules are there in:
   - **a** 8.0 g of oxygen, O_2
   - **b** 22.0 g of carbon dioxide, CO_2
3. How many moles of ions are there in:
   - **a** 15.0 g of carbonate ions, CO_3^{2-}
   - **b** 28.0 g of ammonium ions, NH_4^+
4. 0.25 mol of element X has a mass of 4.0 g. What is the A_r of element X?
5. 2.0 mol of compound Y has a mass of 80.0 g. What is the M_r of compound Y?
6. One of the earlier scales of atomic weight had hydrogen as the standard atom. On that scale, the atomic weight of hydrogen was 1. Suggest why teachers were among the people who argued against changing to a different scale, in which oxygen was the standard atom with an atomic weight of 16.

One mole is the number of atoms in 12 g of ^{12}C. This is what 12 g of carbon looks like.

2.04 The ideal gas equation

OBJECTIVES

already from AS Level, you know

- that equal volumes of gases, at the same temperature and pressure, contain the same number of particles

and after this spread you should

- be able to recall the ideal gas equation
- be able to convert temperatures from °C to K

If you pump too much air into a party balloon, the pressure of the air inside will burst it with a loud bang. Put one into the fridge and it will shrink a bit. You have carried out two simple experiments into the behaviour of gases. Over three centuries ago, scientists were carrying out careful measurements to see what gases do if their pressure or temperature is changed.

Boyle's law

Robert Boyle, the author of *The Sceptical Chymist*, published the results of his experiments on gases in 1661. Boyle discovered that as long as the temperature of a gas stayed the same, the higher the pressure the smaller its volume. **Boyle's law** states that

- The volume of a fixed mass of gas (at a constant temperature) is inversely proportional to its pressure.

This can be shown using the equation:

$$V \propto 1/p$$

where V is the volume and p is the pressure. The $\propto$ sign means 'is proportional to'.

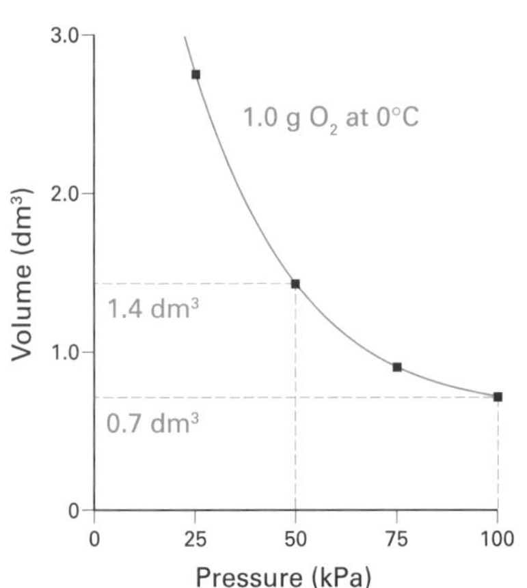

From Boyle's law, it follows that if you double the pressure on a gas its volume halves.

These marshmallows are at normal atmospheric pressure.

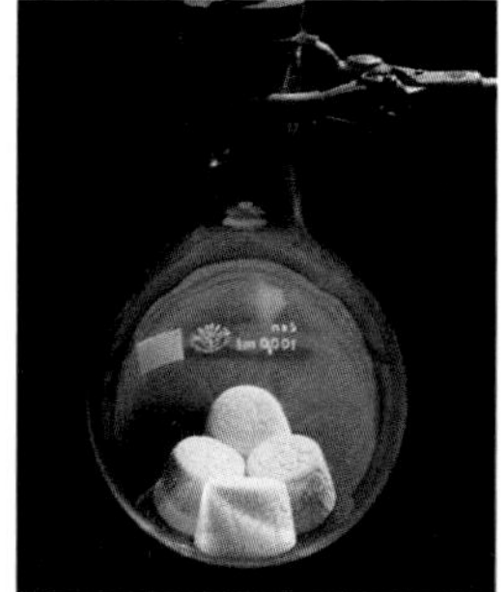

These marshmallows are under reduced pressure. The air bubbles inside them have expanded.

The Magdeburg hemispheres

Robert Boyle's experiments were greatly helped by Otto von Guericke's invention of the vacuum pump in 1650. Boyle could easily change the pressure exerted on a sample of gas using a vacuum pump. But the inventor himself did something quite amazing using such a pump.

Von Guericke carried out a famous experiment in Magdeburg in 1654 in front of the Holy Roman Emperor (Ferdinand III) and a crowd of very impressed people. He pressed two copper hemispheres together and pumped the air out. Two teams of eight horses could not pull them apart. Air pressure pushed the hemispheres tightly together until air was let back in.

Charles's law

Jacques Charles investigated the relationship between the temperature of a gas and its volume. The French scientist released his results in 1787. **Charles's law** states that

- The volume of a fixed mass of gas (at a constant pressure) is proportional to its **absolute temperature**.

This can be shown using the equation:

$$V \propto T$$

where V is the volume and T is the absolute temperature

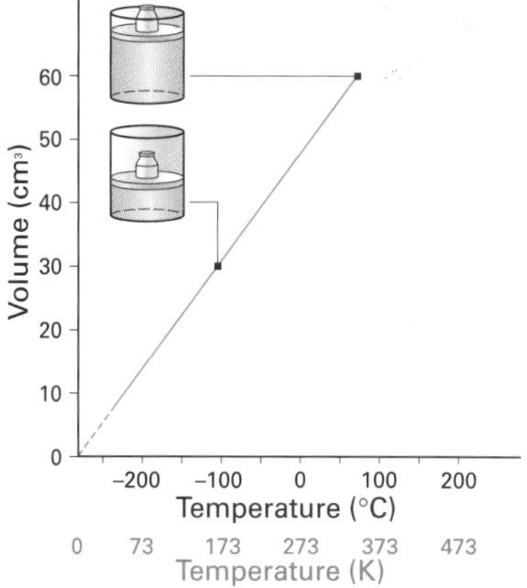

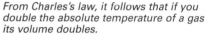

From Charles's law, it follows that if you double the absolute temperature of a gas its volume doubles.

Putting it all together

By 1811 there were three gas laws:

Boyle's law: $V \propto 1/p$

Charles's law: $V \propto T$

Avogadro's principle: $V \propto n$ where n is the number of moles of gas

They can be combined like this:

$$V \propto \frac{nT}{p}$$

You can remove the proportional sign if you multiply by a constant (a number that stays the same):

$$V = \frac{RnT}{p}$$

where R is the **gas constant**

This equation is usually rearranged to form the **ideal gas equation**:

$$pV = nRT$$

You must be able to recall and use this equation. There are some important things to note about it:

- p is the pressure measured in pascals, Pa
- V is the volume measured in cubic metres, m^3
- n is the amount of gas measured in moles, mol
- R is the gas constant. It is $8.31\,J\,K^{-1}\,mol^{-1}$ (joules per kelvin per mole)
- T is the absolute temperature measured in kelvin, K

Check your understanding

1. A balloon is filled with $1.0\,dm^3$ of air at normal atmospheric pressure. If its temperature stays the same, what is the volume of the air at:
   a double atmospheric pressure
   b half atmospheric pressure?
2. A balloon is filled with $1.0\,dm^3$ of air at 300 K. If its pressure stays the same, what is the volume of the air at:
   a 600 K
   b 150 K?
3. State the ideal gas equation and give the units used in it.
4. Convert these temperatures to kelvin, K:
   a 0 °C
   b 150 °C
   c 1535 °C
   d −40 °C

Absolute temperature

If you study the graph of volume against temperature, you will see that there is a temperature at which the gas would have zero volume. This is called **absolute zero.** It is a very chilly −273.15 °C. The absolute temperature scale starts at this temperature. Absolute temperature is measured in **kelvin,** K. The great thing about the absolute temperature is that there are no negative temperatures. This makes any calculations much easier to do. A change of 1 K is the same as a change of 1 °C. Notice that you would say "one kelvin" and not "one degree Kelvin".

Converting to kelvin

If you are given a temperature in °C you must convert it into K for the ideal gas equation. At this level you just add 273. So −100 °C = 173 K, 0 °C = 273K, and 100 °C = 373 K.

2.05 Using the ideal gas equation

OBJECTIVES

already from AS Level, you

- can recall the ideal gas equation
- can convert temperatures from °C to K

and after this spread you should

- be able to use the ideal gas equation, rearranging it if necessary

Using standard index form

If you are comfortable using standard index form, there are much easier ways to do these conversions.

- To convert kPa to Pa, just multiply by 10^3.

  So 0.25 kPa = 0.25×10^3 Pa

- To convert dm^3 to m^3, just multiply by 10^{-3}.

  So $100\,dm^3 = 100 \times 10^{-3}\,m^3$ (or $1 \times 10^{-1}\,m^3$)

- To convert cm^3 to m^3, just multiply by 10^{-6}.

  So $100\,cm^3 = 100 \times 10^{-6}\,m^3$ (or $1 \times 10^{-4}\,m^3$)

Conversions for the ideal gas equation

Remember that the ideal gas equation is $pV = nRT$. You already know how to convert from °C to K for the absolute temperature T. But you may have other conversions to do before you can use the equation in a calculation.

Pressure

Pressure is measured in pascals, Pa. Atmospheric pressure is about 100 000 Pa. So you can see that the pascal is quite a small measure. This means that pressure is often given in kilopascal, kPa. You must convert any pressure given in kPa into Pa before using it in the ideal gas equation.

To do this, you multiply the pressure in kPa by 1000. For example: 1 kPa = 1000 Pa, and 0.25 kPa = 250 Pa.

Volume

Volume is usually measured in cubic decimetres, dm^3, or in cubic centimetres, cm^3. But the ideal gas equation needs volumes measured in cubic metres, m^3. You must convert any volume into m^3 before using it in the ideal gas equation.

To do this

- divide volumes in dm^3 by 1000
- divide volumes in cm^3 by 1 000 000

For example:

- $100\,dm^3 = 100 \div 1000 = 0.1\,m^3$
- $100\,cm^3 = 100 \div 1\,000\,000 = 0.0001\,m^3$

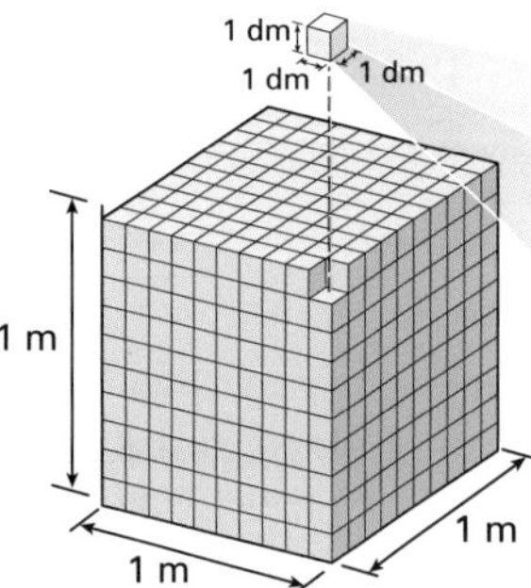

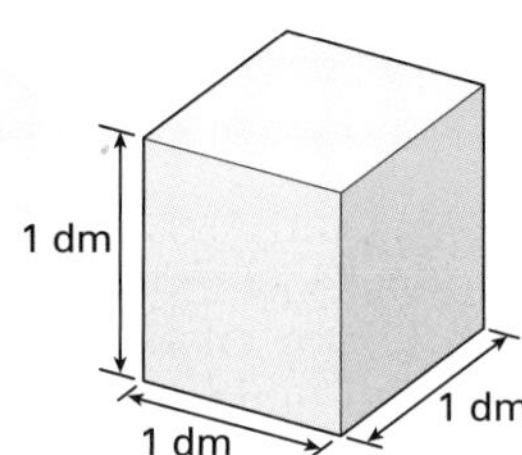

There are one thousand dm^3 in one m^3.

Rearranging equations

You rearrange mathematical equations by doing the same things to both sides. So if you divide one side by V you must also divide the other side by V. The aim is to get the quantity that you have been asked to find on its own on one side.

For example, you are given $a = b \times c$, and asked to find c. You would divide both sides by b. This gives $a/b = b/b \times c$. Since any number divided by itself is 1, this simplifies to $a/b = 1 \times c$, or $a/b = c$. You would then swap it round to get $c = a/b$.

Finding a volume

You can find the volume of a gas if you know its pressure, amount, and absolute temperature. You need to rearrange the ideal gas equation:

$$V = \frac{nRT}{p}$$

Worked example for finding a volume

What is the volume occupied by 0.5 mol of a gas at 400 K and 100 Pa?

Write the rearranged ideal gas equation: $V = \frac{nRT}{p}$

Show your working out:

$$V = \frac{0.5 \times 8.31 \times 400}{100}$$

$$V = \frac{1662}{100}$$

$$= 16.62\,m^3$$

Unless you are told otherwise, show your final answer correct to three significant figures, in this case $16.6\,m^3$.

Finding a pressure

You can find the pressure of a gas if you know its amount, absolute temperature, and volume. You need to rearrange the ideal gas equation:

$$p = \frac{nRT}{V}$$

Worked example for finding a pressure

What is the pressure exerted by 2.0 mol of gas at 25 °C in a 5 dm³ container?

Write the rearranged ideal gas equation: $p = \frac{nRT}{V}$

Remember that 25 °C = 25 + 273 = 298 K.

Also, $5\,dm^3 = 5 \times 10^{-3}\,m^3$ (or, $5 \div 1000 = 0.005\,m^3$)

Show your working out:

$$p = \frac{2.0 \times 8.31 \times 298}{5 \times 10^{-3}}$$

$$p = \frac{4952.76}{5 \times 10^{-3}} = 990\,552\ Pa$$

This is 991 kPa (or 9.91×10^5 Pa) correct to three significant figures.

Finding the temperature of a gas

You can find the temperature of a gas if you know its pressure, volume, and amount. You need to rearrange the ideal gas equation:

$$T = \frac{pV}{nR}$$

Worked example for finding a temperature

What is the temperature of 0.2 mol of gas at 2000 Pa in a 40 dm³ container?

Write the rearranged ideal gas equation: $T = \frac{pV}{nR}$

Remember that $40\,dm^3 = 40 \times 10^{-3}\,m^3$ (or, $40 \div 1000 = 0.04\,m^3$)

Show your working out:

$$T = \frac{2000 \times 40 \times 10^{-3}}{0.2 \times 8.31}$$

$$T = \frac{80}{1.662}$$

$$= 48.1\ K$$

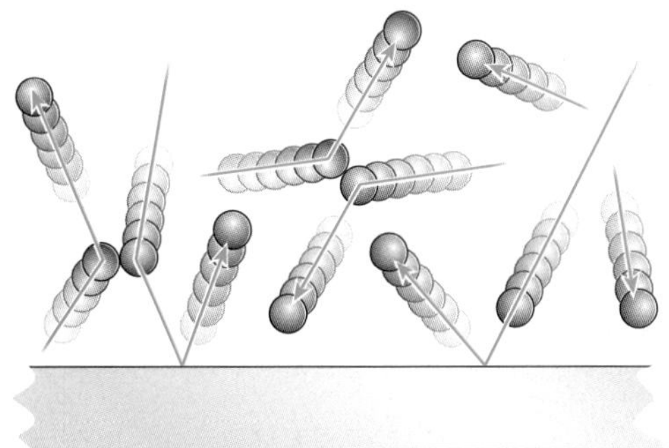

A gas exerts pressure when its molecules collide with the container wall and rebound.

Check your understanding

1. Convert the following pressures to Pa.
   a 2.5 kPa **b** 0.1 kPa
2. Convert the following volumes to m^3.
   a $1250\,dm^3$ **b** $500\,dm^3$ **c** $25\,cm^3$
3. What is the volume occupied by 1.0 mol of a gas at 20 °C and 100 kPa?
4. What is the pressure exerted by 5.0 mol of gas at 600 K in a $2.5\,dm^3$ container?
5. What is the temperature of 2.0 mol of gas at 10 kPa in a $4.0\,dm^3$ container?

2.06 Experiments to find M_r

OBJECTIVES

already from AS Level, you can

- work out the mass of a substance from its relative mass and amount
- use the ideal gas equation, re-arranging it if needed

and after this spread you should

- know how to find the M_r of a volatile liquid or a gas

M_r from the ideal gas equation

You will remember the relationship between mass, M_r, and amount of substance:

$$\text{mass} = M_r \times n$$

You can re-arrange it like this:

$$M_r = \frac{\text{mass}}{n}$$

So you can find the relative molecular mass of a substance from a sample. You just need to know the mass and number of moles of the substance in your sample. A precise balance lets you find the mass, and the ideal gas equation lets you find the number of moles if the sample is a gas.

The ideal gas equation is $pV = nRT$. You need to re-arrange it to find the number of moles of gas:

$$n = \frac{pV}{RT}$$

The M_r of a gas can be found by weighing a sample then measuring its volume, temperature, and pressure. This also works for a **volatile** liquid, a liquid that vaporizes easily.

Worked example for finding M_r

0.260 g of a gas occupies 144 cm³ at 101 kPa and 19 °C. What is its M_r?

Convert the quantities to the correct units:

$$144\,\text{cm}^3 = 144 \times 10^{-6}\,\text{m}^3$$

$$101\,\text{kPa} = 101 \times 10^3\,\text{Pa}$$

$$19\,°\text{C} = 19 + 273 = 292\,\text{K}$$

Write the re-arranged ideal gas equation: $n = \frac{pV}{RT}$

Show your working out:

$$n = \frac{101 \times 10^3 \times 144 \times 10^{-6}}{8.31 \times 292}$$

$$n = \frac{14.544}{2426.52} = 5.99 \times 10^{-3}\,\text{mol}$$

Then:

$$M_r = \frac{\text{mass}}{n}$$

$$= \frac{0.260}{5.99 \times 10^{-3}}$$

$$= 43.4$$

Percentage difference

The gas in the example was actually carbon dioxide. The true M_r of carbon dioxide is 44.0. How close was the experimental M_r?

Difference = experimental value − true value
= 43.4 − 44.0
= −0.6

The negative sign tells you that the experimental value was too small. A positive sign would tell you that the experimental value was too large.

It is often useful to calculate the **percentage difference**, too. Use this equation:

$$\%\ \text{difference} = \frac{\text{experimental value} - \text{true value}}{\text{true value}}$$

So for the worked example:

$$\%\ \text{difference} = \frac{43.4 - 44.0}{44.0} \times 100$$

$$= -1.36\%$$

Carrying out an experiment

M_r of a gas

The volume of a sample of gas is easily measured using a **gas syringe**. This is a glass syringe with a low-friction plunger that moves out smoothly. The barrel of the gas syringe is graduated to show the volume of gas it contains.

Some gases can be made using chemical reactions. For example carbon dioxide can be made using the reaction between powdered calcium carbonate and dilute hydrochloric acid. Or they can be released from a pressurized cylinder. In both cases, it is important to find the mass of the gas released into the gas syringe. This is usually done by weighing the container before and after releasing the gas.

The other measurements needed to find the M_r of the gas are:

- temperature using a thermometer
- pressure using a **barometer**

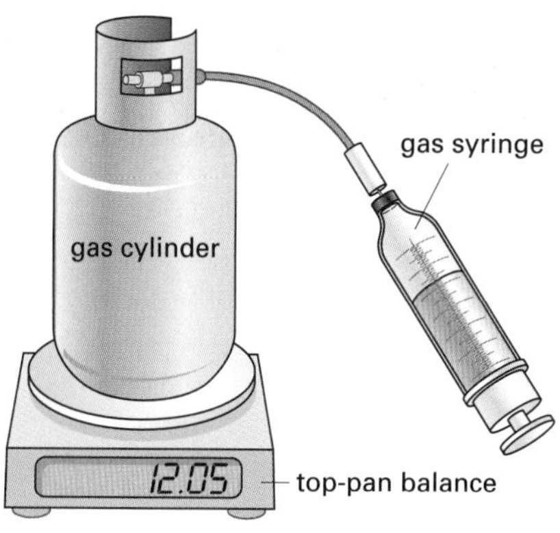

A simple experiment to find the M_r of a gas.

M_r of a volatile liquid

Volatile liquids easily turn into a gas when warmed up. For example, hexane boils at just 69 °C. A gas syringe with a rubber cap fitted is used. It is put in an oven so that liquid will vaporize inside it. A second syringe fitted with a hypodermic needle is filled with the liquid. The liquid is injected though the rubber cap into the gas syringe, where it vaporizes. The gas expands and pushes the plunger out, allowing its volume to be measured.

The mass of injected volatile liquid is found by weighing the hypodermic syringe before and after the injection. As before, the temperature and pressure must also be measured.

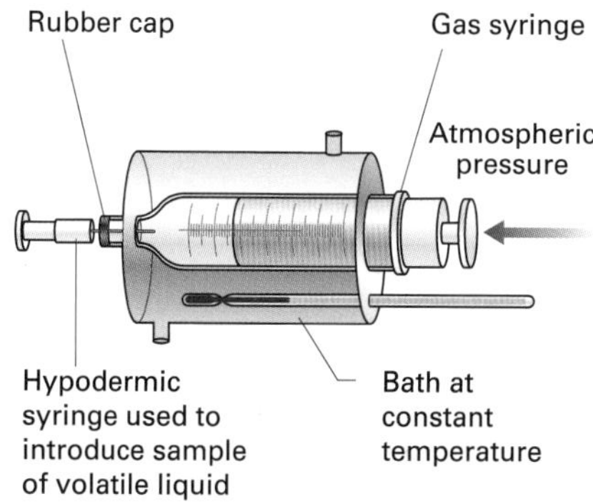

Apparatus needed to find the M_r of a volatile liquid.

Check your understanding

1. A cylinder of gas X weighed 242.487 g. After it was used to release some gas into a gas syringe, the cylinder weighed 242.313 g.
   - a Calculate the mass of gas released.
   - b The reading on the gas syringe was 75 cm^3, the temperature was 27 °C and the atmospheric pressure was 100 kPa. Calculate the number of moles of gas released into the syringe.
   - c Use your answers to parts a) and b) to calculate the M_r of the gas.
2. An experiment was carried out to find the M_r of a volatile liquid, hexane, C_6H_{14}.
   - a Calculate the M_r of hexane from its formula.
   - b The experimental value for the M_r of hexane was 87.2. Calculate the percentage difference between this value and the true value.

2.07 Molecular and empirical formulae

OBJECTIVES

already from GCSE, you know

- that the formula of a compound shows the numbers and types of atoms that are joined together to make the compound

and after this spread you should

- understand molecular and empirical formulae, and the relationship between them
- be able to calculate empirical formulae from data giving percentage composition by mass

Molecular formula

The **molecular formula** of a compound gives the *actual* number of atoms of each element in one molecule of the compound.

This is the same as the number of moles of each element in a mole of the compound. You will be familiar with several molecular formulae from your GCSE studies. For example

- H_2O is the molecular formula of water. It shows that each water molecule contains two hydrogen atoms and one oxygen atom.
- C_2H_5OH is the molecular formula of ethanol (the alcohol in beer, wine, and aftershave). It shows that each ethanol molecule contains two carbon atoms, six hydrogen atoms, and one oxygen atom.

Molecular formulae are used to write chemical equations and in chemical calculations. They are used even if the compound is an ionic substance that contains ions rather than molecules. For example

- NaCl is the molecular formula of sodium chloride (common table salt and an important industrial **raw material**). It shows that each mole of sodium chloride contains one mole of sodium ions and one mole of chloride ions.

Empirical formula

The **empirical formula** of a compound gives the *simplest whole number ratio* of atoms of each element in the compound. For example

- C_2H_6 is the molecular formula of ethane. But this can be simplified by dividing both numbers by two. This gives the empirical formula, which is CH_3.
- $C_6H_{12}O_6$ is the molecular formula of glucose. This can also be simplified, but this time by dividing the numbers by six. So its empirical formula is CH_2O.

Often the molecular and empirical formulae are the same. For example, H_2O cannot be simplified without having half an atom in the empirical formula. So the molecular and empirical formulae of water are the same.

Compounds may also have different molecular formulae but the same empirical formula. For example, the molecular formula of propene is C_3H_6 and the molecular formula of octene is C_8H_{16}. But both have the same empirical formula, CH_2.

Working out an empirical formula

An empirical formula can be worked out if you are given the **percentage composition** by mass of each element in a compound. Study this example.

Worked example for finding an empirical formula (1)

An oxide of sulfur contains 50% by mass of sulfur, and 50% by mass of oxygen. What is its empirical formula?

Step 1 Assume the total mass of the oxide was 100 g. This makes the mathematics easy, because the oxide will contain 50 g of sulfur and 50 g of oxygen.

Step 2 Divide each mass of each element by its relative atomic mass, A_r. This gives you the number of moles of each element.

Step 3 Divide each number of moles by the smallest number.

Step 4 Write out the empirical formula.

	S	O
mass (g)	50	50
n (mol)	50 ÷ 32.1 = 1.558	50 ÷ 16.0 = 3.125
divide by smallest number	1.558 ÷ 1.558 = 1	3.125 ÷ 1.558 = 2
write out the empirical formula	SO_2	

Molecular formulae from empirical formulae

You need to know the relative formula mass, M_r, of a compound to find its molecular formula from its empirical formula.

For example, the empirical formula of butane is C_2H_5 and the M_r of butane is 58.

The M_r of the empirical formula = (2 × 12) + (5 × 1) = 29.

The ratio of the two M_r values is 58 ÷ 29 = 2. So you have to multiply each number in the empirical formula by 2 to get the molecular formula. This is C_4H_{10}.

The method works even if you have more than two elements in the compound. Study this next example.

Worked example for finding an empirical formula (2)

A certain compound contains oxygen, and 43.4% Na and 11.3% C. What is its empirical formula?

Calculate the percentage of oxygen in the compound. This will be 100 – 43.4 – 11.3 = 45.3%.

Now carry on from step 1 as in the first example.

	Na	C	O
mass (g)	43.4	11.3	45.3
n (mol)	43.4 ÷ 23.0 = 1.89	11.3 ÷ 12.0 = 0.94	45.3 ÷ 16.0 = 2.83
divide by smallest number	1.89 ÷ 1.89 = 1.0	0.94 ÷ 1.89 = 0.5	2.83 ÷ 1.89 = 1.5
multiply each number by 2	2.0	1.0	3.0
write out the empirical formula	Na_2CO_3		

Notice how you had to add an extra stage and multiply all three numbers by two. The empirical formula must use only whole numbers. The last calculation step gets rid of decimal fractions. If you had 1.33 after dividing by the smallest number you would multiply by 3.

Check your understanding

1. a What is the difference between an empirical formula and a molecular formula?
   b What information do you need to find a molecular formula from an empirical formula?
2. A certain compound contains 51.5% Cu, 9.7% C, and 38.9% O.
   What is its empirical formula?
3. a A certain compound contains 85.7% C and 14.3% H.
   What is its empirical formula?
   b The M_r of the compound is 42.
   What is the molecular formula of the compound?

2.08 Balancing equations

OBJECTIVES

already from GCSE, you

- balance some simple chemical equations
- use state symbols

and after this spread you should

- be able to balance equations for unfamiliar reactions when you are given the reactants and products

State symbols

State symbols are often used to show the state of the reactants and products. They are four of them:

(s) means solid
(l) means liquid
(g) means gas
(aq) means aqueous solution (dissolved in water)

For example

$Na_2S_2O_3(aq) + 2HCl(aq) \rightarrow$
$2NaCl(aq) + S(s) + H_2O(l) +$
$SO_2(g)$

Take care not to confuse (l) and (aq). When solid sodium hydroxide is added to water, sodium hydroxide solution forms. Even though this mixture is a liquid, the sodium hydroxide itself is not a liquid. It has dissolved to make an aqueous solution.

No atoms are lost or made during chemical reactions. They are just re-arranged. Chemical equations show the formulae of the different substances involved in a reaction. They need to be balanced to show the relative amounts of each substance.

Background knowledge

In general, chemical equations are written like this:

$$\text{reactants} \rightarrow \text{products}$$

The **reactants** are the chemicals that react together, and the **products** are the chemicals made in the reaction. For example

$$HCl + NaOH \rightarrow NaCl + H_2O$$

HCl and NaOH are the reactants, and NaCl and H_2O are the products.

Balancing equations

You balance equations to get the same number of atoms of each element on both sides of the equation. Being able to balance equations is an important skill for all the Units in your chemistry studies. You must make sure that you can do it. Unless you are already good at this, take time to do a lot of practice.

Look at this equation:

$$S + O_2 \rightarrow SO_2$$

It is already balanced because you have one sulfur atom and two oxygen atoms on each side of the arrow.

Look at this equation:

$$Mg + HCl \rightarrow MgCl_2 + H_2$$

It is not balanced. You have one magnesium atom on each side of the arrow. But you have one hydrogen atom and one chlorine atom on the left, and two hydrogen atoms and two chlorine atoms on the right. *Never* change a chemical formula to balance an equation. Instead, adjust the number of molecules by writing numbers just to the left of the formulae.

This is the balanced equation:

$$Mg + 2HCl \rightarrow MgCl_2 + H_2$$

A number 2 has been written next to HCl. This means that there are now two hydrogen atoms and two chlorine atoms on the left of the arrow.

Odd or even?

If you have a more complicated equation to balance, it can be difficult to know where to start. Looking for odd and even numbers on each side of the arrow can help you get started.

Look at this unbalanced equation:

$$Na + H_2O \rightarrow NaOH + H_2$$

There is an even number (two) of hydrogen atoms on the left of the arrow and an odd number (three) of hydrogen atoms on the right. This tells you straightaway that the equation is not balanced. Here is how you can balance it.

Step 1 Make the number of hydrogen atoms on the right even by adding a number 2 to the NaOH.

$$Na + H_2O \rightarrow 2NaOH + H_2$$

Step 2 Count each atom to check if the equation is balanced. There is one sodium atom on the left and two on the right. So adjust the number of sodium atoms on the left by adding a number 2 to the Na.

$$2Na + H_2O \rightarrow 2NaOH + H_2$$

Step 3 Count each atom to check if the equation is balanced. There is one oxygen atom on the left and two on the right. So adjust the number of oxygen atoms on the left by adding a number 2 to the H_2O.

$$2Na + 2H_2O \rightarrow 2NaOH + H_2$$

Step 4 Count each atom to check if the equation is balanced. It is.

Fractions are allowed

It is acceptable to have fractions in your balanced equations. This often helps if the equation seems difficult to balance.

Look at this example:

$$C_2H_6 + 3\frac{1}{2}O_2 \rightarrow 2CO_2 + 3H_2O$$

The 3½ in the equation does not mean that you have three and a half oxygen molecules. The equation tells you that one mole of C_2H_6 would react with three and a half moles of oxygen, forming two moles of carbon dioxide and three moles of water.

Watch for groups of atoms

Sometimes you do not have to count all the atoms.

Look at this example:

$$Al + H_2SO_4 \rightarrow Al_2(SO_4)_3 + H_2$$

Notice how SO_4 appears on the left and on the right. It is much easier to treat it as one unit. Remember that the small 3 outside the brackets in $A1_2(SO_4)_3$ means that there are three SO_4 present. So you just need a 3 next to H_2SO_4 to get three SO_4 on the left, and a 2 next to the A1 to get two A1 on the left. Now see that $3H_2SO_4$ contains $3H_2$, so add a 3 by the H_2 on the right.

$$2Al + 3H_2SO_4 \rightarrow Al_2(SO_4)_3 + 3H_2$$

… balanced!

Check your understanding

1. Balance these equations:
   - a $Mg + O_2 \rightarrow MgO$
   - b $Na + O_2 \rightarrow Na_2O$
   - c $Al + Cl_2 \rightarrow AlCl_3$
   - d $Fe_2O_3 + CO \rightarrow Fe + CO_2$
   - e $C_2H_5OH + O_2 \rightarrow CO_2 + H_2O$
   - f $NH_3 + O_2 \rightarrow NO_2 + H_2O$
   - g $NH_4OH + H_3PO_4 \rightarrow (NH_4)_3PO_4 + H_2O$

2.09 Ionic equations

OBJECTIVES

already from AS Level, you can

- balance equations for unfamiliar reactions when you are given the reactants and products

and after this spread you should

- be able to write balanced ionic equations

Salts from neutralization reactions

You will remember from your GCSE studies that the salt formed in a neutralization reaction depends upon the metal in the alkali and the acid used. For example, potassium hydroxide and nitric acid form potassium nitrate (and water). Now you know why.

This is what you get if you write the ions out separately:

$K^+(aq) + OH^-(aq) + H^+(aq) + NO_3^-(aq) \rightarrow K^+(aq) + NO_3^-(aq) + H_2O(l)$

If you identify the spectator ions, you will see that they make potassium nitrate, KNO_3:

$\mathbf{K^+}(aq) + OH^-(aq) + H^+(aq) + \mathbf{NO_3^-}(aq) \rightarrow \mathbf{K^+}(aq) + \mathbf{NO_3^-}(aq) + H_2O(l)$

The chemical equations you have studied so far are **full equations**. They show the complete formulae for all the reactants and products. But reactions involving ionic compounds can be written as **ionic equations**. These focus on the ions that react together and ignore the ones that do not take part in the reaction.

Spectator ions

Here is the full equation for the reaction between hydrochloric acid and sodium hydroxide solution:

$$HCl(aq) + NaOH(aq) \rightarrow NaCl(aq) + H_2O(l)$$

Water is the only substance here that exists as molecules. The other three are ionic compounds, made from ions. Here is the equation again, this time with all the ions written out separately:

$$H^+(aq) + Cl^-(aq) + Na^+(aq) + OH^-(aq) \rightarrow Na^+(aq) + Cl^-(aq) + H_2O(l)$$

Notice that Na^+ and Cl^- ions appear on both sides of the equation. They separate in solution and do not take part in the reaction. Ions that do this are called **spectator ions**. They "watch" the reaction happening without reacting themselves.

You can leave out the spectator ions by cancelling them from the equation.

$$H^+(aq) + \cancel{Cl}^-(aq) + \cancel{Na}^+(aq) + OH^-(aq) \rightarrow \cancel{Na}^+(aq) + \cancel{Cl}^-(aq) + H_2O(l)$$

This gives you

$$H^+(aq) + OH^-(aq) \rightarrow H_2O(l)$$

This is an ionic equation. Notice that it is simpler than the full equation. It helps you to see which ions have actually taken part in the reaction. You may well have seen this particular equation during your GCSE studies. It is the equation for the reaction between any acid and alkali. This is the reaction that produces the heat in a neutralization reaction.

Using ionic equations

Ionic equations are useful when describing these types of reaction:

- neutralization
- precipitation
- displacement

In all cases, the total charge on each side of the equation must be the same.

Neutralization

You have already seen how to work out the ionic equation for the reaction between hydrochloric acid and sodium hydroxide solution. What about other acids and alkalis?

This is the full equation for the reaction between sulfuric acid and ammonia solution, which contains $NH_4OH(aq)$:

$$H_2SO_4(aq) + 2NH_4OH(aq) \rightarrow (NH_4)_2SO_4(aq) + 2H_2O(l)$$

Here it is with the individual ions written out separately:

$$2H^+(aq) + SO_4^{2-}(aq) + 2NH_4^+(aq) + 2OH^-(aq) \rightarrow$$
$$2NH_4^+(aq) + SO_4^{2-}(aq) + 2H_2O(l)$$

You can leave out the spectator ions by cancelling them from the equation:

$$2H^+(aq) + \cancel{SO_4^{2-}}(aq) + \cancel{2NH_4^+}(aq) + 2OH^-(aq) \rightarrow$$
$$\cancel{2NH_4^+}(aq) + \cancel{SO_4^{2-}}(aq) + 2H_2O(l)$$

This gives you

$$2H^+(aq) + 2OH^-(aq) \rightarrow 2H_2O(l)$$

This simplifies to

$$H^+(aq) + OH^-(aq) \rightarrow H_2O(l)$$

Notice that this is exactly the same ionic equation as the first example.

When lead(II) nitrate solution and potassium iodide solution are mixed, they immediately form a bright yellow precipitate of lead(II) iodide.

Precipitation

You may have come across precipitation reactions during your GCSE studies. They are particularly useful for identifying ions in solution. For example, silver nitrate solution is used to identify **halide ions** in solution (chlorides, bromides, and iodides). Coloured silver compounds form in the reaction. These are insoluble in water and so form a cloudy **precipitate**.

Here is an example of the full equation for a precipitation reaction:

$$AgNO_3(aq) + KCl(aq) \rightarrow AgCl(s) + KNO_3(aq)$$

Here it is with the individual ions written out separately:

$$Ag^+(aq) + NO_3^-(aq) + K^+(aq) + Cl^-(aq) \rightarrow AgCl(s) + K^+(aq) + NO_3^-(aq)$$

You can leave out the spectator ions by cancelling them from the equation:

$$Ag^+(aq) + \cancel{NO_3^-}(aq) + \cancel{K^+}(aq) + Cl^-(aq) \rightarrow$$
$$\cancel{NO_3^-}(aq) + \cancel{K^+}(aq) + AgCl(s)$$

This gives you

$$Ag^+(aq) + Cl^-(aq) \rightarrow AgCl(s)$$

Check your understanding

1. Write the balanced ionic equation corresponding to each of the following full equations:

   a $KOH(aq) + HCl(aq) \rightarrow KCl(aq) + H_2O(l)$

   b $2NaOH(aq) + H_2SO_4(aq) \rightarrow Na_2SO_4(aq) + 2H_2O(l)$

   c $BaCl_2(aq) + CuSO_4(aq) \rightarrow BaSO_4(s) + CuCl_2(aq)$

   d $Cu(s) + 2AgNO_3(aq) \rightarrow 2Ag(s) + Cu(NO_3)_2(aq)$

2. Balance these ionic equations:

   a $Mg^{2+}(aq) + OH^-(aq) \rightarrow Mg(OH)_2(s)$

   b $Al^{3+}(aq) + OH^-(aq) \rightarrow Al(OH)_3(s)$

   c $Fe(s) + Pb^{2+}(aq) \rightarrow Fe^{3+}(aq) + Pb(s)$

***Displacement reactions:** copper displaces silver from silver nitrate solution. Silver crystals coat the copper wire.*

$Cu(s) + 2Ag^+(aq) \rightarrow Cu^{2+}(aq) + 2Ag(s)$

2.10 Reacting masses

OBJECTIVES

already from GCSE, you know

- that in a reaction the mass of the products equals the mass of the reactants

already from AS Level, you can

- balance equations for unfamiliar reactions when you are given the reactants and products
- work out relative formula masses
- work out the mass of a substance from its relative mass and amount

and after this spread you should be able to

- calculate reacting masses from balanced equations

The French chemist Antoine Lavoisier established the *law of conservation of mass* in 1789. It means that the mass of the products made in a chemical reaction is always equal to the mass of the reactants. This is really useful because it lets you calculate the mass of product that could be made in a chemical reaction. In a similar way, you can also calculate the mass of reactants needed to make a certain mass of product. These are called reacting mass calculations.

Reacting masses - an example

Before reaction, the total mass of a flask of silver nitrate solution and a measuring cylinder of potassium chromate(VI) solution is 173.64g.

After mixing the reactants, silver chromate(VI) and potassium nitrate solution form. The total mass is still 173.64 g.

Step 1	Write a balanced equation	Fe_2O_3	+	3C	→	2Fe	+	3CO
Step 2	Write the M_r values underneath	160		12		56		28
Step 3	Multiply by the big numbers in the equation	160		3 × 12		2 × 56		3 × 28
Step 4	Convert to grams	160g		36g		112g		84g
Step 5	*Ignore 3C and 3CO because they were not in the question*							
Step 6	Divide by 160 to get 1 g of iron oxide	1g				$\frac{112}{160}$g		
Step 7	Multiply by 25 to get 25g of iron oxide	25g				$\frac{112}{160} \times 25$ = 17.5g		

The result: *17.5g of iron is formed by reducing 25g of iron(III) oxide.*

This method is simple but it does not explicitly use the idea of the mole. At AS Level you should show that you understand this idea. Also remember to use M_r values expressed to one decimal place.

Using moles

There is an easy way to do reacting mass calculations involving moles:

Step 1 Write the balanced equation.

Step 2 Underline the two substances mentioned in the question.

Step 3 Write down under each underlined substance the information you know about it.

Step 4 You will know two things about one of the underlined substances. Use this information to work out the number of moles.

Step 5 Use the balanced equation to work out the number of moles of the other underlined substance.

Step 6 Use the answer to step 5 to work out the mass of the other underlined substance.

This box shows an example.

What mass of iron(III) oxide is produced by the oxidation of 83.7g of iron?			
Step 1	$4Fe$	$+ \quad 3O_2 \quad \rightarrow$	$2Fe_2O_3$
Step 2	$\underline{4Fe}$	$+ \quad 3O_2 \quad \rightarrow$	$\underline{2Fe_2O_3}$
Step 3	83.7g $A_r = 55.8$		$M_r = 159.6$
Step 4	moles = mass ÷ A_r moles = 83.7 ÷ 55.8 moles = 1.50		
Step 5	4 mol Fe → 2 mol Fe_2O_3 1 mol Fe → 0.50 mol Fe_2O_3 1.50 mol Fe → 0.75 mol Fe_2O_3		
Step 6			mass = M_r × mol mass = (159.6 × 0.75) g mass = 119.7g

The result: *119.7g of iron(III) oxide is formed from 83.7g of iron.*

Kilograms and tonnes

You may be asked to do a reacting mass calculation involving kilograms or tonnes instead of grams. For example, over a million tonnes of ammonia is produced in the UK each year. You could do the calculation as usual without converting the mass from grams, remembering that the numbers in the masses represent kilograms or tonnes instead of grams. This will give you the correct answer but with incorrect working.

It is better to show the masses in grams in standard form. So 25kg is 25×10^3g and 25 tonnes is 25×10^6g. This way, your working out is correct. You can leave the $\times 10^3$ and $\times 10^6$ bits in the calculation right until the end, then convert back to kilograms or tonnes if you wish.

Check your understanding

1. Calcium carbonate is a raw material used in the extraction of iron from iron ore. It decomposes to form calcium oxide and carbon dioxide when heated:

   $CaCO_3(s) \rightarrow CaO(s) + CO_2(g)$

   a What mass of calcium oxide is obtained by the thermal decomposition of 20g of calcium carbonate?

   b What mass of calcium carbonate is needed to produce 2 tonnes of calcium oxide?

2. Sodium reacts vigorously with water to produce sodium hydroxide and hydrogen:

   $2Na(s) + 2H_2O(l) \rightarrow 2NaOH(aq) + H_2(g)$

   a What mass of sodium hydroxide is produced when 1g of sodium reacts completely with water?

   b What mass of hydrogen is produced when 50g of sodium reacts completely with water?

2.11 Percentage yield

OBJECTIVES

already from AS Level, you can

- calculate reacting masses from balanced equations

and after this spread you should be able to

- calculate percentage yield

Some of the product may be left behind during processes such as filtration.

Ammonia is used to produce fertilizers such as ammonium nitrate and ammonium phosphate.

The **yield** of a chemical reaction is the amount of product formed. You can work out the **theoretical yield** using a reacting mass calculation. This is the maximum possible amount of product from the reactants used. In practice you are unlikely to get the theoretical yield. The **actual yield** is usually much less than the theoretical yield. There are several reasons for this, including:

- the reactants may be impure
- the reaction may not go to completion
- some of the product may be left in the container
- it may be difficult to purify the product.

In a laboratory situation this might just be annoying, as you will have to use more of each reactant than you calculated. But in an industrial situation more raw materials and energy will be used, more waste will be produced, and it could cost a lot of money.

Calculating percentage yield

The **percentage yield** tells you how close to the theoretical yield you have got. The higher the percentage yield, the closer you are to the theoretical yield. Here is the equation for calculating percentage yield:

$$\text{percentage yield} = \frac{\text{actual yield}}{\text{theoretical yield}} \times 100$$

The percentage yield would be 100% if all the reactants were converted into products, and there were no losses during processes such as pouring and filtering.

The theoretical yield for a certain method to obtain copper(II) sulfate crystals is 2.0 g. The actual yield obtained was 1.8 g. What was the percentage yield?

$$\text{percentage yield} = \frac{1.8}{2.0} \times 100 = 0.9 \times 100 = 90\%$$

The Haber Process

Ammonia is manufactured from nitrogen and hydrogen using the Haber Process. Over one hundred million tonnes of ammonia is manufactured worldwide each year. About 85% of this is used to make fertilizers. Without the Haber Process it would be very difficult to produce enough food for everyone. Yet under typical conditions the percentage yield of ammonia is only about 15%. The reaction is reversible and does not go to completion:

$$N_2(g) + 3H_2(g) \rightleftharpoons 2NH_3(g)$$

Worked example

If the percentage yield of ammonia in the Haber Process is 15%, what mass of ammonia is produced from 6 tonnes of hydrogen?

$$N_2(g) + \underline{3H_2}(g) \rightleftharpoons \underline{2NH_3}(g)$$

mass of hydrogen = 6×10^6 g

M_r of hydrogen = 2.0

amount of hydrogen = $6 \times 10^6 \div 2.0 = 3 \times 10^6$ mol

mole ratio of $H_2:NH_3$ is 3:2, so theoretical yield of $NH_3 = 2 \times 10^6$ mol

M_r of ammonia = 17.0

theoretical yield of $NH_3 = 2 \times 10^6 \times 17.0 = 34.0 \times 10^6$ g

$$\text{percentage yield} = \frac{\text{actual yield}}{\text{theoretical yield}} \times 100$$

$$\text{rearrange equation: actual yield} = \frac{\text{percentage yield} \times \text{theoretical yield}}{100}$$

$$\text{mass of ammonia produced} = \frac{15 \times 34.0 \times 10^6}{100} = 5.1 \times 10^6 = 5.1\,\text{tonnes}$$

Limiting reactant

The reactants in a chemical reaction are often mixed in different proportions from the ones in the balanced equation. Where there are two reactants, this means that one of them will be in **excess**. It will not all be used up in the reaction. The other reactant is the **limiting reactant**. It will be completely used up in the reaction and so determines the theoretical yield.

For example, iron reacts with sulfur to produce iron(II) sulfide:

$$Fe(s) + S(s) \rightarrow FeS(s)$$

One mole of iron reacts with one mole of sulfur to produce one mole of iron(II) sulfide. If more than one mole of iron is mixed with one mole of sulfur, the iron will be in excess. The sulfur will be the limiting reactant. No matter how much excess iron is added, no more than one mole of iron(II) sulfide can be produced from one mole of sulfur.

Iron and sulfur react vigorously when heated to produce iron(II) sulfide.

Check your understanding

1. Ethanol, CH_3CH_2OH, is produced from glucose, $C_6H_{12}O_6$, by fermentation: $C_6H_{12}O_6 \rightarrow 2CH_3CH_2OH + 2CO_2$
   - a What is the theoretical yield of ethanol from 18 g of glucose?
   - b What is the percentage yield if the actual yield of ethanol is 1.2 g?
2. Calcium carbonate reacts with hydrochloric acid to produce calcium chloride, water, and carbon dioxide.
   - a Write the balanced equation for the reaction between calcium carbonate and hydrochloric acid.
   - b What mass of calcium carbonate is needed to produce 0.22 g of carbon dioxide?
   - c Limestone is impure calcium carbonate. A lump of limestone weighing 0.51 g produces 0.22 g of carbon dioxide when it is added to excess hydrochloric acid. Use your answer to part **b** to calculate the percentage purity of the lump of limestone.

2.12 Atom economy

OBJECTIVES

already from AS Level, you can

- calculate reacting masses and percentage yield from balanced equations

and after this spread you should be able to

- calculate percentage atom economy from a balanced equation

Chemical reactions with a low atom economy produce a lot of waste.

The **atom economy** of a chemical reaction is the proportion of reactants that are converted into useful products rather than waste products. Chemical processes that have a high atom economy are more efficient and produce less waste than processes with a low atom economy. This is important for the environment and for **sustainable development**. Efficient chemical processes help us to meet our needs without reducing the ability of future generations to meet their needs.

Calculating atom economy

The atom economy of a reaction is calculated using this equation:

$$\%\text{atom economy} = \frac{\text{mass of desired product}}{\text{total mass of reactants}} \times 100$$

For example, in a certain process 50 tonnes of reactants produce 30 tonnes of the desired product and 20 tonnes of waste by-product. The atom economy is:

$$\%\text{atom economy} = \frac{30}{50} \times 100 = 60\%$$

This process would be less efficient than one with an atom economy of 80%.

It is also possible to predict the atom economy of a process using the relative molecular masses of the reactants and products:

$$\%\text{atom economy} = \frac{\text{relative mass of desired product}}{\text{total relative masses of all reactants}} \times 100$$

You need to know:

- the balanced equation for the reaction
- the relative atomic masses or relative molecular masses of the products

Iron can be extracted from its ore using carbon monoxide in a blast furnace. What is the atom economy of this process?

Balanced equation is: $Fe_2O_3(s) + 3CO(g) \rightarrow 2Fe(l) + 3CO_2(g)$

A_r of Fe = 55.8

M_r of CO_2 = 44.0

Total relative mass of desired product = 2 × 55.8 = 111.6

Total relative masses of all products = (2 × 55.8) + (3 × 44.0) = 111.6 + 132.0 = 243.6

$$\%\text{atom economy} = \frac{111.6}{243.6} \times 100 = 45.8\%$$

Sometimes the relative masses of the reactants are easier to calculate than the relative masses of the products. Mass is conserved in a chemical reaction, so you can also work out atom economy like this:

$$\%\text{atom economy} = \frac{\text{relative mass of desired product}}{\text{total relative masses of all reactants}} \times 100$$

In the previous example, M_r of Fe_2O_3 is 159.6 and M_r of CO is 28.0.

The total relative masses of reactants = 159.6 + (3 × 28.0)
= 159.6 + 84.0 = 243.6

This is the same as the total relative masses of products.

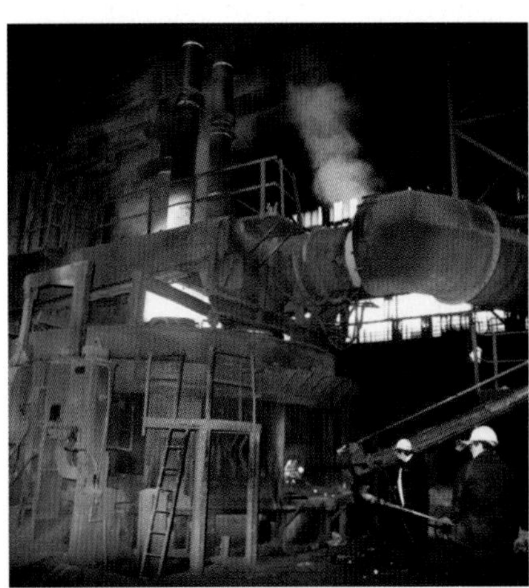

Iron is produced from iron ore in a blast furnace.

Comparing reactions

Atom economy is a useful way to compare different ways of making the same desired product. For example, hydrogen gas is used in the manufacture of margarine and as a fuel. It can be produced by reacting coal with steam:

$$C(s) + H_2O(g) \rightarrow H_2(g) + CO(g)$$

$$\%\text{ atom economy} = \frac{2.0}{2.0 + 28.0} \times 100 = \frac{2.0}{30.0} \times 100 = 6.67\%$$

Hydrogen can also be produced by reacting natural gas with steam:

$$CH_4(g) + H_2O(g) \rightarrow 3H_2(g) + CO(g)$$

$$\%\text{ atom economy} = \frac{(3 \times 2.0)}{(3 \times 2.0) + 28.0} \times 100 = \frac{30.0}{50.0} \times 100 = 17.6\%$$

Both processes produce carbon monoxide as a waste product. But the second process has a higher atom economy. It produces a higher proportion of useful product and is more efficient.

The atom economy of the first process can be improved by reacting the carbon monoxide waste product with steam:

$$CO(g) + H_2O(g) \rightarrow CO_2(g) + H_2(g)$$

This means that overall the equation for the reaction is:

$$C(s) + 2H_2O(g) \rightarrow 2H_2(g) + CO_2(g)$$

$$\%\text{atom economy} = \frac{(2 \times 2.0)}{(2 \times 2.0) + 44.0} \times 100 = \frac{4.0}{48.0} \times 100 = 8.33\%$$

There is an advantage to having carbon dioxide as the waste product instead of carbon monoxide. Carbon monoxide is insoluble and toxic but carbon dioxide dissolves easily in sodium hydroxide solution. This leaves pure hydrogen behind and reduces the cost of purifying the hydrogen. The overall energy required for the process is reduced because the reaction between carbon monoxide and steam is exothermic.

Hydrogen is used to 'harden' vegetable oils to make fats for margarine.

Check your understanding

1. A research chemist synthesizes 24 g of a desired product. If the total mass of products made in the reaction is 120 g, what is the atom economy of the reaction?
2. Ethanol, CH_3CH_2OH, and carbon dioxide are produced from glucose, $C_6H_{12}O_6$, by fermentation.
   - **a** Write the balanced equation for the reaction.
   - **b** Calculate the atom economy for producing ethanol by fermentation.
3. **a** Hydrogen can be produced by the electrolysis of brine, concentrated sodium chloride solution. The overall equation for the process is
   $2NaCl(aq) + 2H_2O(l) \rightarrow 2NaOH(aq) + Cl_2(g) + H_2(g)$
   Calculate the atom economy for producing hydrogen this way.
   - **b** Hydrogen can also be produced by the electrolysis of water. The overall equation for the process is:
     $2H_2O(l) \rightarrow 2H_2(g) + O_2(g)$
     Calculate the atom economy for producing hydrogen this way.
   - **c** Explain which of these two processes is the more efficient for producing hydrogen.
   - **d** Suggest why the reaction between steam and hydrocarbons such as natural gas is the most common large-scale process for the manufacture of hydrogen.

2.13 Sustainable development

OBJECTIVES

already from AS Level, you can

- calculate percentage yield
- calculate percentage atom economy from a balanced equation

and after this spread you should

- understand the importance of developing chemical processes with high percentage yield and atom economy

The human race is living beyond its means. Non-renewable resources such as fossil fuels and many metal ores are being used at increasing rates. Waste pollutes the environment. Land is being cleared to make way for homes, factories, and farms. These activities are putting a huge burden on the Earth. Sustainable development aims to meet our needs without harming the chances of future generations to meet their needs. One way to contribute to sustainable development is to develop chemical processes that use natural resources more effectively and produce less waste.

Waste

Waste is inevitable. Your body is releasing waste carbon dioxide, water vapour, and thermal energy while you are just sitting and reading this book. Waste from industrial processes can be harmful to the environment. Whenever a chemical process produces waste, it means that some of the raw materials have not been converted into useful products. Treating and disposing of waste involves the use of more chemicals, equipment, and energy. Reducing the amount of waste produced in the manufacture of chemicals is important for sustainable development.

Chemical waste must be disposed of carefully.

Percentage yield versus atom economy

A process may have a low percentage yield but a high atom economy. The Haber Process for making ammonia from nitrogen and hydrogen is like this. Only around 15% of the reactants are converted into products. But ammonia is the only product, so the atom economy is 100%. Unreacted reactants are recycled, increasing the overall efficiency.

A process may have a high percentage yield but a low atom economy. A high proportion of the reactants are converted into products, but a high proportion of these are not the desired product. A method used to manufacture epoxyethane in the last century was like that. Its yield of epoxyethane was about 80% but its atom economy was just 25.4%.

Epoxyethane is a raw material for making polyester polymers, widely used in clothing.

Epoxyethane manufacture

Epoxyethane, C_2H_4O, is used for the production of ethane-1,2-diol. This is used as antifreeze for car cooling systems, and to make polyester for fabrics and plastic bottles. Epoxyethane is manufactured from ethene, C_2H_4, a compound obtained from crude oil. It was originally manufactured in two steps:

Step 1 $C_2H_4 + Cl_2 + H_2O \rightarrow CH_2ClCH_2OH + HCl$

Step 2 $CH_2ClCH_2OH + Ca(OH)_2 \rightarrow C_2H_4O + CaCl_2 + 2H_2O$

Overall, this is:

$$C_2H_4 + Cl_2 + Ca(OH)_2 \rightarrow C_2H_4O + CaCl_2 + H_2O$$

The yield of this process was about 80% but its atom economy was just 25.4%. Overall, the mass of waste products was around three times the mass of the epoxyethane manufactured. This way of making epoxythane was abandoned by 1975.

The modern process is a one-step process. It is an addition reaction using silver as a catalyst:

$$C_2H_4 + \tfrac{1}{2}O_2 \rightarrow C_2H_4O$$

Some ethene is oxidized to carbon dioxide and water vapour, reducing the yield of epoxyethane to about 80%. But the atom economy is 100%. There is still some waste but very much less than the original process.

O

$H_2C—CH_2$

The C–O bonds in epoxyethane are strained, making it very reactive.

Different types of reaction

Catalysts allow different reactions to be used to make a particular product, with better yields and atom economies. For example, iron is a catalyst for making ammonia by the Haber Process, and silver is a catalyst for making epoxyethane.

Some types of reaction have better atom economies than others. The Haber Process and the manufacture of epoxyethane involve addition reactions. There is only one product in these reactions. The atom economy is 100% because all the reactant atoms end up in the desired product. Other types of reaction have atom economies of less than 100%. For example:

- *Substitution reactions* involve replacing a group of atoms on a molecule with a different group.
- *Elimination reactions* involve removing a group of atoms from a molecule.
- *Condensation reactions* involve two molecules joining together to make a larger molecule and a smaller one.

Check your understanding

1. Using the Haber Process for manufacturing ammonia as your example, explain why a process with a low percentage yield might be chosen.
2. Compare the two processes used to manufacture epoxyethane. To what extent is the percentage yield an unreliable indicator of the efficiency of these processes?
3. Explain how the use of catalysts in industrial processes can help to achieve sustainable development.

2.14 Reacting volumes

OBJECTIVES

already from AS Level, you can

- calculate reacting masses from balanced equations
- use the ideal gas equation, rearranging it if needed

and after this spread you should be able to

- calculate reacting volumes of gases

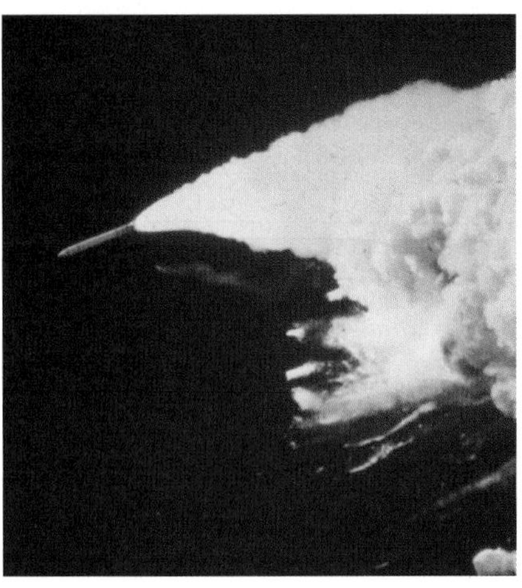

Hydrogen reacts explosively with oxygen.

The law of combining volumes

Joseph Gay-Lussac was a French chemist who proposed the *law of combining volumes* in 1809. This states that at constant temperature and pressure, the volumes of reacting gases and their gaseous products can be simplified as ratios of small whole numbers. For example, in the reaction between hydrogen and oxygen, two volumes of hydrogen react with one volume of oxygen to produce two volumes of water vapour. This is easily seen when you look at the balanced equation:

$$2H_2(g) + O_2(g) \rightarrow 2H_2O(g)$$

So 50 cm^3 of hydrogen and 25 cm^3 of oxygen would react completely to produce 50 cm^3 of water vapour, assuming the temperature and pressure stayed the same. Gay-Lussac's law was used two years later by Avogadro to develop the Avogadro principle: equal volumes of gases, at the same temperature and pressure, contain the same number of particles.

You can use the law of combining volumes to calculate the volumes of gases involved in chemical reactions.

Simple reacting volume calculations

You can do simple calculations if you know:

- the balanced equation
- the volume of at least one gas (reactant or product)
- that the temperature and pressure stay the same

Worked example

What volume of ammonia is produced when nitrogen reacts completely with 75 cm$_3$ of hydrogen?

$$N_2(g) + 3H_2(g) \rightarrow 2NH_3(g)$$

Mole ratio of ammonia to hydrogen is 2:3, so:

$$\text{volume of ammonia} = \frac{2}{3} \times 75 = 50\,\text{cm}^3$$

The limiting reactant can also be identified easily. This is the gas that is present in a volume that is too small to react completely with the other reactant gas.

100 cm^3 of nitrogen and 340 cm^3 of hydrogen react to produce ammonia.

1 *Which gas is the limiting reactant?*

$$N_2(g) + 3H_2(g) \rightarrow 2NH_3(g)$$

100 cm^3 of nitrogen will react with 300 cm^3 of hydrogen. There is 340 cm^3 of hydrogen, so nitrogen is the limiting reactant.

2 *What volume of ammonia will be produced if the nitrogen reacts completely?*

Mole ratio of ammonia to nitrogen is 2:1, so:

$$\text{volume of ammonia} = \frac{2}{1} \times 100 = 200\,\text{cm}^3$$

Molar volume

The molar volume, V_m, is the volume occupied by one mole of gas. It depends on the temperature and pressure, so it is important that these are stated. Unless you are told otherwise, you should assume that V_m is measured at standard temperature and pressure, STP.

- Standard temperature is 273.15 K (assumed to be 273 K at A Level).
- Standard pressure is 1×10^5 Pa.

The ideal gas equation is $pV = nRT$. It can be re-arranged to find volumes:

$$V = \frac{nRT}{p}$$

$$\text{So } V_m \text{ at STP} = \frac{1 \times 8.31 \times 273}{1 \times 10^5} = 0.0227\,\text{m}^3 \text{ or } 22.7\,\text{dm}^3$$

The molar volume can be used to work out the volume of gas produced in a reaction at STP.

Magnesium reacts vigorously with hydrochloric acid to produce hydrogen and magnesium chloride.

Worked example

What volume of hydrogen is produced at STP when 1.7 g of magnesium ribbon reacts completely with excess hydrochloric acid?

$$Mg(s) + 2HCl(aq) \rightarrow MgCl_2(aq) + H_2(g)$$

$$\text{amount of magnesium} = \frac{\text{mass}}{A_r} = \frac{1.7}{24.3} = 0.07\,\text{mol}$$

Mole ratio of $Mg:H_2$ is 1:1, so 0.07 mol of hydrogen is produced.

volume of hydrogen at STP = $0.07 \times 22.7 = 1.59\,\text{dm}^3$

Defining STP

Standard pressure used to be defined by IUPAC as 1 atmosphere. This is 1.01325×10^5 Pa. At this pressure and 273 K, V_m is 22.4 dm³. Many text books and websites still use this value for V_m. But since 1982 IUPAC has recommended that standard pressure should be 1×10^5 Pa.

V_m at STP is 22.7 dm³.

Check your understanding

1. Hydrogen and chlorine react to produce hydrogen chloride gas:

   $$H_2(g) + Cl_2(g) \rightarrow 2HCl(g)$$

   a What volume of chlorine will react completely with 100 cm³ of hydrogen?

   b What volume of hydrogen chloride will produced when 100 cm³ of chlorine reacts completely with hydrogen?

   c What volume of hydrogen chloride will be produced in the reaction between 50 cm³ of hydrogen and 100 cm³ of chlorine?

2. Airbags in cars contain sodium azide, NaN_3. In a crash, this is ignited electrically. It decomposes rapidly to produce sodium and nitrogen:

   $$2NaN_3(s) \rightarrow 2Na(s) + 3N_2(g)$$

   The nitrogen gas inflates the bag within about 20 ms.

   What mass of sodium azide is needed to produce 40 dm³ of nitrogen at STP?

Nitrogen is an ideal choice for airbags because it is non-toxic and non-flammable.

2.15 Concentrations of solutions

OBJECTIVES

already from AS Level, you can

- calculate reacting masses from balanced equations

and after this spread you should be able to

- calculate concentrations of solutions

Concentrations at GCSE

You may have calculated concentrations in your GCSE studies. If you did, you will have written the unit of concentration as mol/dm^3. This has the same meaning as mol dm^{-3}, because dm^{-3} means 1/dm^3. The calculations at A Level are identical to the ones you might have done at GCSE. Remember that the **solute** is the substance that dissolves in the **solvent**.

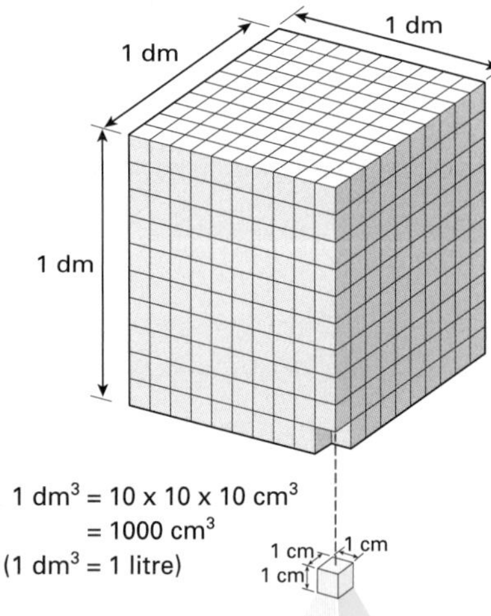

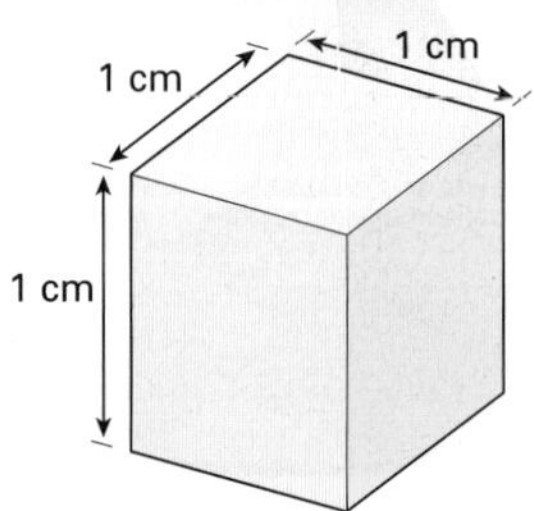

There are 1000 cm^3 in 1 dm^3.

In everyday life the concentrations of solutions may have several different units. Ingredients labels show units such as g per 100 ml, g/l, mg/ml, and % vol. Different units can be confusing and could lead to misunderstanding. In chemistry, the unit of concentration is mole per cubic decimetre, written as mol dm^{-3}. This often shortened to M, which is pronounced as molar. But note that strictly speaking, the word *molar* really means *divided by the number of moles*. The concentration of a solution measured in mol dm^{-3} is its **molarity**.

Calculating concentration

Concentration is calculated using this equation:

$$c = \frac{n}{V}$$

where c = concentration in mol dm^{-3}
n = amount of substance in mol
V = volume in dm^3

Example

What is the concentration of 0.25 dm^3 of a solution containing 0.50 mol of sodium hydroxide?

$$c = \frac{n}{V} = \frac{0.50}{0.25} = 2.0 \text{ mol dm}^{-3}$$

Note that the answer could also be written as 2.0 M.

A **standard solution** is a solution with an accurately known concentration. It is usual to use a **volumetric flask** when you make a standard solution. The most commonly used size is 250 cm^3. It is important to divide this volume by 1000 to convert it into dm^3 before using it in a calculation.

Worked example

0.050 mol of sodium carbonate was dissolved in water and added to a 250 cm^3 volumetric flask. The solution was made up to the mark with water. What was its concentration?

$$250 \text{ cm}^3 = 250 \div 1000 = 0.250 \text{ dm}^3$$

$$c = \frac{n}{V} = \frac{0.05}{0.25} = 0.200 \text{ mol dm}^{-3}$$

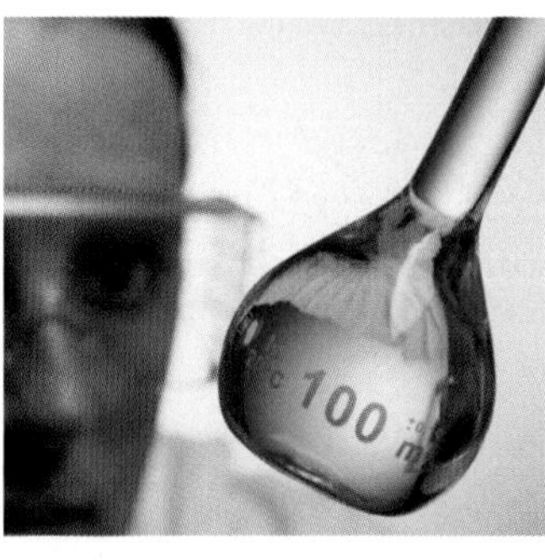

Volumetric flasks are used to make standard solutions.

Using concentrations

Find a required volume

You may need to calculate the volume of solution required in chemical reactions. The equation is rearranged like this:

$$V = \frac{n}{c}$$

Remember that the volume calculated will be in dm^3, not cm^3. You will need to multiply your answer by 1000 to get cm^3.

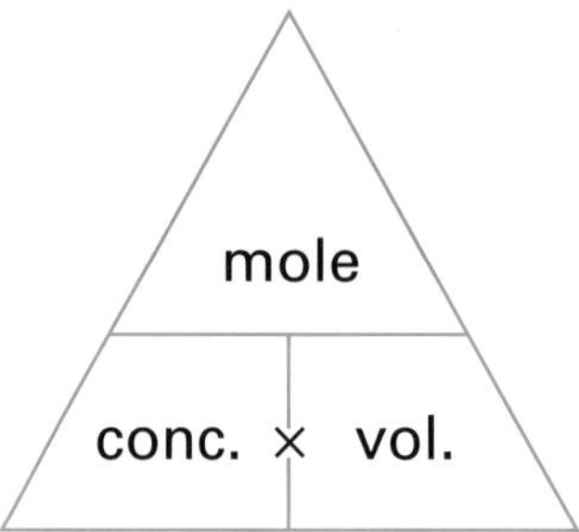

One way to remember how to work out concentration is to think "concentration is moles over voles". This "magic triangle" can help you see how the equation can be rearranged.

Worked example

What volume of 0.10 M hydrochloric acid contains 2.5×10^{-3} mol?

$$V = \frac{n}{c} = \frac{2.5 \times 10^{-3}}{0.10} = 2.5 \times 10^{-2}\,\text{dm}^{-3}$$

Note that 2.5×10^{-2} dm^3 is 25 cm^3 ($2.5 \times 10^{-2} \times 1000$).

Finding a required mass

You need to calculate the mass of solid required when you make up your own solutions. The calculation is in two parts.

1 Calculate the amount of substance needed from the intended volume and concentration

2 Calculate the mass of substance needed from its amount and M_r

Worked example

What mass of sodium hydroxide is needed to make 250 cm^3 of a 0.500 M solution?

1 Re-arrange the equation:

$$n = c \times V$$

$$= 0.500 \times \frac{250}{1000} = 0.125\,\text{mol}$$

2 M_r of NaOH = 23.0 + 16.0 + 1.0 = 40.0

$$\text{mass} = M_r \times n$$

$$= 40.0 \times 0.125$$

$$= 5.00\,\text{g}$$

Check your understanding

1. Calculate the concentration of these solutions:
   a 1.0 mol of solute in 2.0 dm^3 of solution
   b 2.0 mol of solute in 250 cm^3 of solution
   c 2.0 g of sodium hydroxide in 25 cm^3 of solution?
2. How many moles of solute are dissolved in
   a 0.5 dm^3 of a 2.5 mol dm^{-3} solution
   b 250 cm^3 of a 0.05 M solution?
3. What mass of anhydrous sodium carbonate, Na_2CO_3, is needed to make 250 cm^3 of a 0.10 M solution?

2.16 Acid–base titrations

OBJECTIVES

already from AS Level, you can

- calculate concentrations of solutions
- calculate reacting masses from balanced equations

and after this spread you should be able to

- calculate concentrations and volumes for reactions in solutions

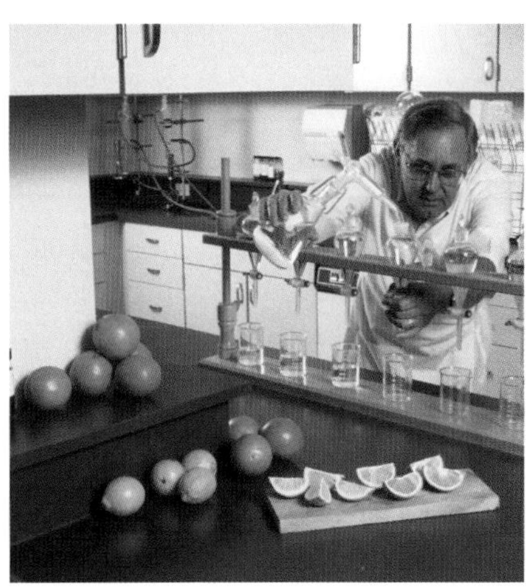

Titration is used to check batches of acidic liquids such as orange juice.

Titration is a method used to find the concentration of a reactant in solution. It is often used to find the concentration of an acid or an alkali. Titration is important for quality testing of vinegar, orange juice, and other acidic liquids. It can even be used to find the M_r of unknown acids and bases. The identity of the unknown substance may be found using this information.

Titration needs:

- a standard solution, a solution with an accurately known concentration
- a graduated **burette** to deliver precise volumes of one of the solutions
- a conical flask to hold a precise volume of the other solution, measured using a **transfer pipette**
- an indicator to reveal the endpoint of the titration

The **titrant** is the solution delivered by the burette. It is added to the other solution until the **equivalence point** is reached. This is the point when the two substances in solution have completely reacted with each other. It is revealed using an indicator in acid–base titrations. The change in the indicator's colour at the equivalence point shows that the endpoint of the titration has been reached. The **titre** is the volume of titrant needed to react completely with the reagent in the flask.

Calculations

You may have carried out titrations in your GCSE studies. If you did, you may also have calculated the amounts involved. The calculations at A Level are identical to the ones you might have done at GCSE. You will either be given the balanced equation, or the question will be about monoprotic acids and bases. A monoprotic acid has a general formula HA. Hydrochloric acid, HCl, and nitric acid, HNO_3, are monoprotic acids.

Indicators

Two acid–base indicators are commonly used in titrations. Phenolphthalein changes from pink in alkaline solution to colourless in acidic solution. This happens in the pH range 10.0 to 8.2, so phenolphthalein is particularly useful when the endpoint is above pH 7. Methyl orange changes from yellow to red in increasingly acidic solution. This happens in the pH range 4.4 to 3.2, so methyl orange is particularly useful when the endpoint is below pH 7.

Titration calculations

Titration calculations rely on the idea that you know only the volume of one of the solutions, but you know both the volume and the concentration of the other solution. With this information, and the balanced equation, you can work out the concentration of the first solution.

These are the steps needed in your calculations.

Step 1 Write the balanced equation for the reaction.

Step 2 Underline the two reacting substances in the titration.

Step 3 Write down under each underlined substance the information you know about it.

Step 4 You will know two things about one of the underlined substances. Use this information to work out the number of moles.

Step 5 Use the balanced equation to work out the number of moles of the other underlined substance.

Step 6 Use the answer to step 5 to work out the concentration of the other underlined substance.

Worked example

$25.0\,cm^3$ of sodium hydroxide solution is exactly neutralized by $22.5\,cm^3$ of $0.10\,mol\,dm^{-3}$ hydrochloric acid. What was the concentration of the sodium hydroxide solution?

Step 1	$HCl(aq)$	+	$NaOH(aq)$	→	$NaCl(aq)$	+	$H_2O(l)$
Step 2	$\underline{HCl}(aq)$	+	$\underline{NaOH}(aq)$	→	$NaCl(aq)$	+	$H_2O(l)$
Step 3	$22.5\ cm^3$ $0.10\ mol\ dm^{-3}$		$25.0\ cm^3$				
Step 4	moles = concentration × volume $n = 0.10 \times \frac{2.25}{1000}$ $n = 2.25 \times 10^{-3}\,mol$						
Step 5	1 mol HCl reacts with 1 mol NaOH $2.25 \times 10^{-3}\,mol$ HCl reacts with $2.25 \times 10^{-3}\,mol$ NaOH						
Step 6	concentration = $\frac{\text{amount}}{\text{volume}}$ $c = \frac{2.25 \times 10^{-3}}{25.0} \times 1000$ $c = 0.09\,mol\,dm^{-3}$						

The concentration of the sodium hydroxide solution was $0.09\,mol\,dm^{-3}$. Note that you do not have to use any M_r values in these calculations.

The factor of 1000

In step 4, you have to divide the volume in cm^3 by 1000 to convert it to dm^3. You then have to do the same thing in step 6. One easy way to tackle this is to remember that wherever the volume in cm^3 appears in these calculations, you need to put 1000 on the opposite side of the line. So if the volume is on the top, the 1000 goes on the bottom. If the volume is on the bottom, the 1000 goes on the top. If you forget the 1000 both times, you will get the right answer but with incorrect working out.

Check your understanding

1. $25.0\,cm^3$ of sodium hydroxide solution is exactly neutralized by $27.5\,cm^3$ of $0.20\,mol\,dm^{-3}$ hydrochloric acid. What was the concentration of the sodium hydroxide solution?
2. $25.0\,cm^3$ of an unknown monoprotic acid HA is exactly neutralized by $15.0\,cm^3$ of $0.50\,mol\,dm^{-3}$ sodium hydroxide solution.
   - **a** Write the equation for the reaction between HA and NaOH.
   - **b** What was the molarity of the unknown acid?
   - **c** The concentration of the acid was $10.95\,g\,dm^{-3}$. Calculate its M_r.
   - **d** Use your answer to part **c**) to suggest the identity of the acid.

2.17 Carrying out a titration

OBJECTIVES

already from AS Level, you can

- calculate concentrations and volumes for reactions in solutions

and after this spread you should be able to

- make up a volumetric solution
- carry out a simple acid–base titration

Making up a volumetric solution involves accurately weighing the solid, dissolving it, and then making the solution to the required volume in a volumetric flask. When you carry out practical work, make sure your equipment is clean and dry before use.

Weighing accurately

Weighing by difference gives you an accurate measure of the mass of solid used. These are the steps you need to take.

1. Use a spatula to transfer roughly the required amount of solid to a small empty beaker. You do not need to use a precise balance at this stage and you do not need to record the reading on the balance.
2. Weigh the beaker of solid on the precise balance and record the reading.
3. Carefully transfer the solid to another, larger beaker. This is the beaker you will use to dissolve the solid. Re-weigh the empty smaller beaker on the precise balance and record the reading. The difference between the two readings is the accurate mass transferred to the larger beaker.

Weighing boats

You might use a small glass sample bottle, plastic weighing boat, or piece of paper or aluminium foil instead of a small beaker.

Work carefully to avoid making a mess around the balance.

Weighing precautions

Take care that you

- record your two precise readings to the precision of the balance and include the unit. For example, you should record a reading from a balance precise to ± 0.01 g as 1.60 g not 1.6 g
- start again if the accurate mass transferred is outside the required range.

Making the solution

1. Use a wash bottle to add de-ionized water to the solid. Wash the sides of the beaker carefully. Make sure all the solid is wet before adding more water. Use a glass rod to stir until all the solid has dissolved.
2. Put a funnel in the volumetric flask and pour the solution through it. Use the wash bottle to wash the beaker and rod. Transfer all the washings to the funnel. Repeat this. Wash the funnel then remove it.
3. Add de-ionized water to the volumetric flask. As you get close to the mark, use a teat pipette to add water dropwise.
4. Stopper the volumetric flask and swirl the flask to mix its contents thoroughly. Label the flask with your name, the date, and the contents.

The bottom of the meniscus should touch the mark when viewed at eye level.

Dissolving precautions

Take care that you
- include all your washings
- make up to the mark on the volumetric flask (you will have to start all over again if you go over)
- mix the contents of the flask thoroughly

The titration

1. It is usual in an acid–base titration to put the acid into the burette. You should rinse the burette with de-ionized water, then with the acid. Make sure there are no bubbles in the tip. There is no need to start at exactly $0.00\,cm^3$, as long as you record the initial volume in your results table.

2. Use a transfer pipette to put an accurate volume of alkali solution into a conical flask. Use a pipette filler to do this safely and make sure that the liquid drains out under gravity. Hold the tip against the inside of the flask for a few seconds and do not blow the liquid out. If you make a mistake, start again, otherwise your titre will be inaccurate.

A pipette filler is used with a pipette to transfer the titrant safely.

3. Remember to add a few drops of indicator to the conical flask. This is usually phenolphthalein or methyl orange, depending on the combination of acid and base used. You can see the colour change more easily when the conical flask is on a white tile.

4. Repeat your titration until your get at least two concordant results. These are usually within $0.10\,cm^3$ of each other.

Watch carefully for the endpoint.

Titration precautions

Take care that you:
- set the burette vertically with the tip just inside the conical flask
- remove the funnel before each run
- swirl the conical flask during the run
- add the titrant dropwise near the endpoint
- wash the inside of the flask with de-ionized water just before the endpoint
- record the initial and final readings to two decimal places, with the last number being 0 or 5. For example, record 24.50 not 24.5, or 24.55 if the meniscus is halfway between 24.50 and 24.60
- repeat until you get at least two concordant results

	Run 1	Run 2	Run 3
Final volume (cm^3)	23.85	47.15	23.55
Initial volume (cm^3)	0.15	23.70	0.00
Titre (cm^3)	23.70	23.45 ✓	23.55 ✓
Mean titre (cm^3)	23.50		

Record your results in a table like this one. Tick the concordant titres used for your mean titre.

3.01 Ionic bonding

OBJECTIVES

already from GCSE, you know

- an ionic compound is a giant structure of ions
- ionic bonding is the strong force of attraction between oppositely charged ions

already from AS Level, you know

- how to represent the electron configurations of atoms and ions from $Z = 1$ to $Z = 36$ in terms of levels and sub-levels

and after this spread you should

- understand that ionic bonding involves attraction between oppositely charged ions in a lattice
- know the structure of the sodium chloride crystal

Chemical **bonds** are forces of attraction. Metallic bonds occur in metals and covalent bonds occur between two non-metals. In general, ionic bonds occur when a metal and a non-metal react to form a compound.

Towards a stable electron configuration

Metal atoms lose electrons to achieve a stable electron configuration. Non-metal atoms may gain electrons to do this. The stable electron configuration is often that of a noble gas. For example:

- A sodium atom loses one electron to form a sodium ion: $Na \rightarrow Na^+ + e^-$. The electron configuration of the sodium ion is $1s^2\ 2s^2\ 2p^6$. This is isoelectronic with neon.
- A chlorine atom gains one electron to form a chloride ion: $Cl + e^- \rightarrow Cl^-$. The electron configuration of the chloride ion is $1s^2\ 2s^2\ 2p^6\ 3s^2\ 3p^6$. This is isoelectronic with argon.

Some d block metals form ions with a noble gas configuration but most of them do not. For example, the electron configuration of the copper(II) ion, Cu^{2+}, is $1s^2\ 2s^2\ 2p^6\ 3s^2\ 3p^6\ 3d^9$.

Dot and cross diagrams

The formation of ions is often shown by a **dot and cross diagram**. Usually, only the electrons in the highest occupied energy level are shown in these diagrams.

For example, sodium chloride is formed when sodium and chlorine react together. Each sodium atom transfers one electron to a chlorine atom. The transferred electron is shown as a dot in the dot and cross diagram.

Cations and anions

Positively charged ions are called **cations**, pronounced 'cat-ions' and not 'cayshuns'. This is because positively charged ions would be attracted to the cathode, the negatively charged electrode, during **electrolysis**.

Negatively charged ions are called **anions**. This is because negatively charged ions would be attracted to the anode, the positively charged electrode, during electrolysis.

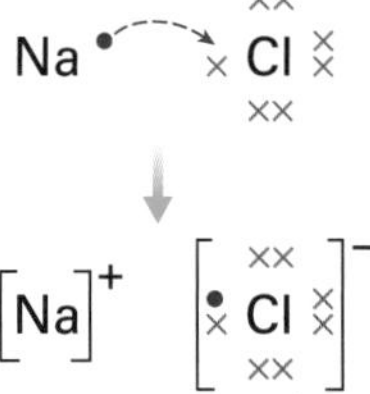

The 3s electron from a sodium atom transfers to the 3p sub-level of a chlorine atom, causing a sodium ion and a chloride ion to form.

The reaction between sodium and chlorine is very vigorous.

Remember that there is no difference between electrons shown with a dot and those shown with a cross. The dots and crosses just help you keep track of where the electrons have come from.

Ionic bonds and lattices

Compounds that are made up of ions are called **ionic compounds**. There is a strong **electrostatic** force of attraction between oppositely charged ions in an ionic compound. These forces are the **ionic bonds**. The ions in an ionic compound form a regular three-dimensional structure, called a **lattice**.

Each ion is surrounded by ions with the opposite charge. This structure is repeated very many times in an ionic compound, so it is a **giant structure**.

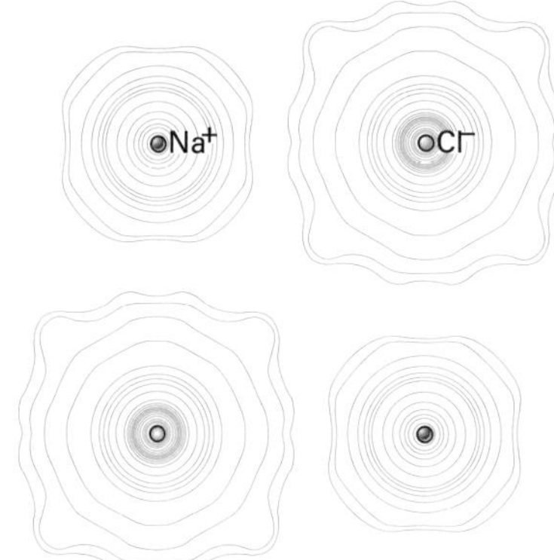

This is an electron density map of sodium chloride. The lines are like contours on a normal map but they show areas of equal electron density instead of equal height. You can see that the ions are distinct, with very little electron density between them.

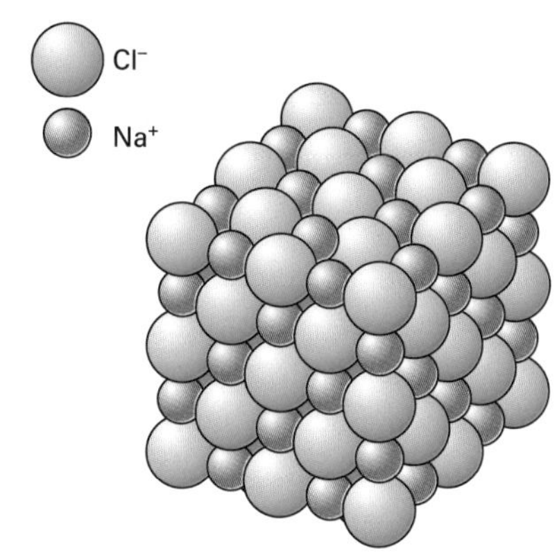

This is a space-filling diagram to show the lattice structure of sodium chloride. It shows the relative sizes of the ions but is difficult to draw.

Each sodium ion in the sodium chloride lattice is surrounded by six chloride ions. Each chloride ion is surrounded by six sodium ions. This is more easily seen if the ions are shown as simple circles with lines between them, as on the right.

Ionic formulae

Ionic compounds do not exist as molecules. So you cannot talk about the molecular formula of an ionic compound. Instead, a compound's ionic formula is the empirical formula of the compound. The formula for sodium chloride is NaCl. It tells you that there are equal numbers of sodium ions and chloride ions in the ionic lattice.

When you work out the formula for an ionic compound, remember that you need equal numbers of positive and negative charges. For example:

- magnesium oxide contains Mg^{2+} and O^{2-} ions, so its formula is MgO
- calcium chloride contains Ca^{2+} and Cl^- ions, so its formula is $CaCl_2$
- aluminium oxide contains Al^{3+} and O^{2-} ions, so its formula is Al_2O_3 (six positive charges and six negative charges)

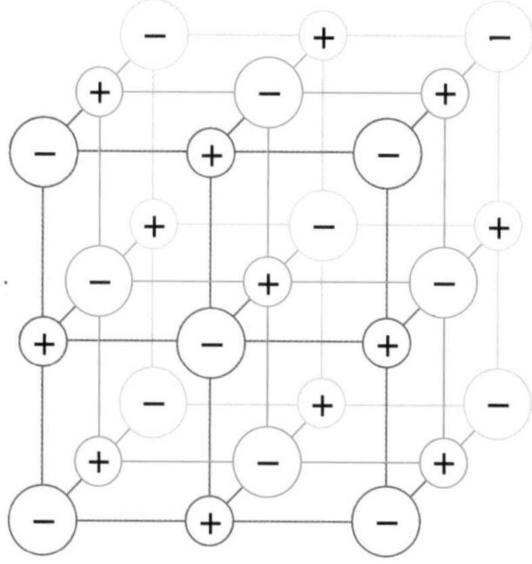

This is the easiest way to show the lattice structure of sodium chloride. You may need to draw or complete a diagram like this in the examination.

Check your understanding

1. Draw dot and cross diagrams for the formation of these ionic compounds:
   a NaCl **b** MgO **c** $MgCl_2$
2. What is an:
   a ionic bond **b** ionic compound **c** ionic lattice?
3. The diagram below represents part of the sodium chloride crystal. One sodium ion is shown by a + sign.
   a Copy and complete the diagram to show the correct positions of chloride ions with a – sign.
   b How many nearest sodium ions surround each chloride ion?

3.02 Metallic bonding

OBJECTIVES

already from GCSE, you know

- that atoms in metals are regularly arranged

and after this spread you should

- understand that metallic bonding involves a lattice of positive ions surrounded by delocalized electrons
- know the structure of magnesium

Metals consist of a lattice of positively charged metal ions surrounded by a 'sea' of delocalized electrons. The overall structure is electrically neutral.

Ionic lattices and metallic lattices

Take great care not to confuse ionic lattices and metallic lattices. In both cases there are charged particles, arranged in a regular way. In an ionic lattice the charged particles are oppositely charged ions. In a metallic lattice, the charged particles are metal cations surrounded by a 'sea' of delocalized electrons. You may lose marks if you mention the word 'ionic' in an answer to a question about metallic bonding.

Metallic bonds occur in metals. Depending on your GCSE examination board for science, you may already understand metallic bonding. But you probably do not know the structure of magnesium. You need to at AS Level.

A sea of electrons

Metal atoms are closely packed in the solid state. The highest occupied energy level of one metal atom overlaps with those from neighbouring atoms. As a result, electrons in the highest occupied energy level become **delocalized**. Delocalized electrons are not associated with any particular atom and they are free to move through a structure.

The atoms in metals are effectively ionized to form positive ions or cations. The electron configuration for magnesium atoms is $1s^2\ 2s^2\ 2p^6\ 3s^2$. The two 3s electrons are delocalized in each atom, leaving magnesium ions Mg^{2+} ($1s^2\ 2s^2\ 2p^6$). The metal cations are regularly arranged to form a lattice.

You would expect the positively charged metal ions to repel each other because they have like charges. But there are strong electrostatic forces of attraction between the positively charged metal ions and the negatively charged delocalized electrons. These forces are the metallic bonds.

Metal crystals

Metals are usually **malleable**. This means that they can be pressed or hammered into shape. Metals are usually **ductile**, too. This means that they can be stretched out to make a wire. Gold is the most malleable and ductile metal of all. Just 1 g of gold could be hammered into a sheet just 0.02 mm thick or stretched into a wire over 2 km long. Metals behave like this because layers of metal cations can slide over each other if sufficient force is applied. The metallic bonds stop the structure being completely disrupted unless large forces are applied.

If enough force is applied to a metal, layers of metal ions can slide past each other. Metals are malleable and ductile as a result.

You can think of metals as existing as crystals because the metal cations form regular lattice structures. Usually the whole of a piece of metal is not a single crystal. Instead, it is broken up into many crystals that are set at various angles to each other. This makes it more difficult for the layers of metal cations to slide over each other.

The properties of a metal can be deliberately altered by adding other metals to it to form an **alloy**. Some alloys are much tougher than the individual metals they contain. Commercial bronze is an alloy of copper and zinc. It resists stretching better than copper or zinc alone. Other alloys are more easily shaped than the pure metal.

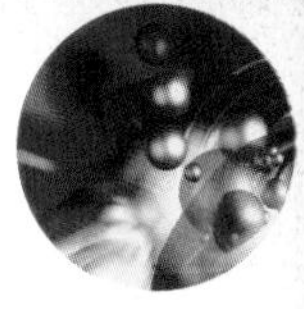

Drinks cans are made of aluminium alloyed with magnesium and manganese. This is more easily pressed into shape than pure aluminium. Slightly different alloys are used for the can and its top.

The structure of magnesium

Many metals have a structure called hexagonal close-packed. Magnesium and zinc have this structure. The hexagonal close-packed structure is one of the most efficient ways to arrange particles. 74.1% of the volume is filled with particles in the hexagonal close-packed structure, but just 52.4% of the volume is filled if you stack particles on top of each other in a simple cubic structure.

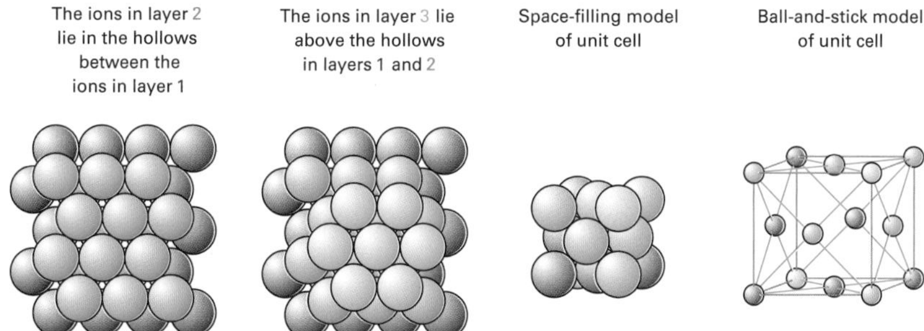
The ions in layer 2 lie in the hollows between the ions in layer 1
The ions in layer 3 lie above the hollows in layers 1 and 2
Space-filling model of unit cell
Ball-and-stick model of unit cell

The hexagonal close-packed structure of magnesium. The alternating layers of magnesium cations have been coloured differently so that you can see the structure more clearly.

There are two alternating layers in the hexagonal close-packed structure. You can build this using marbles in a box. Start the first layer with the marbles arranged as in the photograph. The second layer of marbles goes in the gaps between the marbles in the first layer. The marbles in the third layer are placed exactly over the marbles in the first layer.

You can use marbles in a tray to investigate the hexagonal close-packed structure.

Face-centred cubic

You get a different structure if the third layer of marbles is placed over the hollows in the first two layers. This is called cubic close-packed or face-centred cubic. Gold, copper, and aluminium have this structure. So does sodium chloride.

Check your understanding

1. Describe the bonding in metals.
2. Describe the structure of the magnesium crystal.
3. Explain why metals are usually malleable and ductile.

3.03 Covalent bonding

OBJECTIVES

already from GCSE, you know

- when atoms share pairs of electrons, they form strong covalent bonds
- some covalently bonded substances consist of simple molecules
- others, such as diamond, have giant covalent structures

and after this spread you should

- know that a covalent bond involves a shared pair of electrons
- know the structure of diamond

A shared pair of electrons

A **covalent bond** is a shared pair of electrons. The shared electrons are from the highest occupied energy level of each of the two atoms involved in the bond. Both nuclei are attracted to the shared pair of electrons. A lot of energy is needed to overcome this electrostatic force of attraction. This means that covalent bonds are strong. Substances containing covalent bonds form molecules.

Simple molecules and elements

Simple molecules are each made from a small number of atoms, joined by covalent bonds. Hydrogen has the smallest molecules like this. Each hydrogen molecule consists of two hydrogen atoms with a single covalent bond between them. A single hydrogen atom has the electron configuration $1s^1$. Remember that an s orbital may hold a maximum of two electrons. By sharing its electron, each hydrogen atom in a hydrogen molecule has the electron configuration $1s^2$. This is isoelectronic with helium, a noble gas.

You can use dot and cross diagrams to show the bonding in molecules. You can also show a covalent bond as a straight line between the two atoms. This sort of diagram is called a **displayed formula** or graphical formula.

H :

This dot and cross diagram shows the bonding in a hydrogen molecule, H_2.

H—H

This is the displayed formula for a hydrogen molecule, H_2.

Only electrons from the highest occupied energy level are involved in covalent bonding. For example, a single iodine atom has the electron configuration $1s^2\ 2s^2\ 2p^6\ 3s^2\ 3p^6\ 3d^{10}\ 4s^2\ 4p^6\ 4d^{10}\ 5s^2\ 5p^5$. This looks very complicated, but you only need to look at energy level 5. Notice that the 5p sub-level is not full – one more electron is needed to fill it. So iodine atoms can each form a covalent bond. An iodine molecule consists of two iodine atoms with a single covalent bond between them.

: I : I :

This dot and cross diagram shows the bonding in an iodine molecule, I_2. Notice that only the electrons in energy level 5 are shown.

I — I

This is the displayed formula for an iodine molecule, I_2.

Covalent compounds

Covalent compounds have covalent bonds between atoms of different elements. For example, hydrogen chloride molecules each consist of a hydrogen atom and a chlorine atom, joined by a single covalent bond. You can show the bonding in a covalent compound using dot and cross diagrams and displayed formulae, as on the left.

Notice that there are three pairs of electrons in the highest occupied energy level of the chlorine atom that are not involved in covalent bonds. These are called **lone pairs** of electrons. They are important for determining the shape of the molecule, and for some of the chemical properties of the compound.

This dot and cross diagram shows the bonding in a hydrogen chloride molecule, HCl.

H—Cl

This is the displayed formula for a hydrogen chloride molecule, HCl.

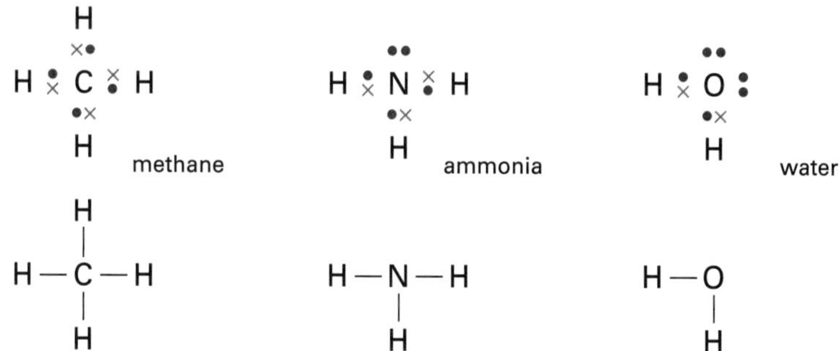

The methane molecule does not have a lone pair of electrons. The ammonia molecule has one lone pair of electrons and the water molecule has two.

Diamond

Diamond and graphite are **allotropes** of carbon. Allotropes of an element are in the same state but have different structures. The electron configuration of carbon is $1s^2\ 2s^2\ 2p^2$. Notice that the 2p sub-level is not full – four more electrons are needed to fill it. So carbon atoms can each form four covalent bonds. In diamond, each carbon atom is covalently bonded to four other carbon atoms. The four atoms that surround a particular carbon atom form the corners of regular four-sided shape, called a tetrahedron. This arrangement is a called a **tetrahedral** arrangement. This is repeated very many times, so diamond has a **giant covalent** structure, also called a **macromolecular** structure.

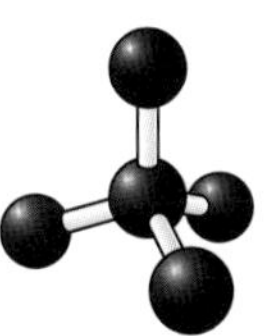

Carbon atoms have a tetrahedral arrangement in diamond.

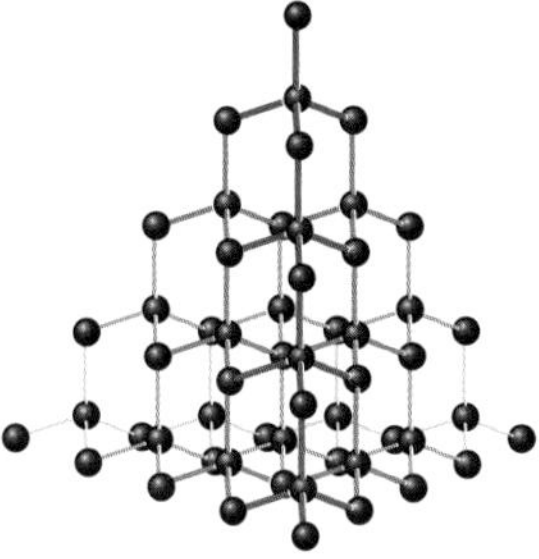

The structure of diamond. If you need to draw a diagram like this in the exam, draw at least five carbon atoms.

Cutting diamonds

Diamond is the hardest natural substance because of its structure and bonding. Industrial diamonds have flaws that make them useless for jewellery. They might be too small, oddly shaped, or coloured. But they make very tough cutting tools such as drills to cut through rock to search for oil, and tools to cut glass. If diamond is so hard, how are diamonds cut to make sparkling gemstones for jewellery?

Diamonds can be cut into shapes with many flat faces, designed to reflect light in an attractive way.

Diamond is stronger in some directions than in others. Expert diamond cutters can identify four crystal faces where diamond can be split in two or 'cleaved'.

If they get it wrong, the diamond could shatter when hit with the cutter's tool. Diamond-tipped tools are also used to saw diamonds and to polish them.

Double and triple bonds

Some atoms can share four electrons, forming a **double covalent bond**. Oxygen and carbon atoms can do this. Double bonds are represented by two parallel lines in displayed formulae. The displayed formula for O_2 is O=O, and the displayed formula for CO_2 is O=C=O.

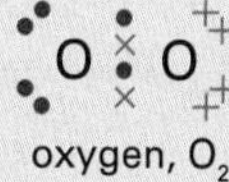

oxygen, O_2

carbon dioxide, CO_2

Dot and cross diagrams for oxygen and carbon dioxide.

Some atoms can share six electrons, forming a **triple covalent bond**. Carbon and nitrogen atoms can do this. Triple bonds are represented by three parallel lines in displayed formulae. The displayed formula for N_2 is N≡N.

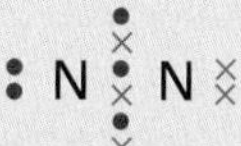

Dot and cross diagram for N_2.

Check your understanding

1. What is a covalent bond?
2. Draw dot and cross diagrams to show the bonding in:
   a Cl_2
   b CCl_4
   c ethane, $H_3C—CH_3$
   d ethene, $H_2C{=}CH_2$
3. Describe the structure and bonding in diamond. Include a diagram in your answer.

3.04 Co-ordinate bonding

OBJECTIVES

already from AS Level, you know

- that a covalent bond involves a shared pair of electrons

and after this spread you should

- be able to explain and recognize co-ordinate bonding in compounds and ions

You will recall that a covalent bond is a shared pair of electrons. Usually, the two atoms involved in a covalent bond each contribute one electron. Something different happens in a **co-ordinate bond**, also called a **dative covalent bond**.

Into a vacant orbital

In a co-ordinate bond, one of the two atoms contributes *both* of the shared electrons. The other atom does not contribute any electrons to the bond. Once formed, a co-ordinate bond behaves just the same as other covalent bonds.

For a co-ordinate bond to form, there must be

- a lone pair of electrons on one of the atoms and
- a vacant orbital on the other atom.

For example, a co-ordinate bond forms between ammonia and boron trifluoride. The ammonia molecule, NH_3, has a lone pair of electrons on its nitrogen atom. This is because nitrogen has five electrons in its highest occupied energy level, but only three of these are involved in covalent bonds with the hydrogen atoms. The boron trifluoride molecule, BF_3, has a vacant orbital on its boron atom. This is because boron has three electrons in its highest occupied energy level and all three are involved in covalent bonds with the fluorine atoms. The nitrogen atom contributes both electrons to form a co-ordinate bond with the boron atom.

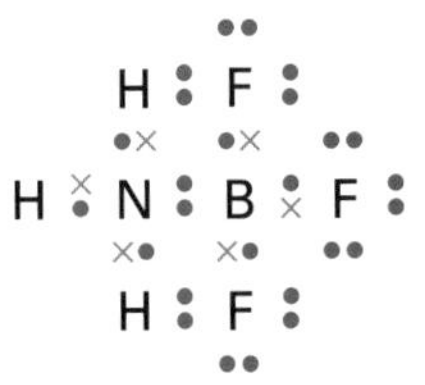

This dot and cross diagram shows the formation of a co-ordinate bond between ammonia and boron trifluoride.

When you draw a displayed formula involving a co-ordinate bond, the bonding line is replaced by an arrow. This points from the atom that donates the lone pair of electrons towards the atom with the vacant orbital.

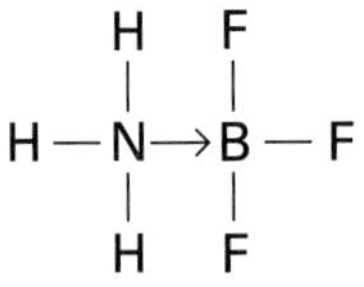

This is the displayed formula for the compound formed when ammonia reacts with boron trifluoride. You can also write the formula as $H_3N{\rightarrow}BF_3$.

The carbon monoxide molecule, CO, has a triple covalent bond between its two atoms. Two of these are covalent bonds in which the carbon atom and oxygen atom each contribute one electron. The other bond is a co-ordinate bond in which the oxygen atom contributes both electrons.

Aluminium chloride

You might expect aluminium chloride to be an ionic substance, as it consists of a metal and a non-metal. Instead, it is a covalent compound. Aluminium has three electrons in its highest occupied energy level, so it can form three covalent bonds. Again, you might expect the formula of aluminium chloride to be $AlCl_3$. But measurements of the relative molecular mass of solid aluminium chloride show that it exists as Al_2Cl_6. How can this be?

The chlorine atoms in $AlCl_3$ each have three lone pairs of electrons. The aluminium atom has a vacant orbital. A chlorine atom in one molecule of $AlCl_3$ contributes both electrons to form a co-ordinate bond with the aluminium atom in second molecule of $AlCl_3$. A chlorine atom from this molecule forms a co-ordinate bond with the first molecule of $AlCl_3$. A **dimer** forms, containing two co-ordinate bonds.

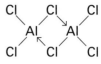

Group 7 and Group 3

Fluorine and chlorine are in Group 7 and both can donate a lone pair of electrons to form a co-ordinate bond. Boron and aluminium are in Group 3 and both can have vacant orbitals in their compounds. As a result, these elements are frequently involved in examination questions about co-ordinate bonding. For example, $H_3N{\rightarrow}BCl_3$, $H_3N{\rightarrow}AlCl_3$ and $H_3N{\rightarrow}AlF_3$ would have identical dot and cross diagrams to $H_3N{\rightarrow}BF_3$ (apart from the atom symbols of course).

The bonding in the aluminium chloride dimer.

Polyatomic ions

A **polyatomic ion** is a charged particle consisting of two or more atoms joined together by covalent bonds. The ammonium ion, NH_4^+, is a polyatomic ion formed from an ammonia molecule and a hydrogen ion, H^+. It contains four covalent bonds, and one of these is a co-ordinate bond. The ammonia molecule has a lone pair of electrons on its nitrogen atom. The hydrogen ion has a vacant orbital. This is because it is formed when a hydrogen atom loses its only electron. The nitrogen atom contributes both electrons to form a co-ordinate bond with the hydrogen ion.

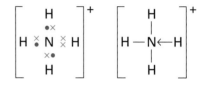

The bonding in an ammonium ion.

Ions from hydrogen

The electron configuration of the hydrogen atom is $1s^1$. A hydrogen atom can lose this electron to form a hydrogen ion, H^+. This is identical to a proton, and the terms *hydrogen ion* and *proton* are often used in similar ways. But note that it would be wrong to say that a nucleus contained hydrogen ions.

A hydrogen atom can also gain an electron to form a hydride ion, H^-. The electron configuration of the hydride ion is $1s^2$. These two electrons let the hydride ion form a co-ordinate bond with atoms that have a vacant orbital. For example, the compound $LiAlH_4$ contains the ion AlH_4^-. This has four covalent bonds with hydrogen atoms. One of these is a co-ordinate bond involving a hydride ion.

Check your understanding

1. What is a co-ordinate bond?
2. a Draw a dot and cross diagram to show the bonding in $H_3N{\rightarrow}AlCl_3$.
   b Explain what the arrow in the formula means.
3. The AlH_4^- ion contains a co-ordinate bond.
   a Draw a dot and cross diagram to show the bonding in this ion.
   b Draw its displayed formula.

3.05 Shapes of molecules

OBJECTIVES

already from AS Level, you know

- that a covalent bond is a shared pair of electrons

and after this spread you should

- understand the concept of bonding pairs of electrons as charge clouds
- be able to predict the shapes of simple molecules, and their bond angles

You can work out the shape of a simple molecule by counting the number of pairs of electrons around the central atom. Each pair of electrons acts as a cloud of negative charge. Pairs of electrons repel each other so that they are as far apart as possible. This keeps the force of repulsion to a minimum and gives each molecule a characteristic shape.

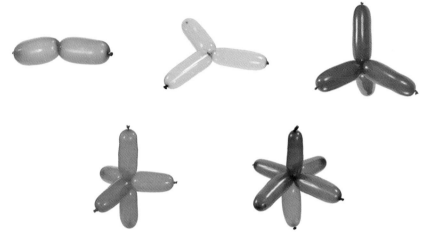

You can use balloons to model the five main molecular shapes.

Cl Be Cl

The beryllium chloride molecule is linear.

180°
Cl—Be—Cl

The bond angle is 180° in beryllium chloride and other linear molecules.

Linear molecules

The atoms in a **linear molecule** lie in a straight line. The beryllium chloride molecule $BeCl_2$ is linear. The two bonding pairs of electrons around the central beryllium atom repel each other as far as possible. This is achieved when the three atoms lie in a straight line, with a **bond angle** of 180°. Note that all diatomic molecules are linear.

Some molecules containing multiple bonds may also be linear. You can treat the double or triple bond as a single bond when working out the shape and bond angle. The carbon dioxide molecule O=C=O is linear. Each double bond acts as a single bond, so effectively there are only two charge clouds around the central carbon atom. Hydrogen cyanide molecules H—C≡N and ethyne molecules H—C≡C—H are two examples of linear molecules containing a triple bond.

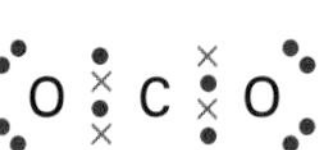

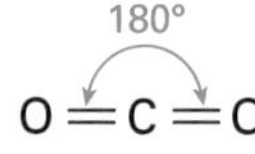

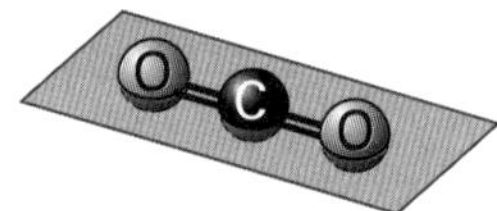

The carbon dioxide molecule is linear.

Trigonal planar molecules

Molecules in which the central atom is surrounded by three bonding pairs of electrons have a **trigonal planar** shape. The boron trichloride molecule BCl_3 is trigonal planar. The three chlorine atoms lie on a flat plane at the points of an equilateral triangle, with the boron atom at the centre. The bond angles are all 120°.

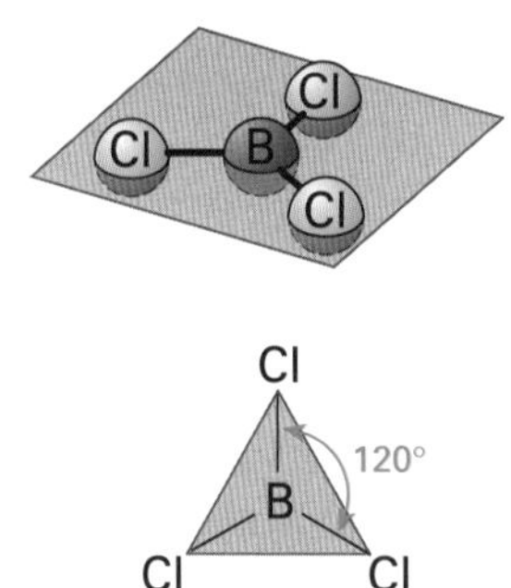

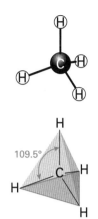

The boron trichloride molecule is trigonal planar.

The methane molecule is tetrahedral.

Tetrahedral molecules

Molecules in which the central atom is surrounded by four bonding pairs of electrons have a tetrahedral shape. The methane molecule CH_4 is tetrahedral. The carbon atom lies at the centre of a regular four-sided shape called a tetrahedron. You have already met this shape in the structure of diamond. The hydrogen atoms each lie at one of the corners. The bond angles are all 109.5° (109° 28′ to be precise).

Trigonal bipyramidal molecules

Molecules in which the central atom is surrounded by five bonding pairs of electrons have a **trigonal bipyramidal** shape. The phosphorus pentafluoride molecule PF_5 is trigonal bipyramidal. This is a more complex shape than the previous ones because there are two different bond angles.

A trigonal pyramid is a pyramid with a triangular base and three sloping sides. The trigonal bipyramidal shape consists of two of these pyramids joined base to base. The atoms around the middle are called **equatorial atoms**. They lie on a plane and have a bond angle of 120°, just as in a trigonal planar molecule. The atoms at each of the two points are called **axial atoms**. They lie in a straight line with the central atom. The bond angle between an equatorial atom and an axial atom is 90°.

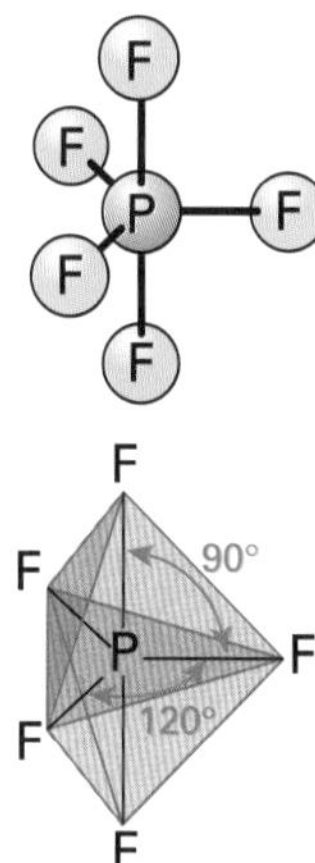

The phosphorus pentafluoride molecule is trigonal bipyramidal.

Octahedral molecules

Molecules in which the central atom is surrounded by six bonding pairs of electrons have an **octahedral** shape. The sulfur hexafluoride molecule SF_6 is an octahedral molecule. This looks like two classic Egyptian pyramids joined base to base. All the bond angle are 90°.

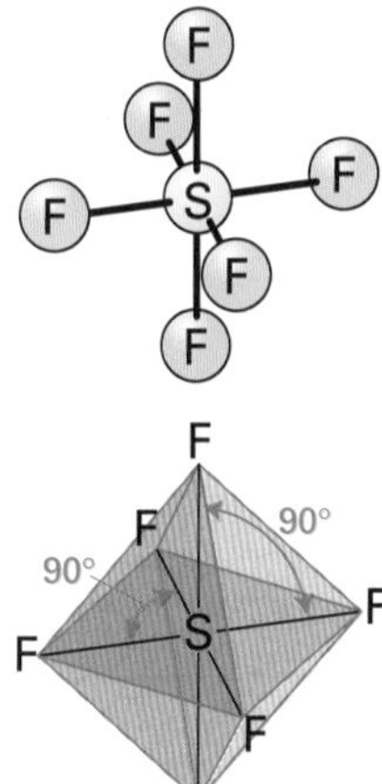

The sulfur hexafluoride molecule is octahedral.

Check your understanding

1. There are two bonding pairs of electrons around the central atom in F_2O. Predict the shape of this molecule and its bond angle.
2. Tetrachloromethane has the formula CCl_4. Predict its shape and bond angle. Hint: draw a dot and cross diagram first to see how many pairs of electrons there are around the central atom.
3. ICl_5 molecules are trigonal bipyramidal. What bond angles are present?
4. Predict the shape and bond angle in aluminium chloride, $AlCl_3$.
5. Give an example of an octahedral molecule and state its bond angle.

3.06 More shapes of molecules

OBJECTIVES

already from AS Level, you

- understand the concept of bonding pairs of electrons as charge clouds
- can predict the shapes of simple molecules and their bond angles

and after this spread you should

- be able to predict the shapes of simple molecules with lone pairs of electrons around the central atom, and their bond angles

There may be one or more lone pairs of electrons around a central atom in a molecule. These affect the shape and bond angles of the molecule.

Strength of repulsion

You will recall that each pair of electrons acts as a cloud of negative charge. Pairs of electrons repel each other so that they are as far apart as possible. Lone pairs of electrons are more compact than bonding pairs, so they have a greater repulsion. They reduce the bond angles in molecules.

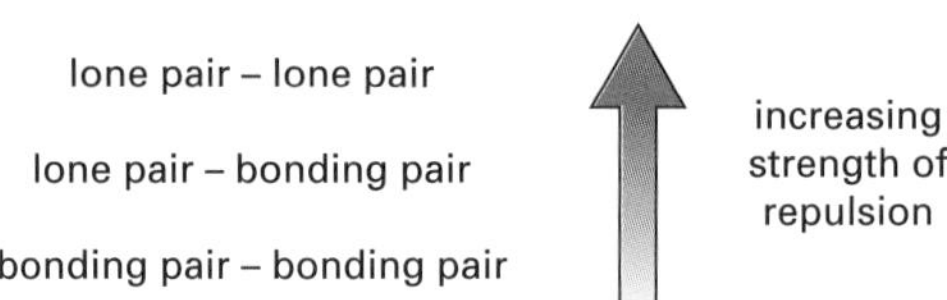

A comparison of the strength of repulsion by pairs of electrons.

Drawing molecules

Linear molecules and trigonal planar molecules are flat. It is easy to draw them on a sheet of paper. Other molecules extend in three dimensions. The bonds in a drawing can have one of three styles to give a sense of depth.

bond in the plane of the paper — bond coming out of the paper — bond going into of the paper

Different bond styles show if a bond comes towards you or away.

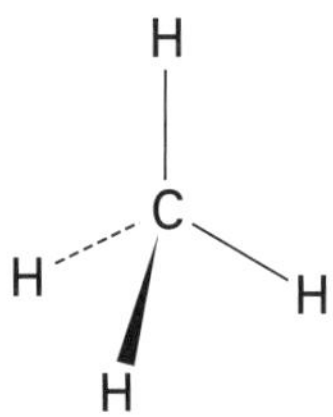

A diagram of a methane molecule. Two bonds are in the plane of the paper, the middle one comes out towards you and the left hand bond goes away from you.

Tetrahedral arrangements

The central carbon atom in methane has four bonding pairs of electrons. These repel each other equally to give a bond angle of 109.5°. Ammonia and water both contain lone pairs of electrons that reduce this angle.

Ammonia

Ammonia, NH_3, has four pairs of electrons around the central nitrogen atom. Three of these are bonding pairs and one is a lone pair. The lone pair of electrons repels the bonding pairs more strongly than the bonding pairs repel each other. This reduces the bond angle to 107°. The arrangement of the pairs of electrons is tetrahedral, but the molecule is **trigonal pyramidal**. It is the shape of a pyramid with a triangular base.

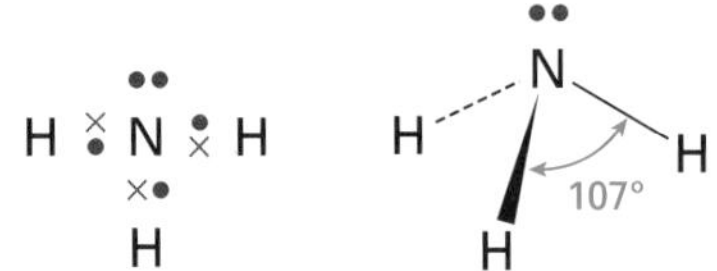

The bond angles in ammonia are 107° because of the presence of a lone pair of electrons.

Water

Water, H_2O, has four pairs of electrons around the central oxygen atom. Two of these are bonding pairs and two are lone pairs. The lone pairs of electrons repel the bonding pairs even more strongly than the one lone pair in ammonia does. This reduces the bond angle further to 104.5°. The arrangement of the pairs of electrons is tetrahedral, but the shape of the molecule is a **bent line**.

The bond angle in water is 104.5° because of the presence of two lone pairs of electrons.

Trigonal bipyramidal arrangements

The central phosphorus atom in phosphorus pentafluoride has five bonding pairs of electrons. These repel each other equally to give bond angles of 120° between equatorial atoms. Sulfur tetrafluoride and chlorine trifluoride both contain lone pairs of electrons. These can reduce this angle and alter the shape of the molecule.

Sulfur tetrafluoride

Sulfur tetrafluoride, SF_4, has five pairs of electrons around the central sulfur atom. Four of these are bonding pairs and one is a lone pair. The lone pair of electrons takes an equatorial position in the molecule. It reduces the bond angle to about 118° and alters the shape of the molecule. The arrangement of the pairs of electrons is trigonal bipyramidal, but the shape of the molecule is not.

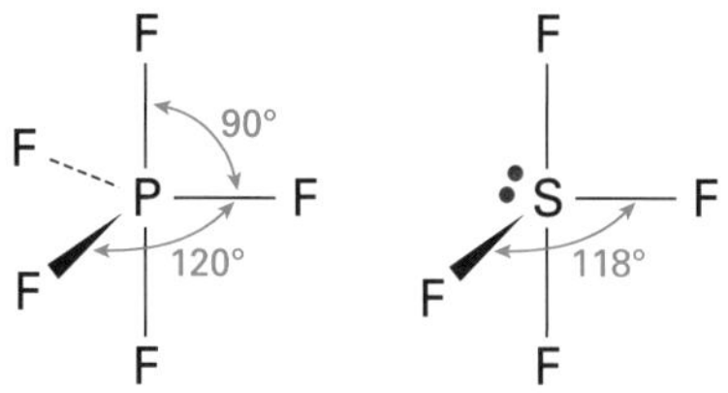

Phosphorus pentafluoride and sulfur tetrafluoride molecules.

Chlorine trifluoride

Chlorine trifluoride, ClF_3, has five pairs of electrons around the central chlorine atom. Three of these are bonding pairs and two are lone pairs. The lone pairs of electrons take equatorial positions in the molecule and alter its shape. The arrangement of the pairs of electrons is trigonal bipyramidal, but the molecule is T-shaped.

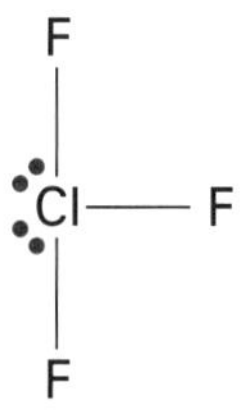

The chlorine trifluoride molecule.

Octahedral arrangements

The central sulfur atom in sulfur hexafluoride has six bonding pairs of electrons. These repel each other equally to give bond angles of 90°. Iodine pentafluoride contains lone pairs of electrons, which alter the shape of the molecule.

Iodine pentafluoride, IF_5, has six pairs of electrons around the central iodine atom. Five of these are bonding pairs and one is a lone pair. You can imagine that the molecule is 'missing' one of its atoms. The arrangement of the pairs of electrons is octahedral, but the shape of the molecule is not. It is **square pyramidal**, the shape of a classic Eqyptian pyramid.

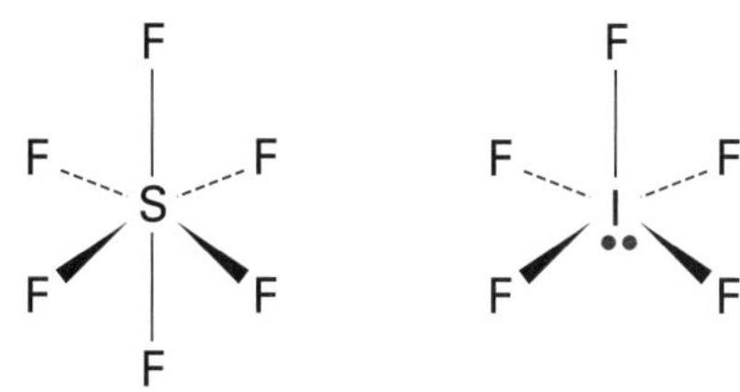

Sulfur hexafluoride and iodine pentafluoride molecules.

Check your understanding

1. Explain the differences between the bond angle in methane and those in ammonia and water.
2. Why do lone pairs of electrons repel more strongly than bonding pairs?
3. There are four pairs of electrons around the central phosphorus atom in phosphine, PH_3. Predict the shape and bond angle in phosphine.
4. There are five pairs of electrons around the central bromine atom in BrF_3. Draw the shape of this molecule and predict its bond angles.
5. Explain why NF_3 and BF_3 have different shapes, even though their formulae are similar.

3.07 Shapes of ions

OBJECTIVES

already from AS Level, you

- understand the concept of bonding pairs of electrons as charge clouds
- understand the effect of lone pairs of electrons on bond angles
- can predict the shapes of simple molecules and their bond angles

and after this spread you should

- be able to predict the shapes of ions and their bond angles

You will recall that a polyatomic ion is a charged particle consisting of two or more atoms joined together by covalent bonds. For example, the ammonium ion NH_4^+ is a polyatomic ion. Just as molecules have shapes, polyatomic ions have shapes, too. Working out their shapes is more complicated because you have to take their charges into account.

A general method for working out shapes

If the central atom is only bonded to atoms that can form single bonds, for example hydrogen and the elements of Group 7, there is a simple method for working out the shape.

1. Find the group number of the central atom.

   Add the number of atoms around the central atom to the group number.
2. If the ion is negatively charged, add the number of charges to the number from step 1.

   If the ion is positively charged, subtract the number of charges from the number from step 1.
3. Divide the number from step 2 by 2. This gives the number of electron pairs around the central atom, letting you work out the arrangement of the electron pairs:
   - 2 pairs = linear
   - 3 pairs = trigonal planar
   - 4 pairs = tetrahedral
   - 5 pairs = trigonal bipyramidal
   - 6 pairs = octahedral
4. Lone pairs alter the basic shape and bond angles. The number of lone pairs is the number from step 3 minus the number of atoms around the central atom.

This works for molecules with single bonds, too. Just remember not to add or subtract anything at step 2.

Here are three examples.

The NH_4^+ ion

1. Group number of central atom N, plus the number of atoms around it

   $5 + 4 = 9$
2. Adjust for single positive charge

   $9 - 1 = 8$
3. Divide by 2

   $8 \div 2 = 4$

   4 pairs give a *tetrahedral* arrangement of electron pairs.
4. The number of lone pairs

   $4 - 4 = 0$

There are no lone pairs, so the NH_4^+ ion has a *tetrahedral* arrangement of atoms, with bond angles of 109.5°.

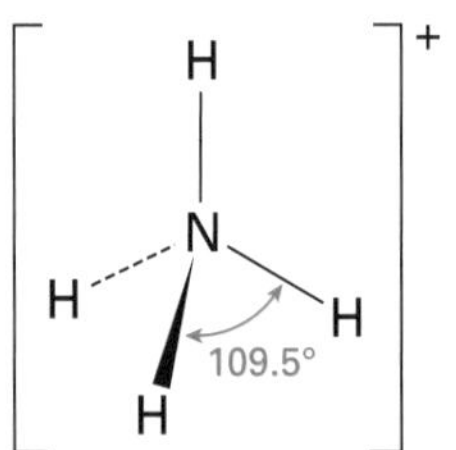

The bond angles in the ammonium ion are all 109.5°.

The I_3^- ion

This one is slightly sneaky. Treat it as II_2^-.

1. Group number of central atom I, plus the number of atoms around it $7 + 2 = 9$
2. Adjust for single negative charge $9 + 1 = 10$
3. Divide by 2 $10 \div 2 = 5$

   5 pairs give a *trigonal bipyramidal* arrangement of electron pairs.

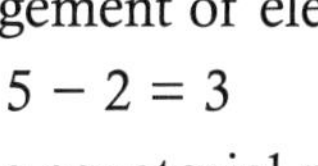

4. The number of lone pairs $5 - 2 = 3$

There are three lone pairs. These occupy the equatorial positions, so the I_3^- ion has a *linear* arrangement of atoms, with bond angles of 180°.

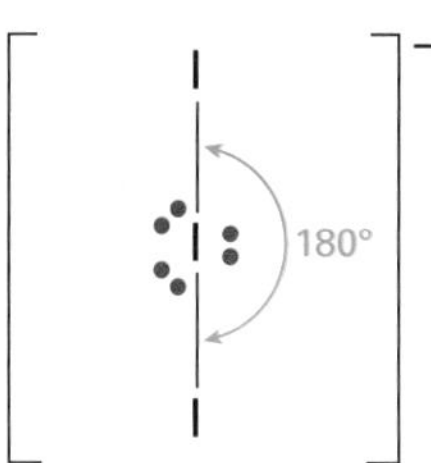

The bond angle in the I_3^- ion is 180°.

The ICl_4^- ion

1. Group number of central atom I, plus the number of atoms around it $7 + 4 = 11$
2. Adjust for single negative charge $11 + 1 = 12$
3. Divide by 2 $12 \div 2 = 6$

   6 pairs give an *octahedral* arrangement of electron pairs.

4. The number of lone pairs $6 - 4 = 2$

There are two lone pairs. The ICl_4^- ion has a *square planar* arrangement of atoms, with bond angles of 90°.

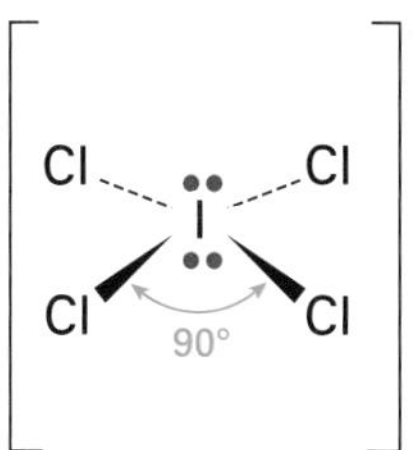

The bond angles in the ICl_4^- ion are all 90°.

Multiple bonds

If the central atom is bonded to oxygen or another atom that can form multiple bonds, you will have to draw a dot and cross diagram. This lets you work out the number of bonding pairs and lone pairs around the central atom. Once you know this, you can predict the shape and bond angles of the ion. You can assume that a multiple bond behaves like a single bond.

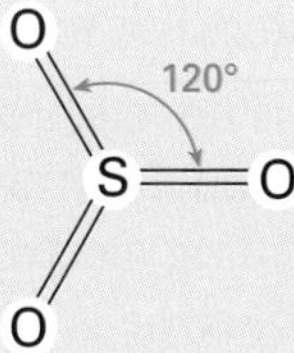

The bond angles are all 120° in the sulfur trioxide molecule.

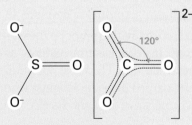

Delocalization of electrons from the C=O bond makes the bond angles all 120° in the carbonate ion.

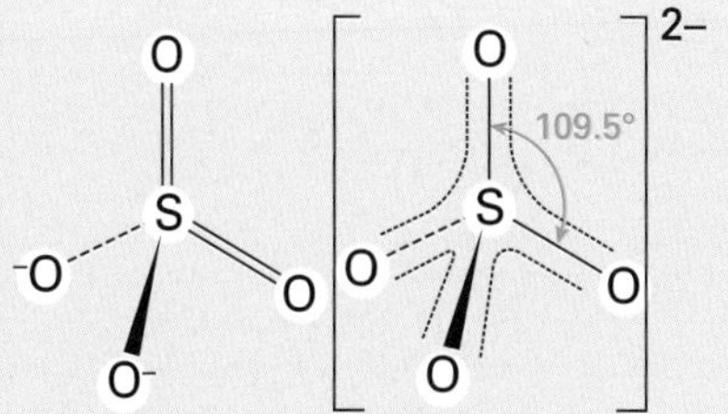

Delocalization of electrons in the S=O bonds makes the bond angles all 109.5° in the sulfate ion.

Check your understanding

1. Predict the shape of the NH_2^- ion and the bond angle.
2. Predict the arrangement of electrons pairs around the central atom in the ICl_4^+ ion and the bond angles.
3. Predict the shape of the PF_6^- ion and the bond angle.
4. Explain why the NH_4^+ ion and the BF_4^- ion have the same shape, even though they have opposite charges.

Science @ Work

VSEPR

Working out the shapes of molecules and ions using the idea of electron pairs repelling each other as far as possible is called the **VSEPR** theory. This stands for *valence shell electron pair repulsion*. The **valence shell** of an atom refers to the energy level containing the electrons that take part in covalent bonding.

3.08 Electronegativity

OBJECTIVES

already from AS Level, you know

- that a covalent bond involves a shared pair of electrons

and after this spread you should

- know the definition of electronegativity
- understand that the electron distribution in a covalent bond may not be symmetrical

A covalent bond is a shared pair of electrons, but are these electrons shared equally between the two bonded atoms? The answer is yes if the two atoms are of the same element. In molecules such as H_2 and Cl_2 the bonding electrons are shared equally. But not if the two bonded atoms are different elements. One of the bonded atoms has more of a share of the bonding electrons than the other. This happens if the two atoms have a different **electronegativity**.

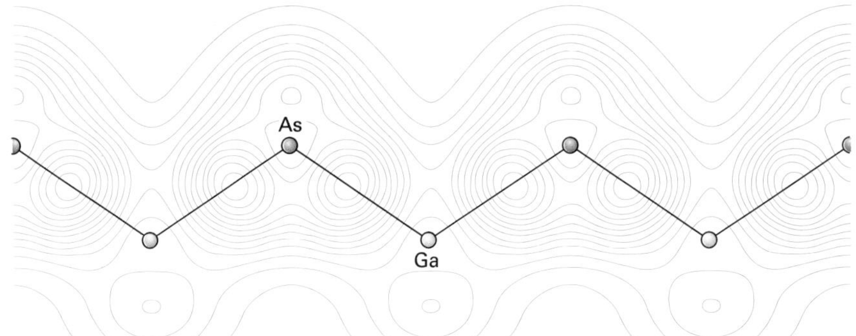

This is an electron density map of gallium arsenide. You can see that the peak of electron density in the As–Ga bond is slightly closer to the arsenic atom. The bonding pair of electrons is not shared equally between the two different atoms.

Electronegativity

Electronegativity is the power of an atom to withdraw electron density from a covalent bond. If the two bonded atoms have identical electronegativities, the bonding electrons will be shared equally. This is what happens in molecules such as H_2 and Cl_2. But if the two bonded atoms have different electronegativities, the more electronegative element will have a greater share of the bonding electrons. This is what happens in gallium arsenide. It happens in many other molecules, too.

Electronegativity cannot be measured directly: there is no 'electronegativity meter'. Instead, electronegativity must be calculated, and there are several ways to do this. The most common way gives the **Pauling electronegativity scale**, named after the scientist who first devised it. Fluorine is the most electronegative element. Its electronegativity value is 4.0 on the Pauling scale.

H 2.2							He –
Li 1.0	Be 1.6	B 2.0	C 2.5	N 3.0	O 3.4	F 4.0	Ne –
Na 0.9	Mg 1.3	Al 1.6	Si 1.9	P 2.2	S 2.6	Cl 3.2	Ar –
K 0.8	Ca 1.0					Br 3.0	Kr 3.0
Rb 0.8						I 2.7	Xe 2.6

In general, electronegativity increases from left to right across a period, and as you go up each group.

Notice that most of the Group 0 elements do not have electronegativity values. This is because they do not form covalent bonds with other atoms. Enough compounds of krypton and xenon have been discovered to make it possible to work out their electronegativity values.

Polar bonds

When the bonding pair of electrons is not shared equally, the bond is a **polar bond**. The more electronegative atom gains a partial negative charge, shown by the symbol δ−. The less electronegative atom gains a partial positive charge, shown by the symbol δ+.

For example, the hydrogen chloride molecule contains a polar bond. Chlorine is more electronegative than hydrogen, so it withdraws electron density from the covalent bond. Chlorine has a partial negative charge δ−, and hydrogen has a partial positive charge δ+. You can write the displayed formula of hydrogen chloride like this: $H^{\delta+}$—$Cl^{\delta-}$. The molecule has a **dipole**, a pair of separated charges of opposite signs.

δ− and δ+

The symbol δ is the Greek lowercase letter 'delta'. Upper case (capital) delta is written as Δ.

Electronegativity scales

The Pauling electronegativity scale is named after Linus Pauling, one of the last century's prominent scientists. Pauling was awarded the 1954 Nobel Prize in Chemistry for his research into the nature of chemical bonds and the structure of complex molecules. He also won the 1962 Nobel Peace Prize for his efforts to stop nuclear weapons testing.

Linus Pauling (1901–1994).

Pauling wrote a scientific paper in 1932 that showed a way to calculate electronegativity values. His method relied on the strengths of different covalent bonds. He looked at the strength of the bonds in two molecules X—X and Y—Y (X and Y represent the two different elements). He noticed that the mean strength of these two bonds was always less than the strength of the X—Y bond. For example, the bond strength for H—H is 436 kJ mol–1 and for Cl—Cl it is 243 kJ mol–1. The mean of these two values is 339.5 kJ mol^{-1} but the bond strength for H—Cl is 432 kJ mol^{-1}.

Pauling argued that the extra strength came from electrostatic attraction between the partially charged atoms. He had a way to calculate *differences* in electronegativity values. To get a scale of individual values, he needed a reference element. Fluorine, the most electronegative element, was given an electronegativity value of 4. This number is big enough to avoid any negative values.

Pauling's scale has been amended over the years to take into account more accurate measurements, and developments such as the discovery of argon and krypton compounds. The electronegativity of hydrogen was revised from 2.1 to 2.2 in 1961. The electronegativity of fluorine is now 3.98 on the scale, but 4.0 will do for most purposes.

Pauling's scale does not seem to work so well for complex molecules or where there is a big difference in electronegativity. Other electronegativity scales have been developed. For example, Robert Mulliken developed one in 1934. The Mulliken electronegativity scale contains values worked out by comparing the energy needed for an atom to lose an electron (its ionization energy) with the energy released when it gains an electron (its electron affinity). It has electronegativity values for all the Group 0 elements. But unlike the Pauling scale, its values have units instead of just being numbers.

Check your understanding

1. Define electronegativity.
2. a Which is the most electronegative element?
   b What is a polar bond?
   c Explain the meaning of the symbols δ– and δ+.
3. Describe the basis of the Pauling electronegativity scale.

3.09 Polar molecules and polarized ions

OBJECTIVES

already from AS Level, you know

- the definition of electronegativity
- that the electron distribution in a covalent bond may not be symmetrical

and after this spread you should

- know that some molecules may be polar
- know that covalent bonds between different elements will be polar to different extents

If a molecule contains polar bonds it may or may not be a polar molecule.

Non-polar molecules

Carbon dioxide contains two polar bonds yet it is not a polar molecule. It is a **non-polar** molecule. You can write its displayed formula as shown in the diagram. Notice that all three atoms lie in a straight line. This means that the two dipoles cancel each other out because they act in opposite directions. There is no net dipole and the molecule is non-polar.

$$\overset{\delta-}{O}=\overset{2\delta+}{C}=\overset{\delta-}{O}$$

The displayed formula for carbon dioxide showing its partial charges.

A similar situation happens in boron trifluoride, BF_3. The three fluorine atoms are arranged at 120° to each other around a central boron atom. Fluorine is more electronegative than boron, so it withdraws electron density from the covalent bond. The B—F bond is polar and you can write the displayed formula as shown in the diagram. The three dipoles cancel each other out. There is no net dipole and the molecule is non-polar.

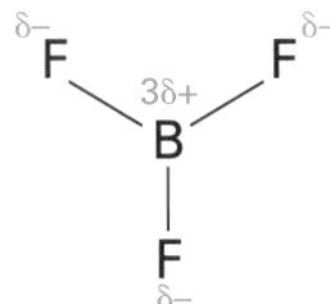

The displayed formula for boron trifluoride showing its partial charges.

Polar molecules

A molecule will be polar if

- it contains at least one polar bond and
- it has a net dipole.

The water molecule is polar because it has two polar bonds and a net dipole. It has a net dipole, while carbon dioxide does not, because its atoms do not lie in a straight line. This means that the two dipoles do not cancel each other out and there is a net dipole on the molecule.

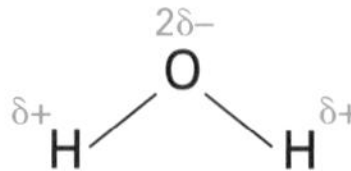

The displayed formula for water showing its partial charges.

Bending water

Polar molecules behave differently to non-polar molecules in an electric field. This is easily shown using a balloon or a plastic rod, charged with static electricity using a piece of cloth. If you hold the charged object near to a stream of water, the water is deflected. A non-polar liquid, such as hexane, is unaffected.

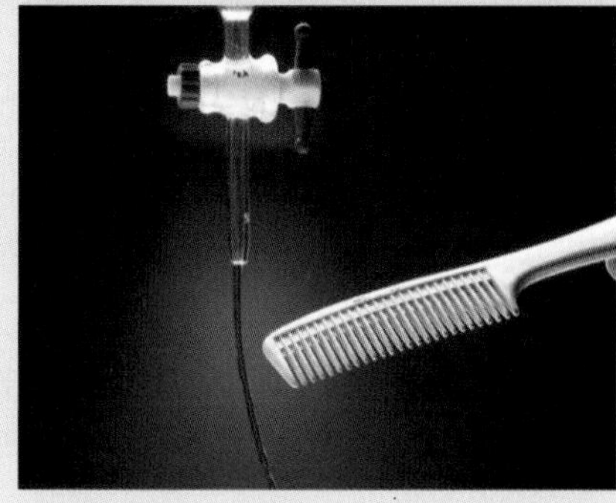

This simple experiment shows that water molecules are polar.

Electronegativity and bond character

The bigger the difference in electronegativity between two atoms in a covalent bond, the bigger the dipole and the more polar the bond.

You should see from the table that:

- H—H, C—C, and Cl—Cl bonds are non-polar because there are no differences in electronegativity
- the C—H bond is slightly polar with a difference of 0.3
- the C—Cl bond is more polar with a difference of 0.7
- the H—Cl bond is the most polar with a difference of 1.0

element	Pauling electronegativity
H	2.2
C	2.5
Cl	3.2

As the difference in electronegativity increases, the covalent bond takes on some ionic character. If there is a very large difference, the bonding is ionic rather than covalent. For example, the electronegativity of sodium is 0.9. So if there were an Na—Cl covalent bond, the difference in electronegativity would be 2.3. This is so large that sodium chloride exists as an ionic compound, not a covalent compound.

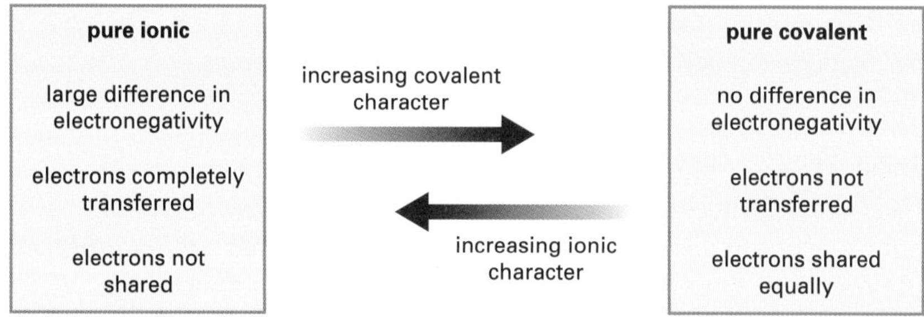

Ionic and covalent bonding can be viewed as opposite extremes of a range of bonding types.

Ionic bonds can show polarity, just as covalent bonds can. The electron cloud around a negative ion can be distorted and drawn towards a positive ion. When this happens, the negative ion is **polarized** by the positive ion. If the negative ion is sufficiently polarized, the ionic bond takes on some covalent character. How much this happens depends on Fajans' rules, developed by Kasimir Fajans.

- Negative ions are most easily polarized if they are large and have a high charge.
- Positive ions have the most polarizing power if they are small and have a high charge (they have a high **charge density**).

Sodium ions have a low charge density. They do not polarize chloride ions, so sodium chloride has ionic bonding. Magnesium ions have a higher charge density. They can polarize chloride ions to some extent, so magnesium chloride has ionic bonding with some covalent character. Aluminium ions have a high charge density. They can polarize chloride ions so much that the bonding becomes covalent, rather than ionic.

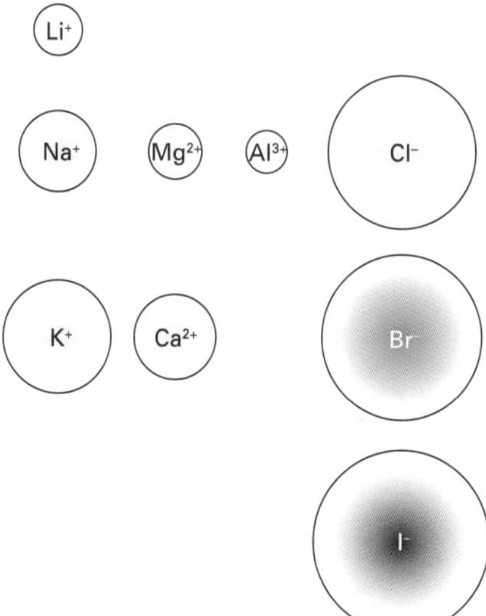

*Aluminium ions have a small **ionic radius** and a high charge. They have a high charge density and are very polarizing. Chloride, bromide, and iodide ions are easily polarized.*

Check your understanding

1. Explain why boron trifluoride molecules are non-polar but water molecules are polar.
2. The electronegativity of sodium is 0.9 and that of sulfur is 2.6 – predict the type of bonding in sodium sulfide and explain your reasoning.
3. Explain why aluminium chloride is a covalent compound, even though it contains a metal and a non-metal.
4. **a** Use Fajans' rules to explain why lithium iodide is an ionic compound with some covalent character .

   b Suggest why Pauling might have viewed it as a covalent compound with some ionic character.

3.10 Temporary dipole–induced dipole forces

OBJECTIVES

already from AS Level, you know

- that covalent substances exist as molecules

and after this spread you should

- understand qualitatively how molecules may interact by temporary dipole–induced dipole forces (van der Waals' forces)
- know the structure of iodine

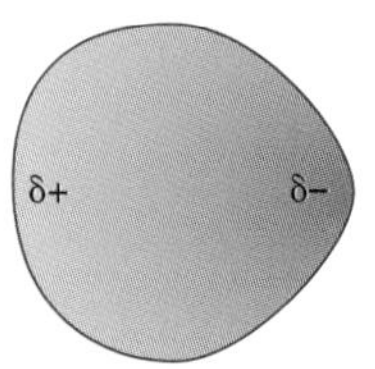

δ+ δ−

Electron cloud slightly distorted to the right, causing partial charges and a temporary dipole

Partial negative charge of temporary dipole repels the charge cloud of a neighbouring molecule, causing an induced dipole

Temporary dipoles and induced dipoles in van der Waals' forces between molecules.

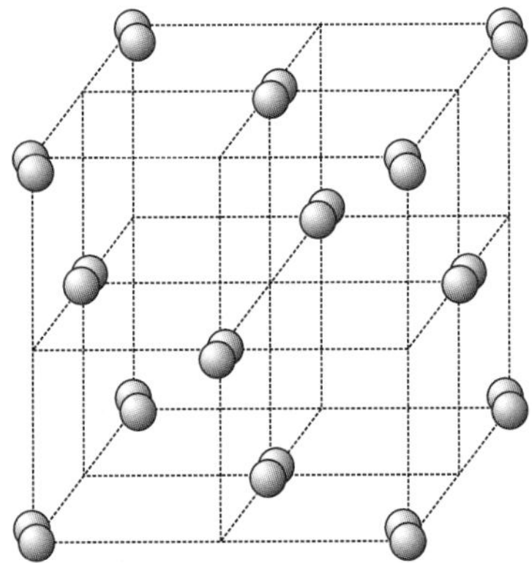

Iodine molecules form a cubic face-centred arrangement in iodine crystals.

Iodine melts at 114°C and boils at 184°C. Its crystals are soft and shiny purple-black. They readily form a purple vapour.

There are forces between molecules, called **intermolecular forces**. These attract molecules towards each other. It is these forces that must be overcome if a simple molecular substance is to melt or boil. The stronger these forces are, the more energy is needed to overcome them, and the higher the melting point and boiling point of the substance. The weakest intermolecular forces, often called **van der Waals' forces**, exist between all molecules.

Van der Waals' forces

Van der Waals' forces ('Waals' is pronounced 'varls') are forces of attraction between a **temporary dipole** on one molecule, and an **induced dipole** on another molecule. They occur between both polar molecules and non-polar molecules, but they are more important in non-polar molecules.

Temporary dipoles

The electron cloud around an atom or a non-polar molecule is not static. On average it is distributed evenly, but small changes in its distribution happen all the time. Changes like these cause a temporary dipole. This is very short-lived and may disappear in the next instant. Temporary dipoles are different from permanent dipoles, which exist continually because of a difference in electronegativity between the two atoms in a covalent bond.

Induced dipoles

The appearance of a temporary dipole in one molecule can cause a dipole to form in a neighbouring molecule. This is called an induced dipole. It is opposite in direction to the temporary dipole that induced it. The opposite partial charges attract each other, causing an attractive intermolecular force.

The structure of iodine

Iodine exists as diatomic molecules, I_2. The two iodine atoms in each molecule are joined together by a covalent bond. The molecules in solid iodine are regularly arranged to form a **molecular crystal**. Van der Waals' forces exist between the molecules in an iodine crystal. It is these forces, not the covalent bonds, which must be overcome for an iodine crystal to change state.

Relatively little energy is needed to overcome the van der Waals' forces between iodine molecules. So the melting point and boiling point of iodine are low. Iodine can **sublime** or turn directly from a solid to a vapour. When an iodine crystal is warmed, it readily forms a purple vapour consisting of individual iodine molecules.

The strength of van der Waals' forces

The strength of van der Waals' forces increases if the following increase:

- the size of the atom or molecule
- the area of contact between molecules

Size of the atom or molecule

The elements in Group 0 are all gases. As you go down the group, the atomic number increases and so does the number of electrons in each atom. A xenon atom near the bottom of the group is larger than a

helium atom at the top. It is more easily polarized. Temporary dipoles form more readily in xenon, and dipoles are more easily induced in neighbouring xenon atoms. The van der Waals' forces between xenon atoms are stronger than those between helium atoms.

You can see a similar trend in the boiling points of the Group 7 elements. These exist as diatomic molecules, but as the atomic number increases so does the boiling point.

You studied alkanes in your GCSE studies. These are molecules containing carbon and hydrogen atoms only. The larger these molecules are, the higher their boiling point. Again this is because the higher of the van der Waals' forces increases.

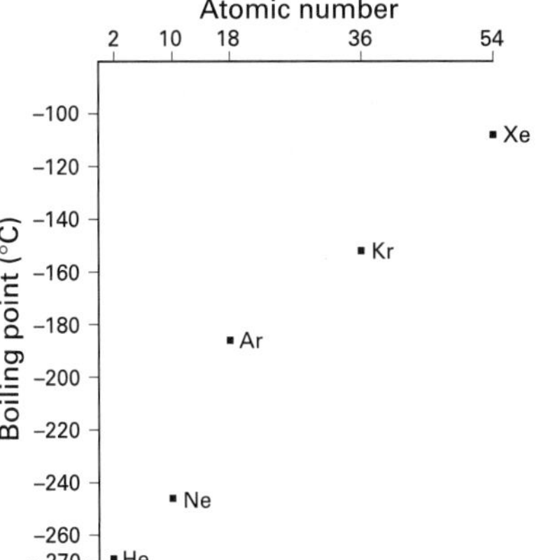

The strength of the van der Waals' forces between noble gas atoms increases as their atomic number increases. The boiling points of the elements increase as a result.

Area of contact between molecules

The greater the area of contact between the molecules, the stronger the van der Waals' forces become. You can see this if you study the effect of having a branch on alkane molecules. In general, the more branches the molecule has, the smaller its area of contact with other molecules, and the weaker the van der Waals' forces become.

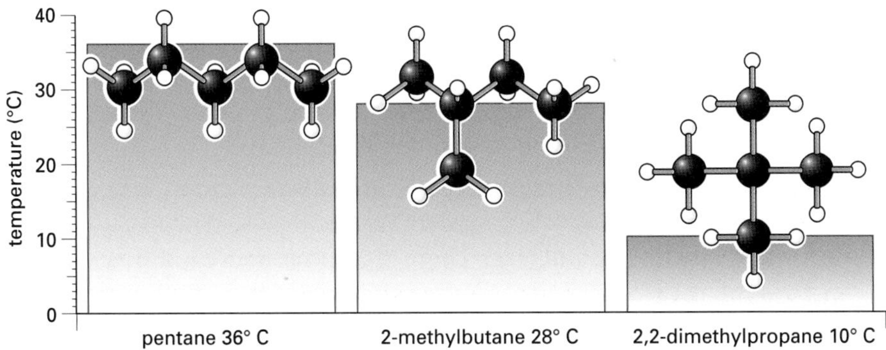

These three alkanes have the same molecular formula, C_5H_{12}. The unbranched alkane has a higher boiling point than the branched alkanes with the same number of carbon atoms.

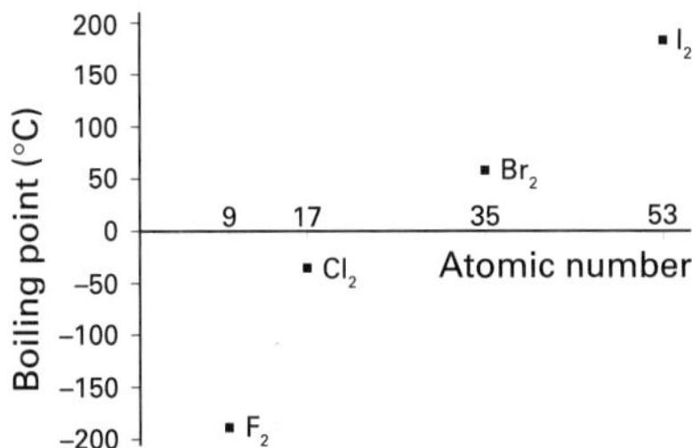

The strength of the van der Waals' forces between Group 7 molecules increases as the atomic number of their atoms increases.

Johannes van der Waals (1837–1923)

Van der Waals' forces are named after Johannes van der Waals, a Dutch scientist. He won the 1910 Nobel Prize in Physics but he needed a lot of determination to get his scientific career started. In the nineteenth century you could only take examinations at university if you had studied classical languages. Van der Waals had not. So he studied in his spare time and was later able to take examinations when the rules changed. He was awarded a doctoral degree in 1873 for his research into liquids and gases. Van der Waals' was the first scientist to suggest the existence of intermolecular forces, which is why they are often named after him.

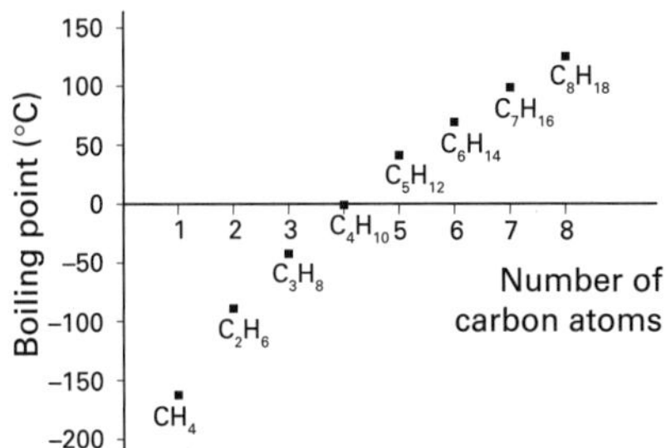

The strength of the van der Waals' forces between alkane molecules increases as the number of carbon atoms in the molecule increases.

Check your understanding

1. Explain how temporary dipoles can cause intermolecular forces.
2. Describe the structure and bonding of solid iodine.
3. Explain why iodine has a higher boiling point than chlorine.
4. Suggest why the boiling point of 3-methylpentane is lower than that of hexane, even though they both have the same molecular formula.

3.11 Permanent dipole–dipole bonding

OBJECTIVES

already from AS Level, you know

- how molecules may interact by temporary dipole–induced dipole forces (van der Waals' forces)

and after this spread you should

- understand qualitatively how molecules may interact by permanent dipole–dipole forces and hydrogen bonding
- understand the importance of hydrogen bonding in determining the boiling points of compounds and the structure of ice

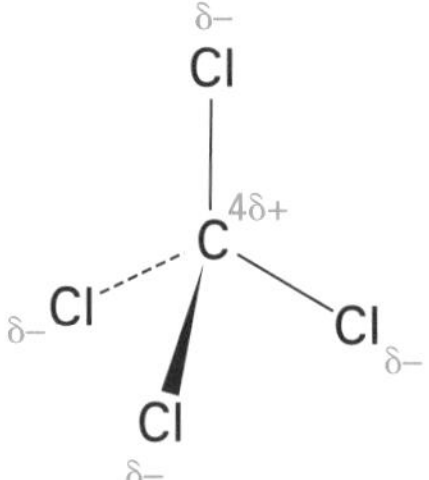

Tetrachloromethane contains four C—Cl bonds. Chlorine is more electronegative than carbon, so these bonds are polar. But the molecule itself is non-polar because the four dipoles cancel each other out. Tetrachloromethane does not have permanent dipole–dipole forces between its molecules.

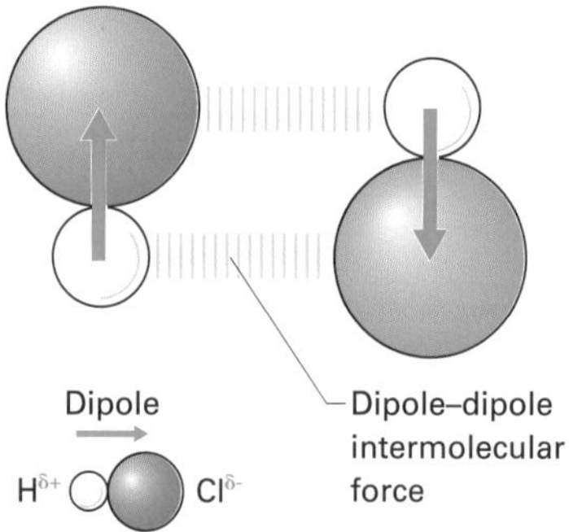

Permanent dipole–dipole forces exist between hydrogen chloride molecules.

Van der Waals' forces exist between all molecules, whether the molecule is polar or non-polar. Additional intermolecular forces exist between polar molecules. These are **permanent dipole–dipole forces**.

Permanent dipole–dipole forces

A permanent dipole occurs when the two atoms in a covalent bond have different electronegativities. It is there all the time and does not rely on the electron cloud being distorted temporarily, as in van der Waals' forces. Molecules containing a permanent dipole may have permanent dipole–dipole forces between them, but only if they are polar molecules.

Molecules in which the dipoles cancel each other out are non-polar. Even though they contain polar bonds, they will not have permanent dipole–dipole forces between them. Carbon dioxide and boron trifluoride are like this. They contain polar bonds but are non-polar molecules. Their dipoles cancel each other out so they do not have a net dipole. They do not have permanent dipole–dipole forces between them, although they still have van der Waals' forces.

Chlorine is more electronegative than hydrogen, so the H—Cl bond in hydrogen chloride is polar. As there is no other polar bond that might cancel out the dipole, hydrogen chloride molecules are polar, too. This means that permanent dipole–dipole forces exist between hydrogen chloride molecules, in addition to van der Waals' forces.

Permanent dipole–dipole forces increase the boiling point of hydrogen chloride by a large amount. Argon and fluorine contain the same number of electrons as hydrogen chloride. There are no permanent dipole–dipole forces between these molecules. Argon boils at −186°C and fluorine boils at −188°C. Hydrogen chloride, which does have permanent dipole–dipole forces, boils at a much higher −85°C.

Hydrogen bonds

Hydrogen bonds are permanent dipole–dipole forces that happen in particular circumstances. They are the strongest type of intermolecular force and are about 10% of the strength of a covalent bond.

Hydrogen bonds will form if

- a molecule contains a hydrogen atom covalently bonded to a nitrogen, oxygen or fluorine atom, and
- there is a lone pair of electrons on the nitrogen, oxygen or fluorine atom.

Nitrogen, oxygen, and fluorine are the three of the four most electronegative elements. They form a very polar covalent bond with hydrogen. The single electron in the hydrogen atom is drawn away. The hydrogen nucleus is strongly attracted to the lone pair of electrons in the nitrogen, oxygen, or fluorine atom in another molecule. A **hydrogen bond** is formed.

$$\overset{\delta+}{H}—\overset{\delta-}{F}\!:\cdots\overset{\delta+}{H}—\overset{\delta-}{F}$$

There are hydrogen bonds between hydrogen fluoride molecules. It helps to show the hydrogen bond as a dashed line that ends at a lone pair of electrons on the electronegative element N, O, or F.

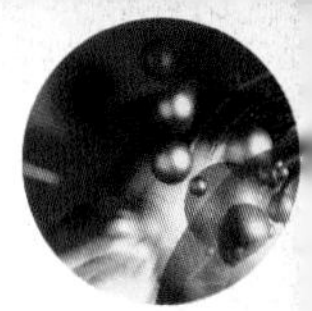

Evidence for hydrogen bonding

The presence of hydrogen bonds considerably increases the boiling point of a substance. A plot of the boiling points of Group 4 hydrides shows a steady increase as you go down the group from CH_4 to SnH_4. This is because the strength of the van der Waals' forces increases. If you do the same thing for Group 5 hydrides, you see the same trend, except that ammonia NH_3 has a much higher boiling point than expected. This is because of the presence of hydrogen bonds. The boiling point of water compared to the other Group 6 hydrides is also much higher than expected. Water has hydrogen bonds, too.

Hydrogen bonding between water molecules.

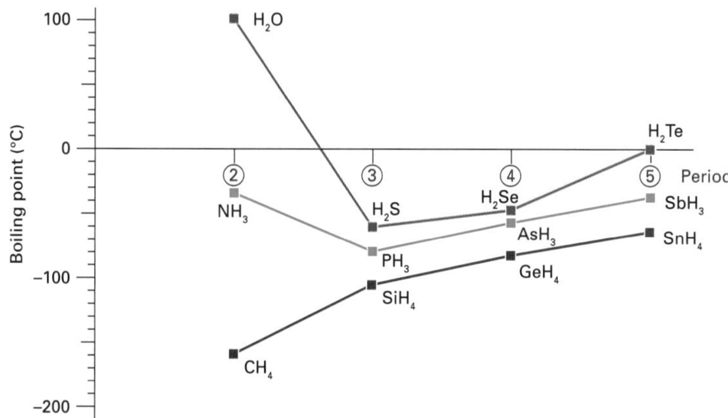

Water and ammonia have anomalously high boiling points. This is due to the presence of hydrogen bonds between their molecules.

More about hydrogen bonds

Chlorine is more electronegative than nitrogen. But it is too large to get close enough to the hydrogen atom to form a hydrogen bond.

Remember that the presence of hydrogen in a molecule may not be enough for hydrogen bonds to form. For example, methane CH_4 does not form hydrogen bonds, because it does not contain one of the electronegative elements N, O, or F.

The structure of ice

The water molecule contains two hydrogen atoms covalently bonded to an oxygen atom. The oxygen atom has two lone pairs of electrons, so water can form hydrogen bonds. These form along the same axis as the H–O bond on the neighbouring water molecule. As a result, ice has a regular open lattice structure. The water molecules are actually further apart than they are in liquid water, so ice has a lower density than liquid water. This is an unusual property, and explains why icebergs float instead of sink.

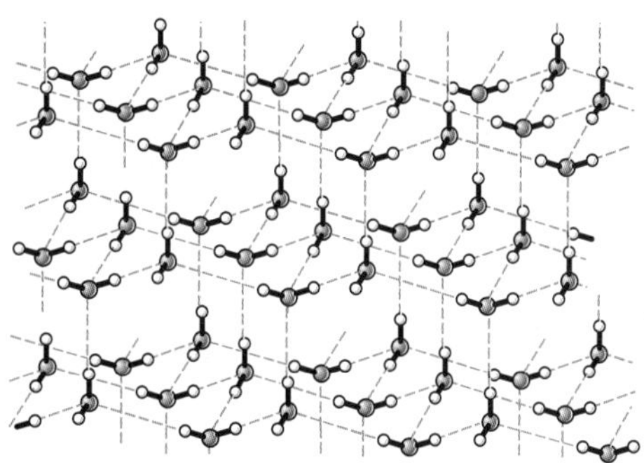

In ice, the water molecules are held in a regular open lattice by hydrogen bonds.

Check your understanding

1. Why do hydrogen chloride molecules have permanent dipole–dipole forces between them but boron trifluoride molecules do not?
2. Why do ammonia molecules have hydrogen bonds between them but hydrogen chloride molecules do not?
3. The table shows the boiling points of hydrogen halides.

hydrogen halide	HF	HCl	HBr	HI
period	2	3	4	5
boiling point (°C)	20	−85	−67	−35

   a Plot a graph of boiling point against period number.
   b Explain why HF has a higher boiling point than expected.

4. Why is ice less dense than liquid water?

3.12 Change of state

OBJECTIVES

already from AS Level, you know

- the structure and bonding present in ionic, metallic, giant covalent, and molecular substances

and after this spread you should

- be able to explain the energy changes associated with changes of state
- be able to relate the melting point and boiling point of a substance to its structure and bonding

States of matter

The three **states of matter** are solid, liquid, and gas. Each state has different physical properties.

state	fixed shape	fixed volume	flows	easily compressed
solid	✓	✓		
liquid		✓	✓	
gas			✓	✓

A summary of some of the typical physical properties of solids, liquids, and gases.

Particles

Several physical properties of solids, liquids, and gases are explained by studying the arrangement and movement of the particles they contain.

Solids

The particles in a solid are close together. They are held together in regular arrangements by strong forces, and are only able to vibrate about fixed positions.

Liquids

The particles in a liquid are close together but there are fewer forces attracting them together. The particles are able to move randomly around each other.

Gases

The particles in a gas are far apart. They have no forces attracting them together and they are able to move randomly in any direction.

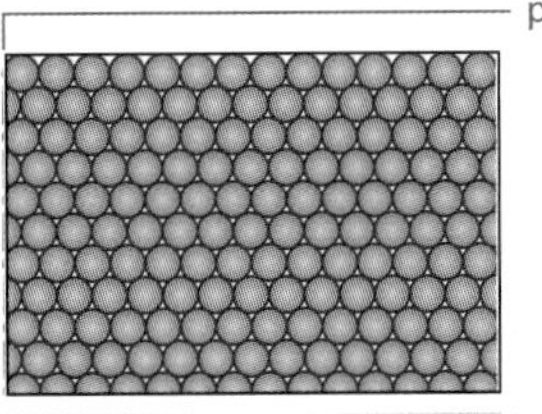

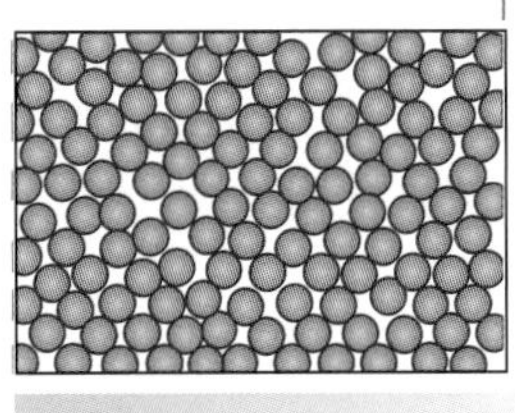

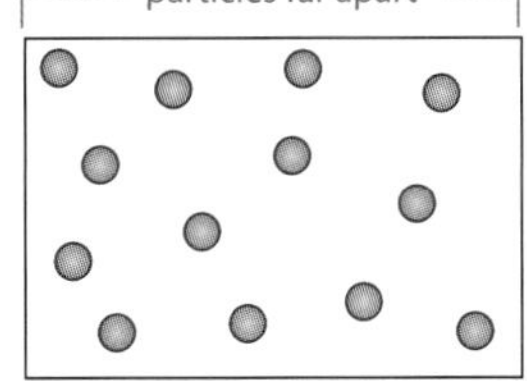

The arrangement of particles in each state of matter is different.

Changing state

Melting

When a solid is heated, its particles gain energy. They vibrate increasingly vigorously and the temperature of the solid increases. At the melting point, T_m, the supplied energy is used to break some of the bonds between the particles. The structure of the solid breaks down and the particles become free to move around each other. A liquid has formed.

Evaporating

When a liquid is heated, its particles gain energy. They move around each other increasingly vigorously and the temperature of the liquid increases. Some particles at the surface of the liquid may have enough energy to break all the bonds keeping them in the liquid. They will escape from the liquid as separate particles, forming a gas. Evaporation can happen at temperatures below the boiling point, T_b.

A solid has fixed shape and volume. It does not flow and it cannot be compressed easily.

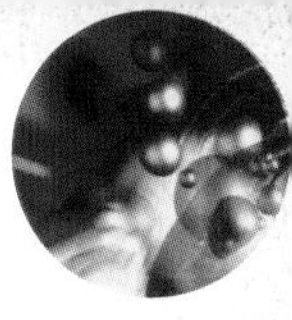

Boiling

At the boiling point, bubbles of vapour form inside the liquid, not just at the surface. The supplied energy is used to break all of the bonds between the particles in the liquid. The liquid is evaporating as fast as it can. When the liquid has boiled, the temperature of the gas rises if it is heated.

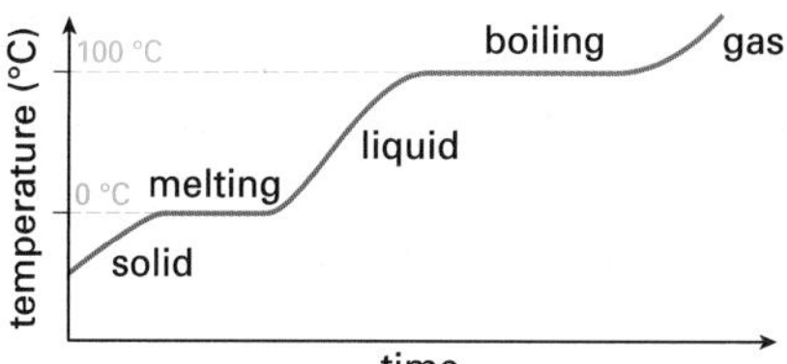

The temperature stays the same while a substance changes state.

Comparing different substances

Ionic, metallic, and giant covalent substances have high melting and boiling points. Molecular substances have low melting and boiling points.

Ionic substances

Ionic substances contain ions. Oppositely charged ions are attracted to each other by electrostatic forces. These ionic bonds are strong, and there are very many of them. A lot of energy is needed to break some of them so that an ionic substance can melt. Even more energy is needed to break all the ionic bonds so that an ionic substance can boil.

Metallic substances

Metals consist of a regular lattice of positively charged metal ions, surrounded by delocalized electrons. The metal ions are attracted to the delocalized electrons by electrostatic forces. These metallic bonds are strong and occur throughout the lattice. A lot of energy is needed to break some of them so that a metal can melt. Even more energy is needed to break all the metallic bonds so that a metal can boil.

Giant covalent substances

Giant covalent substances consist of atoms covalently bonded together. These bonds are strong and there are very many of them. A lot of energy is needed to break them, so the melting and boiling points of giant covalent substances is very high.

Molecular substances

Simple molecular substances consist of molecules attracted to each other by weak intermolecular forces. It is these forces that are overcome when a simple molecular substance melts or boils. Relatively little energy is needed to break some of them, so the melting points of simple molecular substances are low. For example, oxygen melts at −218°C and water at 0°C.

The strong covalent bonds between the individual atoms in a molecule are not broken when a simple molecular substance melts or boils. More energy is needed to break all the intermolecular forces in a liquid molecular substance. This is still relatively low, so the boiling points of simple molecular substances are low. For example, oxygen boils at −183°C and water at 100°C. Remember that substances that contain permanent dipole–dipole forces, or hydrogen bonds, are likely to have higher melting and boiling points than those with just van der Waals' forces.

Boiling points and melting points

- Sodium chloride is an ionic substance. It melts at 801°C and boils at 1413°C.
- Magnesium melts at 650°C and boils at 1090°C. Some metals melt at a very high temperature. Tungsten melts at 3422°C.
- Diamond melts at 3550°C and boils at 4827°C.
- Iodine melts at 114°C and boils at 184°C.

Check your understanding

1. Diamond and iodine both contain covalent bonds. Explain why the melting point of diamond is much higher than the melting point of iodine.
2. Explain why sodium chloride has a much higher melting point than oxygen.
3. a Why do metals have high melting points?
   b Explain which metal, magnesium or tungsten, contains the stronger metallic bonds.
4. Ammonia and methane are simple covalent molecules. They have the same number of electrons and an almost identical relative molecular mass. Explain why ammonia boils at –33°C but methane boils at just –162°C.

3.13 Conducting electricity

OBJECTIVES

already from AS Level, you know

- the structure and bonding present in ionic, metallic, giant covalent, and molecular substances

and after this spread you should

- be able to relate electrical conductivity of a substance to its structure and bonding

An electric current will flow if electric charges move from one place to another. No current will flow if a substance does not have charged particles, or if its charged particles are unable to move through the substance.

Metals

The structure of metals consists of closely packed metal cations surrounded by delocalized electrons. These electrons will move through the metal if a potential difference is applied across it. Electrons carry a negative charge, so an electric current flows through the metal. Metals are good conductors of electricity.

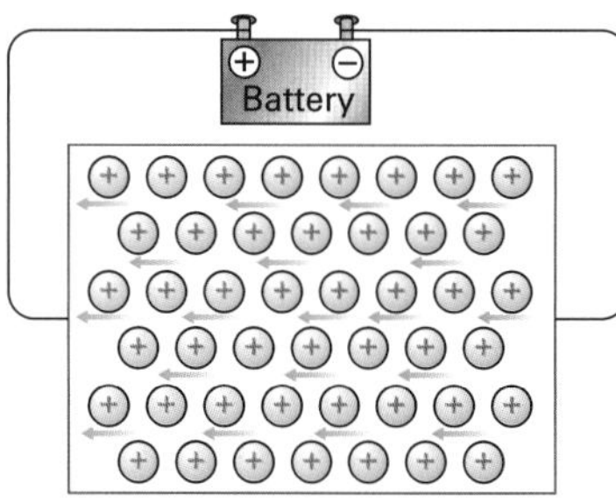

Delocalized electrons flow through a metal from a region of negative potential to a region of positive potential. A battery is a simple way to apply the potential difference.

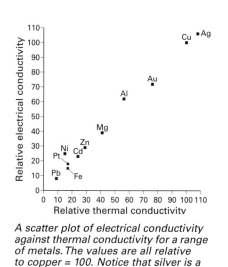

Thermal conductivity

Metals are good conductors of heat. There is good positive correlation between the electrical conductivity of metals and their thermal conductivity.

Electrons are the main conductors of heat in metals. The metal cations in a hot piece of metal have a lot of vibrational energy, and the delocalized electrons have a lot of kinetic energy. The metal cations cannot move from place to place, but the delocalized electrons can. They move through the metal and collide with electrons from colder parts. They transfer some of their kinetic energy to these 'colder' electrons that have less kinetic energy.

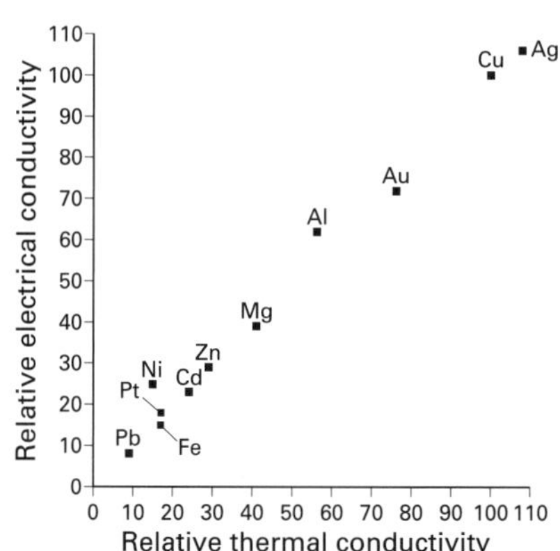

A scatter plot of electrical conductivity against thermal conductivity for a range of metals. The values are all relative to copper = 100. Notice that silver is a better conductor than copper but is too expensive to use for everyday wiring.

Ionic substances

Ionic solids contain ions held in a giant lattice structure by strong ionic bonds between oppositely charged ions. These are charged particles, but they are not free to move from place to place in ionic solids. This means that ionic solids cannot conduct electricity.

Ionic substances can conduct electricity when they are molten (in the liquid state) or dissolved in water. This is because the ions are free to move from place to place. This is the basis of **electrolysis**, the process in which compounds are broken down into simpler substances using electricity.

Aluminium is extracted from its molten ore using electrolysis. Electrolysis is also important in the chloralkali industry. An electric current is passed through concentrated sodium chloride solution, producing sodium hydroxide solution, hydrogen, and chlorine. These are important raw materials for manufacturing products as diverse as soap, margarine, and plastic.

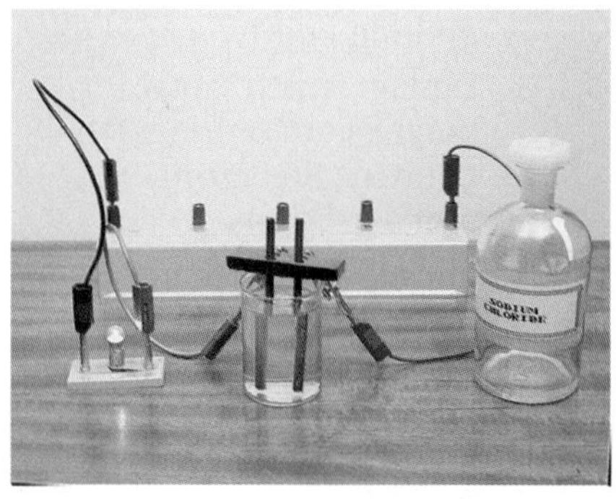
Sodium chloride conducts electricity in aqueous solution but not when it is solid.

Simple molecules

Simple molecules do not conduct electricity very well, if at all. Iodine forms a molecular crystal of iodine molecules. These molecules are not free to move from place to place, as their positions in the crystal are maintained by van der Waals' forces. Even if they could move, the molecules are neutral. Iodine does not conduct electricity.

Giant covalent substances

Diamond and silica

Whether a giant covalent substance conducts electricity depends upon whether it contains delocalized electrons. Diamond does not conduct electricity. Each carbon atom is covalently bonded to four other carbon atoms, so the electrons from the highest occupied energy level are all involved in bonding. There are no delocalized electrons in diamond.

Silica consists of silicon and oxygen atoms covalently bonded to form a giant covalent structure, similar to that of diamond. Silica does not conduct electricity either.

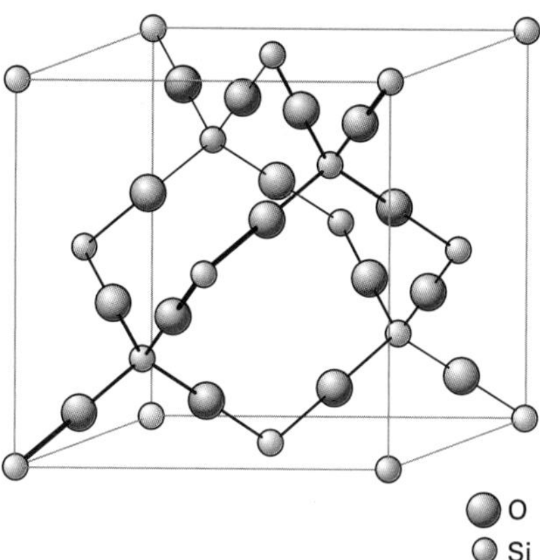

Silica contains two oxygen atoms for every silicon atom. It is often called silicon dioxide as its empirical formula is SiO_2, but silica has a giant covalent structure and is not a simple molecule.

Graphite

Graphite is an allotrope of carbon. Like diamond, it has a giant covalent structure. But unlike diamond, each carbon atom is covalently bonded to three other carbon atoms, not four. Graphite consists of flat layers of carbon atoms arranged in interlocking hexagons, rather like the appearance of chicken wire.

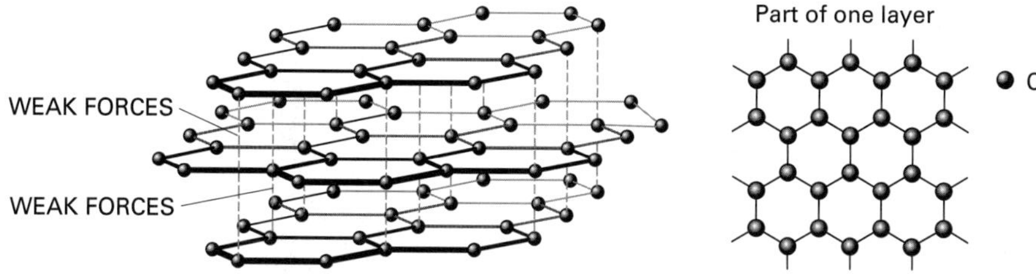

The structure of graphite. Note that the distance between layers is much greater than the distance between atoms in a layer.

Three of the four electrons in the highest occupied energy level of each carbon atom are involved in forming covalent bonds. The remaining electron is delocalized. Graphite can conduct electricity because its delocalized electrons can move through the structure. There are weak van der Waals' forces between layers. These allow the layers to slide over each other, so graphite is soft and slippery.

Ultrapure water

Water consists of simple molecules. It is a poor conductor of electricity. A very small proportion of water molecules ionize to form H^+ and OH^- ions. Impure water also contains ions from dissolved ionic substances. These charged particles are free to move through the liquid, so water does conduct electricity to some extent. The purity of water is often quoted by its electrical conductivity. The lower its conductivity, the poorer water is at conducting electricity and the purer it is. The conductivity of ultrapure water used in manufacturing microprocessor chips is about ten thousand times less than the conductivity of tap water.

Check your understanding

1. Copper and iodine are both solid at room temperature. Explain why copper conducts electricity but iodine does not.
2. Lead bromide conducts electricity when it is molten but not when it is solid.
   a What type of bonding is present in lead bromide?
   b Explain why lead bromide conducts electricity when it is molten.
3. Diamond and graphite are both allotropes of carbon. Explain why one of them conducts electricity but the other does not.

3.14 Period 3 elements

OBJECTIVES

already from AS Level, you know

- the trend in first ionization energy across period 3
- the structure and bonding present in ionic, metallic, giant covalent, and molecular substances

and after this spread you should

- recognize the elements in period 3
- be able to describe and explain the trend atomic radius of these elements

A period is a row in the periodic table. Period 3 contains eight elements, from sodium to argon.

Overview of period 3

Sodium and magnesium are in the s block of the periodic table. The other six elements are in the p block.

Sodium, magnesium, and aluminium are metals. They are good conductors of electricity because they have delocalized electrons in their structure. Phosphorus, sulfur, chlorine, and argon are non-metals. They are poor conductors of electricity. Silicon, near the middle of the period, is a **metalloid**. This means that its properties are intermediate between the properties of a metal and a non-metal. Silicon is a **semiconductor**. Its electrical conductivity is less than the typical conductivity of a metal, but more than the typical conductivity of a non-metal.

element	symbol	electron configuration	electrical conductivity (Cu = 100)
sodium	Na	[Ne] $3s^1$	37
magnesium	Mg	[Ne] $3s^2$	38
aluminium	Al	[Ne] $3s^2\ 3p^1$	64
silicon	Si	[Ne] $3s^2\ 3p^2$	16
phosphorus	P	[Ne] $3s^2\ 3p^3$	1×10^{-16}
sulfur	S	[Ne] $3s^2\ 3p^4$	1×10^{-22}
chlorine	Cl	[Ne] $3s^2\ 3p^5$	negligible
argon	Ar	[Ne] $3s^2\ 3p^6$	negligible

Abbreviated electron configurations and electrical conductivities of the period 3 elements.

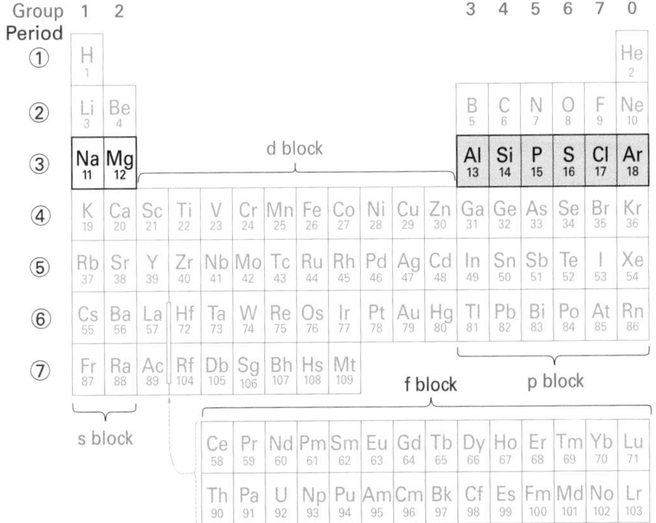

The position of period 3 in the periodic table.

Trend in first ionization energy

You have already learnt about the trend in first ionization energy across period 3 in *Evidence for energy sub-levels*. You will recall that there is a general increase in first ionization energy. As you go across the period, the nuclear charge increases so the force of attraction between the nucleus and the outer electron increases. As a result, the amount of energy needed to remove the outer electron increases. There are two small drops in first ionization energy as you go across the period.

Between magnesium and aluminium

The outer electron in aluminium is in the 3p sub-level whereas the outer electron in magnesium is in the 3s sub-level. The 3p sub-level is higher in energy than the 3s sub-level, so less energy is needed to remove an electron from it.

Between phosphorus and sulfur

The 3p electrons in phosphorus are all unpaired, but in sulfur two of the 3p electrons are paired. There is some repulsion between paired electrons in the same sub-level. This partly counteracts the force of attraction to the nucleus. So less energy is needed to remove one of the paired electrons from a sulfur atom.

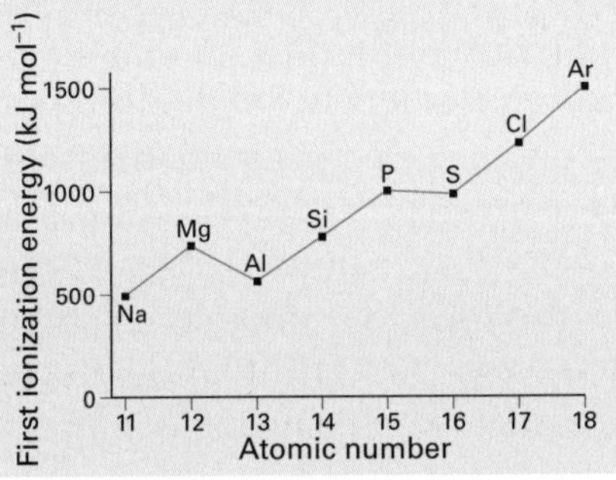

First ionization energy plotted against atomic number for the elements in period 3. (The full-size version of this graph is on spread 1.11.)

Semiconductors

A small proportion of the electrons in silicon are delocalized. The proportion increases as the temperature is increased, which increases the electrical conductivity of silicon. But a better way to increase its conductivity is to add tiny amounts of certain elements from group 3 or 5. This is called 'doping'.

Holes

Boron and gallium are in group 3. They can form three covalent bonds in the silicon giant covalent structure. This means that a neighbouring silicon atom has an unpaired electron while the group 3 atom has a 'hole' that could accept this electron. When a potential difference is applied to the material, electrons move from silicon atoms to fill these holes. This makes other holes and the process continues. The holes are positively charged, so this type of semiconductor material is called p-type.

Extra electrons

Phosphorus and arsenic are in group 5. They have five electrons in their highest occupied energy level. When these atoms form four covalent bonds with neighbouring silicon atoms, they are each left with an unpaired electron. This becomes delocalized, increasing the electrical conductivity of the material. Electrons are negatively charged, so this type of semiconductor material is called n-type. Combinations of the two types make the tiny transistors in microprocessors.

A typical microprocessor in a computer contains hundreds of millions of transistors. This is one of them, seen through an electron microscope.

Trend in atomic radius

If you look at the electron configurations of the period 3 elements, you might expect the size of the atoms to increase as you go across the period. But the reverse is true. The **atomic radius** decreases as you go across the period.

The electrons in the first and second energy levels partly **shield** the electrons in the third energy level from the attraction of the nucleus. As you go across period 3, the nucleus of each successive element has one more proton. This means that the positive charge of the nucleus increases, and so does the force of attraction between the nucleus and the electrons. The amount of shielding does not change across the period, so the electrons in the third energy level are more strongly attracted to the nucleus. As a result, the atomic radius decreases.

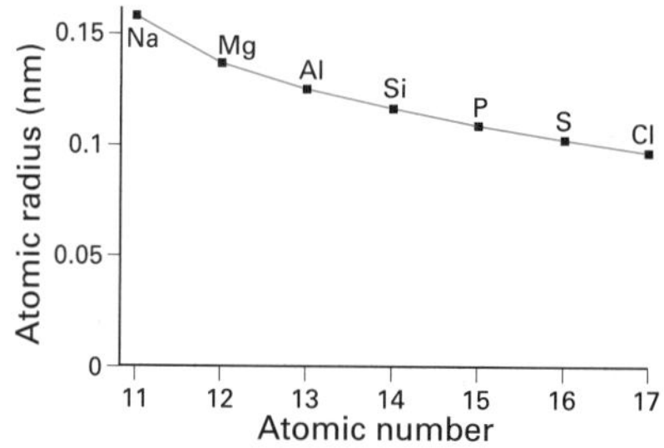

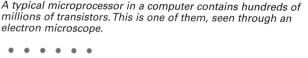

The atomic radius decreases as you go across period 3.

Check your understanding

1. Why are sodium and magnesium placed in the s block of the periodic table?
2. Why is aluminium placed in period 3 of the periodic table?
3. Suggest why aluminium is a better conductor of electricity than sodium and magnesium.
4. a What is the trend in atomic radius across period 3?
   b Explain the reasons for this trend.

3.15 More trends in period 3

OBJECTIVES

already from AS Level, you

- know the structure and bonding present in ionic, metallic, giant covalent, and molecular substances
- can relate the melting point and boiling point of a substance to its structure and bonding

and after this spread you should

- be able to describe the trends in the melting and boiling points of the elements in period 3
- understand the reasons for these trends

Change of state

Bonds are broken when a substance melts or boils. The melting and boiling points of a substance depend on its structure and bonding. Substances with high melting and boiling points have strong bonds. A lot of energy is needed to break these bonds. When a substance melts, some of the bonds between the particles are broken. This lets the particles move freely around each other while still being close together. When a substance boils, all the remaining bonds are broken. This lets the particles move freely in all directions.

Structure and bonding in period 3

The structure and bonding in the elements of period 3 change as you go across the period. The first three elements are metals. Silicon has a giant covalent structure, similar to the structure of diamond. Phosphorus, sulfur, and chlorine exist as simple molecules, while argon exists as separate atoms. Different types of bond must be broken for the elements in period 3 to melt or boil.

element	Na	Mg	Al	Si	P_4	S_8	Cl_2	Ar
structure		metallic		giant covalent		simple molecular		monatomic
bonds broken		metallic		covalent		van der Waals' forces		

The type of structure and bonding changes across period 3.

Trends in melting point and boiling point

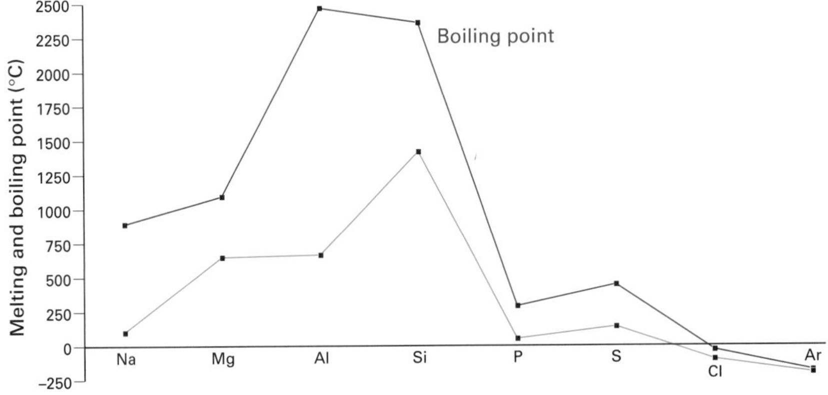

Trends in the melting and boiling points across period 3.

From the graph, you can see that the melting points generally increase going from sodium to silicon, then decrease going to argon. There is a small increase at sulfur. The trend in the boiling points is slightly different. They generally increase going from sodium to aluminium, not silicon, then decrease to argon. Again there is a small increase at sulfur. To explain these trends it helps to look at the three types of bonding present in the period 3 elements.

Sodium, magnesium, and aluminium

These elements are metals. They have metallic bonding in which positive metal ions are attracted to delocalized electrons. Metallic bonds are strong, so a lot of energy is needed to break them. This means that metals tend to have high melting points and boiling points. Notice that the melting and boiling points increase as you go from sodium to aluminium.

The charge on the metal ions increases from +1 for sodium, to +2 for magnesium, and to +3 for aluminium. The size of the ions also decreases from sodium to aluminium, so the charge density increases. The number of delocalized electrons in the structure increases. As a result, the strength of the metallic bonding increases, and so do the melting and boiling points.

Silicon

Silicon is a metalloid. It has a giant covalent structure. Each silicon atom is covalently bonded to four other silicon atoms in a tetrahedral arrangement, just like the carbon atoms in diamond. There are very many covalent bonds and these are strong. A lot of energy is needed to break them, so silicon has high melting and boiling points.

Phosphorus, sulfur, chlorine, and argon

These are all non-metals. Phosphorus, sulfur, and chlorine exist as simple molecules, with strong covalent bonds between their atoms. Argon is monatomic – it exists as separate atoms. These four elements have low melting and boiling points because van der Waals' forces must be overcome to melt or boil them. These forces are very weak, so relatively little energy is needed to do this.

Aluminium and silicon

Silicon has a higher melting point than aluminium, but it has a lower boiling point than aluminium. More energy is needed to break enough bonds to melt silicon than is needed to melt aluminium. A larger increase in temperature is needed to break all the bonds to boil aluminium than is needed to boil silicon. This suggests that the metallic bonds in liquid aluminium are stronger than the remaining covalent bonds in liquid silicon.

	aluminium	silicon
melting point (°C)	660	1410
boiling point (°C)	2467	2355
difference (K)	1807	945

Aluminium is a liquid over a greater range of temperatures than silicon.

The sulfur increase

Sulfur has a higher melting and boiling point than phosphorus, chlorine, and argon. This is because phosphorus exists as P_4 molecules, sulfur as S_8 molecules, chlorine as Cl_2 molecules, and argon as individual atoms. Sulfur molecules have more electrons than the other three, so it has the strongest van der Waals' forces.

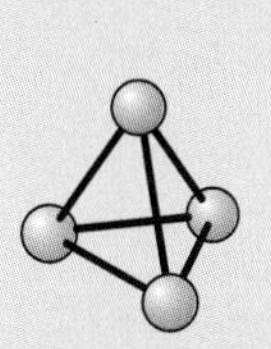

Ball-and-stick model

Space-filling model

Phosphorus exists as P_4 molecules.

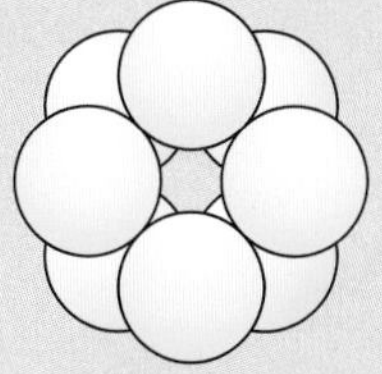

Top view, space-filling model

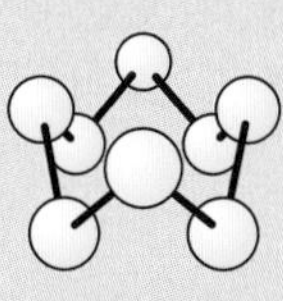

Side view, ball-and-stick model

Sulfur exists as S_8 molecules.

Properties	S_8	P_4	Cl_2	Ar
relative mass, M_r or A_r	256.8	124.0	70.0	39.9
electrons per molecule	128	60	34	18
boiling point (°C)	445	277	−34	−186

The positions of phosphorus and sulfur have been swapped in this table. This makes it easier to see that the boiling point decreases as the size of the molecules decreases.

Check your understanding

1. a Describe the trend in melting point across period 3.
   b Describe the trend in boiling point across period 3.
2. Explain why magnesium has a higher boiling point than sodium.
3. Explain why silicon has a high melting point.
4. Explain why sulfur has a higher melting point than phosphorus.

4.01 An introduction to alkanes

OBJECTIVES

already from GCSE, you know

- that alkanes are hydrocarbons with the general formula C_nH_{2n+2}

already from AS Level, you

- understand the terms molecular formula and empirical formula, and the relationship between them

and after this spread you should

- understand the terms structural formula and homologous series
- be able to apply IUPAC rules for naming unbranched alkanes with up to six carbon atoms

prefix	number of C atoms
meth	1
eth	2
prop	3
but	4
pent	5
hex	6

The alkanes

Alkanes are **hydrocarbons**. This means that they consist of hydrogen and carbon atoms only. Carbon atoms are unusual in that they can form chains, rings, and branches with other carbon atoms. In alkanes with two or more carbon atoms, the carbon atoms are joined with single covalent bonds.

The alkanes form a **homologous series**. They are a 'family' of compounds with the same general formula. This is C_nH_{2n+2} for the alkanes. So the molecular formula for methane is CH_4 ($n = 1$) and for hexane it is C_6H_{14} ($n = 6$).

Naming alkanes

The names of alkanes tell you how many carbon atoms they contain. For unbranched alkanes, this information is found in the first part of the name, the *prefix*. The table shows some common prefixes and their meanings.

The names of alkanes end in *ane*. So *methane* is an alkane containing one carbon atom. You know that the general formula for alkanes is C_nH_{2n+2}, so you can easily work out that the molecular formula of methane is CH_4. In a similar way, *hexane* is an alkane containing six carbon atoms, and it has the molecular formula C_6H_{14}.

Structural formulae of alkanes

A **structural formula** shows the number and type of each atom in a molecule, and it shows how they are joined together. The clearest way to do this is to draw the **displayed formula**.

Displayed formulae

You can draw the displayed formula of an alkane by setting the carbon atoms out as a line of Cs, then drawing single bonds between them. Draw enough extra bonds so that each carbon atom has four bonds coming from it. Complete the displayed formula by adding Hs to the 'empty sticks', bonds that end without an atom symbol.

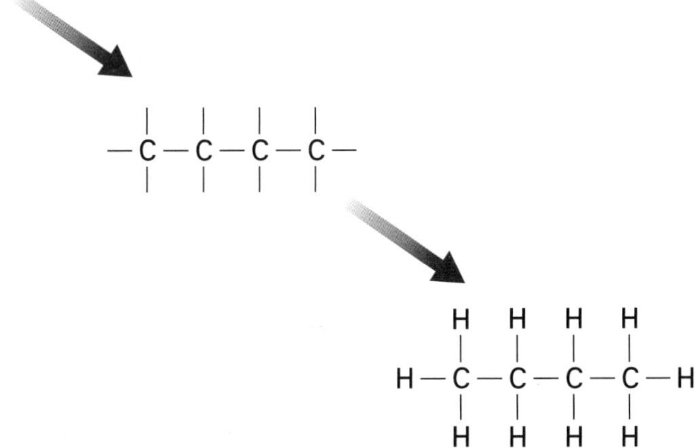

Drawing the displayed formula of butane, C_4H_{10}

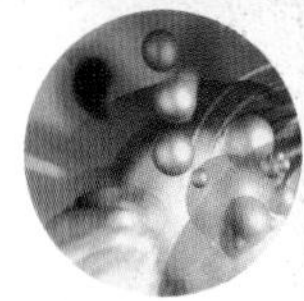

Shortened structural formulae

Displayed formulae are difficult to type, so the **shortened structural formula** may be used instead. For butane this is $CH_3CH_2CH_2CH_3$. Notice that you do not draw any bonds in this type of formula. If you have to write the shortened structural formula of an alkane with many carbon atoms, you can simplify it by putting the CH_2 groups in brackets. Butane becomes $CH_3(CH_2)_2CH_3$ and hexane becomes $CH_3(CH_2)_4CH_3$.

Skeletal formulae

Showing the displayed formulae with the bonds at 90° is simple and clear, but it does not reflect the true shape of the molecule. You will recall that the four bonds of a carbon atom have a tetrahedral arrangement with bond angles of 109.5°. This means that the carbon atoms in alkanes do not really lie on a straight line. They form a zig-zag chain instead.

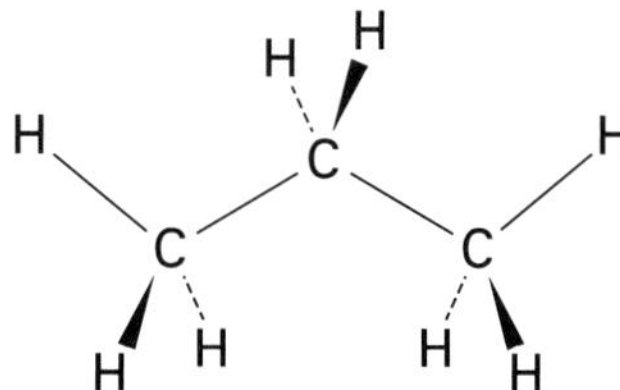

The carbon atoms in propane do not lie on a straight line.

The **skeletal formula** is another shortened way to show the structural formula of an alkane. When you draw a skeletal formula.

- show each covalent bond between two carbon atoms as a single line
- draw the lines so that they show the underlying shape of the molecule
- do not write the symbols for carbon atoms
- write the symbols for other elements if they are present, except for hydrogen where it is joined to a carbon atom

*Skeletal formulae for hexane and cyclohexane (a ring or **cyclic** compound).*

Skeletal formulae are useful if there are a lot of carbon atoms in the molecule. But you can get caught out if you forget that hydrogen atoms are attached at the bends and ends.

Check your understanding

1. What is the molecular formula for eicosane, an alkane containing 20 carbon atoms?
2. Name the compound C_5H_{12}.
3. Compare the advantages and disadvantages of displayed formulae and shortened structural formulae.
4. Write the shortened structural formula for octane, C_5H_{18}.

IUPAC

IUPAC is The *International Union of Pure and Applied Chemistry*. IUPAC is an association of many national scientific societies, including the UK's *Royal Society of Chemistry*. Its aim is to advance the worldwide aspects of chemistry and to help the use of chemistry for everyone's benefit. Around a thousand chemists around the world work voluntarily for IUPAC. They help to develop and establish standardized the naming systems and measurements. For example, IUPAC standardized the spellings of the elements Al and S in 1990. They recommended that *aluminum* should be replaced by the British spelling, *aluminium*, and that *sulphur* should be replaced by the American spelling, *sulfur*. Standardized nomenclature (naming systems) makes chemistry a truly international science.

4.02 Chain isomers of alkanes

OBJECTIVES

already from AS Level, you

- understand the terms structural formula and homologous series
- can apply IUPAC rules for naming unbranched alkanes with up to six carbon atoms

and after this spread you should

- know and understand the meaning of the term structural isomerism
- be able to draw the structures of alkane chain isomers
- be able to apply IUPAC rules for naming branched alkanes with up to six carbon atoms

number of C atoms	side chain	name
1	$-CH_3$	methyl
2	$-CH_2CH_3$	ethyl
3	$-CH_2CH_2CH_3$	propyl

Side chains are named after the number of carbon atoms they contain.

Structural isomerism

Structural isomerism occurs when two or more organic compounds have the same molecular formulae, but different structures. These differences often give the molecules different chemical and physical properties. **Chain isomerism** is a particular type of structural isomerism. It occurs when compounds have identical molecular formulae but their carbon atoms are joined together in different arrangements. Chain **isomers** involve branched and unbranched carbon chains.

If you think of butane, you probably imagine four carbon atoms in a line. This is the unbranched isomer. Now imagine taking one of the carbon atoms off the end and rejoining it to the chain, but not at the end. This gives you a branched alkane – another chain isomer of butane. Note that butane itself is a chain isomer. There is no sense in which it is the 'real' sort of butane.

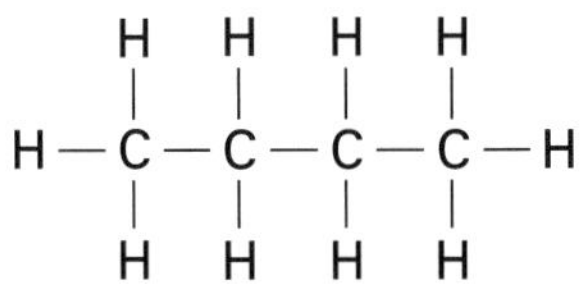

butane

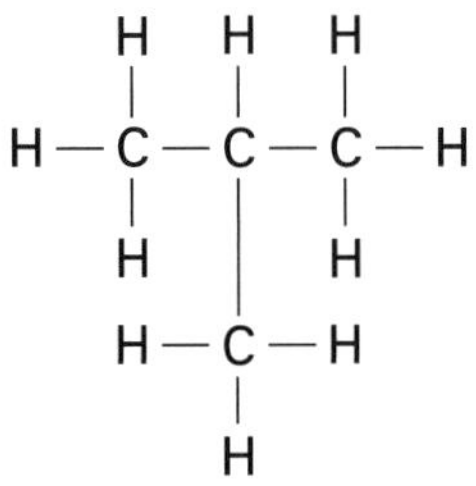

methylpropane

Chain isomers of butane. They have the same molecular formula but their carbon atoms are arranged differently.

Naming branched alkanes

You will have noticed that the branched isomer of butane is called methylpropane. The *methyl* part of the name tells you that there is a branch with one carbon atom. The *propane* part of the name tells you that this is the longest part of the chain – the **main chain**. The branches are called **side chains**.

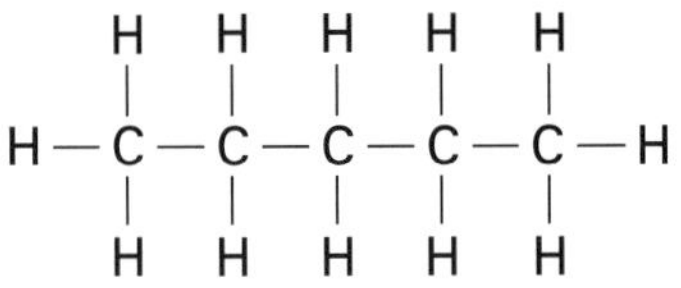

pentane

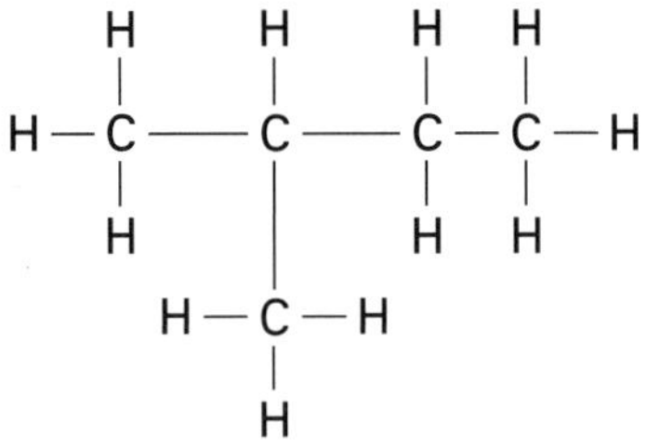

2-methylbutane

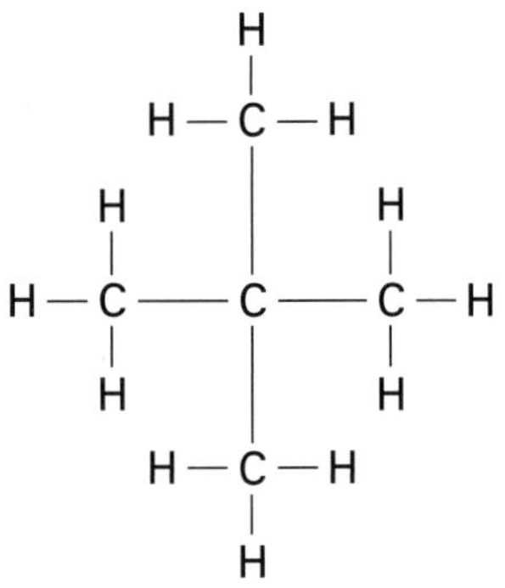

2,2-dimethylpropane

Chain isomers of pentane.

There are three chain isomers of pentane. Notice that numbers are used to show which carbon atom the side chain is attached to. The numbering is always done so you get the lowest total number. So it is 2-methylbutane and not 3-methylbutane. A number and a word are

always separated by a dash. If you have two identical side chains the prefix *di* is put in front of their name. Notice that each side chain has its own number, and that numbers are separated by commas.

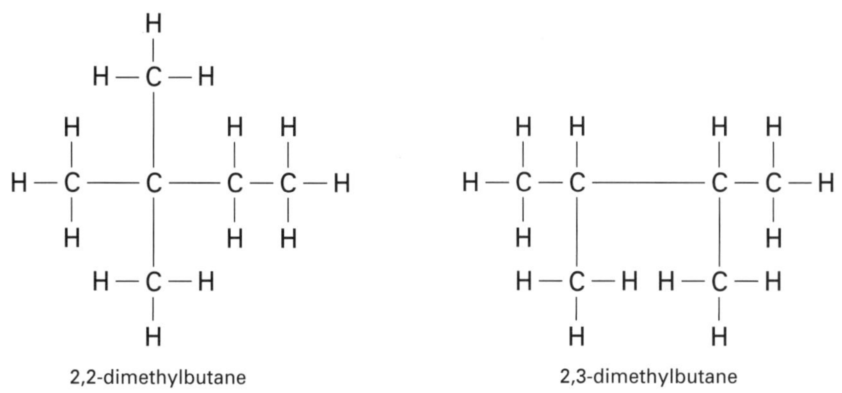

2,2-dimethylbutane and 2,3-dimethylbutane.

number of identical side chains	prefix
2	di
3	tri
4	tetra

Some prefixes for multiple identical side chains.

The lowest total number rule is very important when there are two or more side chains and several possible places where they could be substituted. This means that it must be 2,2-dimethylbutane and not 3,3-dimethylbutane.

Working out chain isomers

Start with the unbranched isomer. It is easiest to leave the hydrogen atoms and their bonds to begin with. These can be added later. Next, draw the unbranched isomer with one less carbon atom. See if you can join this to the main chain anywhere other than the end. If you can do that, see if there are other unique positions on the main chain where it could go.

Take care not to produce mirror images or bent chains – they are just the same chain isomer but drawn differently. Repeat this method, removing one more carbon atom from the main chain each time, until there are no more unique positions.

```
C—C—C—C—C—C                                   no side chain

C—C—C—C—C        C—C—C—C—C
  |                  |                         one side chain
  C                  C

                     C
                     |
C—C—C—C           C—C—C—C                     two side chains
  | |                |
  C C                C
```

Working out the chain isomers of hexane. Remember to add the remaining bonds and hydrogen atoms later.

```
                                              H
                                              |
                                            H—C—H
  H  H  H  H  H               H  H  H         |
  |  |  |  |  |               |  |  |         |
H—C—C—C—C—C—H               H—C—C—C—————C—H
  |  |  |  |  |               |  |  |         |
  H  H  H  H  H               H  H  H         H
```

A bent chain is not a chain isomer of a straight chain.

How many chain isomers are there?

Chemists have been keen to find out how many chain isomers exist for different alkanes for over a hundred years. One way is to draw them all out, but it becomes increasingly difficult to tell if a structure is really different, or if it just a repeat of another structure drawn in a different way. Various mathematical methods may be used. Some are more successful than others and all are complex.

Computers can generate and store all the possible chain isomers for a given molecular formula. They can then count all the unique ones. The numbers rapidly mount up. There are three chain isomers of C_5H_{12} but 75 for $C_{10}H_{22}$. There are 60,523 chain isomers of $C_{18}H_{38}$ and over 900,000 for $C_{21}H_{44}$!

Check your understanding

1. What is *structural isomerism?*
2. What are *chain isomers*?
3. Name this compound

```
          H           H
          |           |
        H—C—H       H—C—H
  H       |     H     |     H
  |       |     |     |     |
H—C———C———C———C———C—H
  |       |     |     |     |
  H       |     H     H     H
        H—C—H
          |
          H
```

4. There are nine chain isomers with the formula C_7H_{16}. Find them and name them.

4.03 Haloalkanes and alkenes

OBJECTIVES

already from GCSE, you know

- that alkenes are hydrocarbons with a double bond and the general formula C_nH_{2n}

already from AS Level, you

- understand the terms structural formula and homologous series
- know and understand the meaning of the term structural isomerism
- can apply IUPAC rules for naming alkanes with up to six carbon atoms

and after this spread you should

- understand the term 'functional group'
- be able to apply IUPAC rules for naming haloalkanes and alkenes with up to six carbon atoms
- be able to draw the structures of position isomers of haloalkanes and alkenes

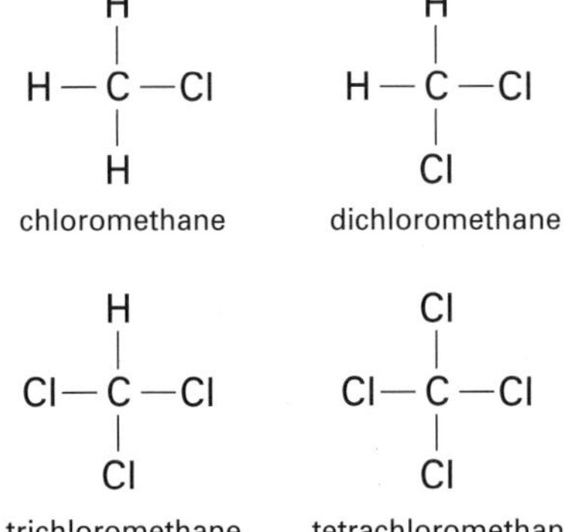

Displayed formulae of some chloroalkanes.

A **functional group** is a particular atom, or a group of atoms, in a molecule that is responsible for the how the molecule reacts. The members of a homologous series will contain the same functional group. For example, haloalkanes contain halogen atoms, and alkenes contain a C=C group.

Haloalkanes

The haloalkanes are alkanes in which at least one hydrogen atom is substituted by a halogen atom. For example, CH_3Cl is a haloalkane in which one of the hydrogen atoms in methane is substituted by a chlorine atom. It is the simplest chloroalkane, a homologous series with the general formula $C_nH_{2n+1}Cl$.

homologous series	functional group
fluoroalkanes	fluorine atom
chloroalkanes	chlorine atom
bromoalkanes	bromine atom
iodoalkanes	iodine atom

The haloalkanes form different homologous series, named after the halogen they contain.

The haloalkanes are named after the halogens they contain. The prefixes fluoro, chloro, bromo, and iodo are added to the name of the parent alkane.

name	formula
fluoromethane	CH_3F
chloromethane	CH_3Cl
bromomethane	CH_3Br
iodomethane	CH_3I

Four haloalkanes based on methane CH_4.

If a haloalkane molecule contains more than one halogen atom, the prefixes di, tri, and tetra are used. So CH_2Cl_2 is dichloromethane, $CHCl_3$ is trichloromethane, and CCl_4 is tetrachloromethane. Notice that CCl_4 is named after methane, even though it does not contain any hydrogen atoms.

Position isomers

Position isomerism is a particular type of structural isomerism. It occurs when the functional group can be in different positions on the same carbon chain. Position isomers are named in a similar way to chain isomers of alkanes. For example, in 1-bromopropane $CH_3CH_2CH_2Br$ the bromine atom is joined to the end of the carbon chain. In 2-bromopropane $CH_3CHBrCH_3$ it is joined to the central carbon atom. You keep the total number as small as possible, so 3-bromopropane would be incorrect for $CH_3CH_2CH_2Br$.

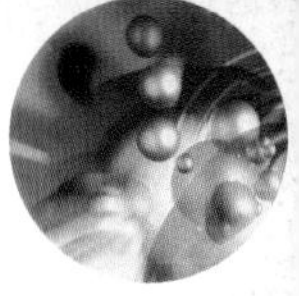

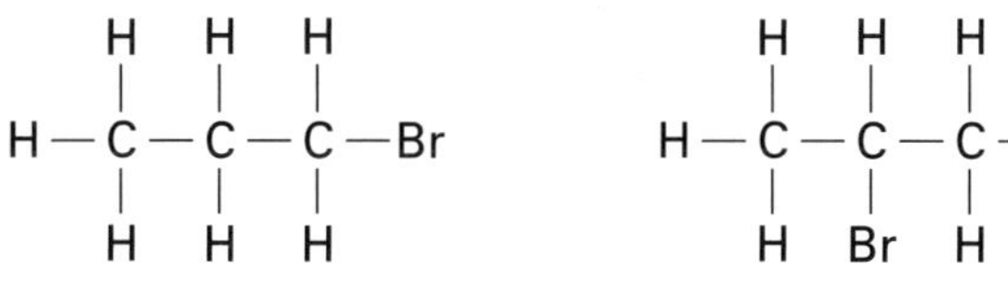

1-bromopropane 2-bromopropane

Position isomers of bromopropane.

```
      H   H   F
      |   |   |
  H — C — C — C — H
      |   |   |
      Cl  I   Br
```

This is 1-bromo-3-chloro-1-fluoro-2-iodopropane. It is not 3-bromo-1-chloro-3-fluoro-2-iodopropane because this gives a total number of 9, which is larger than 7.

Haloalkanes can contain different halogens. When this happens, you name the functional groups in alphabetical order.

Alkenes

The alkenes are a homologous series of hydrocarbons with the functional group >C=C<. Their general formula is C_nH_{2n}. Alkenes are named after the number of carbon atoms they contain, and their names end in ene. The shortened structural formula of ethene is $CH_2{=}CH_2$ and the shortened structural formula of propene is $CH_3CH{=}CH_2$.

Alkenes with four or more carbon atoms may have position isomers. There are two position isomers of butene, C_4H_8. The position of the double bond is shown using a number, just like the position of the halogen atom in a haloalkane. The only difference is that the number goes in the middle of the name with a dash each side. So it would be but-1-ene and not 1-butene.

```
      H   H   H   H              H   H   H   H
      |   |   |   |              |   |   |   |
  H — C — C — C = C          H — C — C = C — C — H
      |   |       |              |           |
      H   H       H              H           H

         but-1-ene                  but-2-ene
```

Position isomers of butene. You keep the total number as small as possible. For example, but-1-ene is correct but not but-3-ene or but-4-ene.

There are chain isomers of alkenes, too. These are named after the parent alkene.

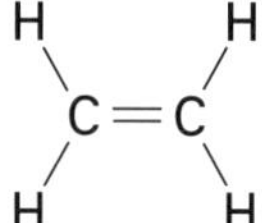

The displayed formula of ethene, the simplest alkene.

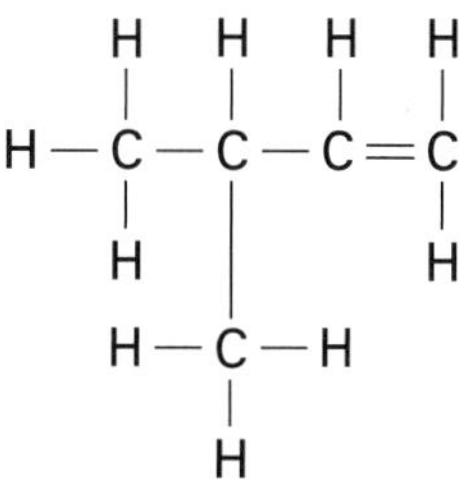

3-methylbut-1-ene, $CH_3CH(CH_3)CH{=}CH_2$ is a chain isomer of pent-1-ene. Notice that it is not called 2-methylbut-3-ene or 2-methylbut-4-ene.

Check your understanding

1. Name these haloalkanes:
   - **a** CH_3CH_2F
   - **b** $CH_3CH_2CHBrCH_2CH_3$
   - **c** $CH_3CFBrCH_2I$
2. Draw the displayed formulae of these haloalkanes:
   - **a** 1-iodopropane
   - **b** 1-bromo-4-fluoro-2-iodobutane
   - **c** 2-chloro-2-methylpropane
3. Name these alkenes:
   - **a** $CH_3CH{=}CH_2$
   - **b** $CH_3CH{=}CHCH_3$
   - **c** $CH_3CH(CH_3)CH{=}CHCH_3$
4. Draw the displayed formulae of these alkenes:
   - **a** pent-2-ene
   - **b** 2-methylpropene
   - **c** buta-1, 3-diene

Bond angles in ethene

The ethene molecule has a trigonal planar shape. You might expect the bond angle H—C—H to be 120°, and many text books suggest that it is. (Using this value in questions is OK!) In fact it is about 118°. The reduction in bond angle is not caused by a lone pair of electrons. Instead it happens because the two bonding pairs of electrons in the C=C bond repel the C—H bonds more than a C—C would do. There is a greater electron density in a C=C bond than in a C—C bond.

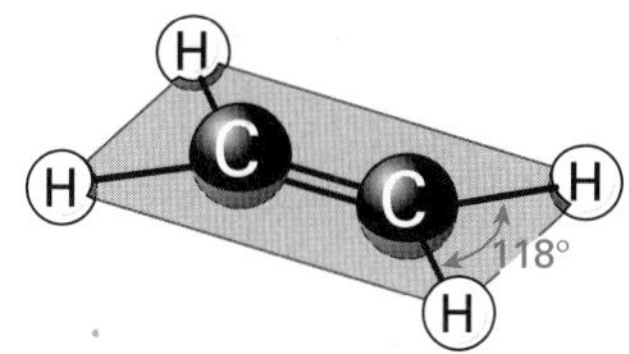

The ethene molecule is planar.

4.04 Petroleum and fractional distillation

OBJECTIVES

already from GCSE, you know

- that crude oil contains hydrocarbons, mainly alkanes
- that alkanes are saturated hydrocarbons
- that crude oil can be separated into fractions by fractional distillation

already from AS Level, you

- understand how molecules may interact by van der Waals' forces
- can relate the boiling point of a substance to its structure and bonding

and after this spread you should

- understand that different petroleum fractions can be drawn off at different levels in a fractionating column because of the temperature gradient

Petroleum is a mixture of hydrocarbons. These are mostly alkanes. The alkanes form a homologous series. They are **saturated** hydrocarbons because they only contain single covalent bonds. Like other homologous series, the alkanes show a trend in their physical properties. As the number of carbon atoms in each unbranched alkane molecule increases, so does the boiling point of the alkane. Larger molecules have stronger van der Waals' forces between them, and more energy is needed to overcome force back.

Fractional distillation relies on differences in boiling points. It lets us separate the different components in petroleum. These components are called **fractions** because they are only part of the original petroleum. A fraction contains hydrocarbon molecules with a similar boiling point and number of carbon atoms per molecule.

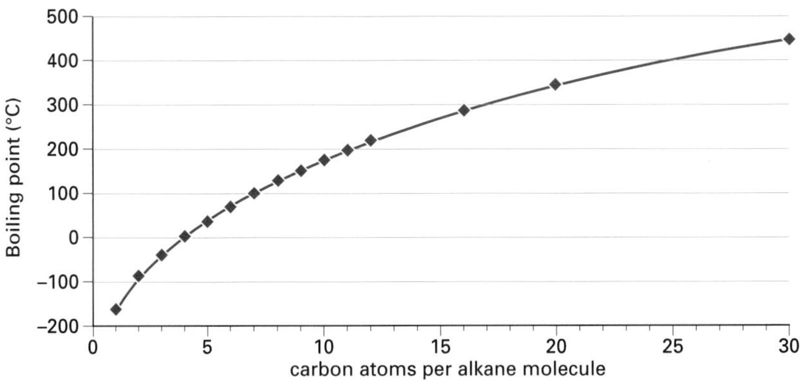

The boiling point of the alkanes increases as the chain length increases.

Vacuum distillation

The residue produced by the *primary distillation* of petroleum contains the largest hydrocarbon molecules. Very high temperatures would normally be needed to separate its components by fractional distillation. This would make the hydrocarbon molecules decompose. To get around this problem, the residue is distilled at reduced pressure, a process called *vacuum distillation*. The fractions in the residue boil at lower temperatures when the pressure is reduced. This *secondary distillation* produces mineral oil, fuel oil, waxes, and bitumen. The bitumen is used to make roads and to make roofs waterproof.

Fractional distillation

The basic needs for fractional distillation are a mixture of liquids with different boiling points, a heat source, a fractionating column, and a condenser. The mixture of liquids is heated and vapour from it rises up the fractionating column. The top of the column is colder than the bottom. Vapour condenses and falls back as a liquid when it reaches a part of the column that is cool enough. The liquid with lowest boiling point escapes from the top as a vapour. This is cooled and condensed to form a liquid using a condenser.

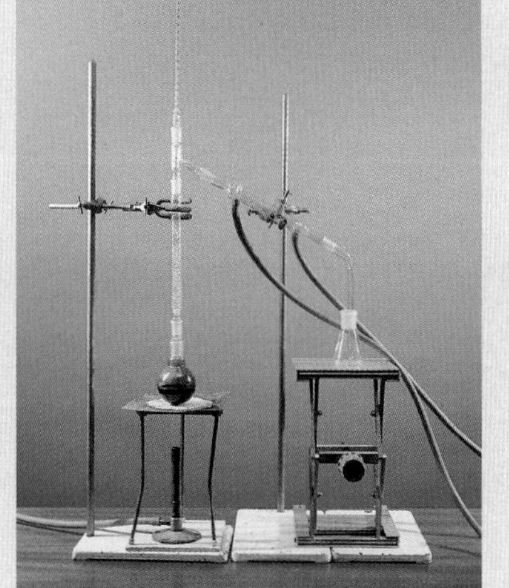

This is a typical laboratory-scale fractional distillation apparatus.

The fractionating column

Over 130 cubic metres of petroleum products are consumed every second worldwide. So the industrial fractional distillation of petroleum takes place on a very large scale. The petroleum is vaporized by heating it to around 350°C in a furnace. The vapours are passed into the fractionating column.

There is a negative temperature gradient inside the column, from about 340°C at the bottom to about 60°C at the top. The largest hydrocarbons have high boiling points. They stay liquid and fall to the bottom of the fractionating column as residue. The smallest hydrocarbons have low boiling points. These leave the top of the column as gases. The remaining hydrocarbons rise until they reach a part of the column that is cold enough for them to condense. They are led out of the column as liquids.

Most of the fractions obtained from petroleum are used as fuels. Naphtha and kerosene are used as **feedstock** for the chemical industry. They are the starting materials for making very many products such as paints, dyes, medicines, polymers, and detergents.

Oil refineries process petroleum to produce a wide range of useful products.

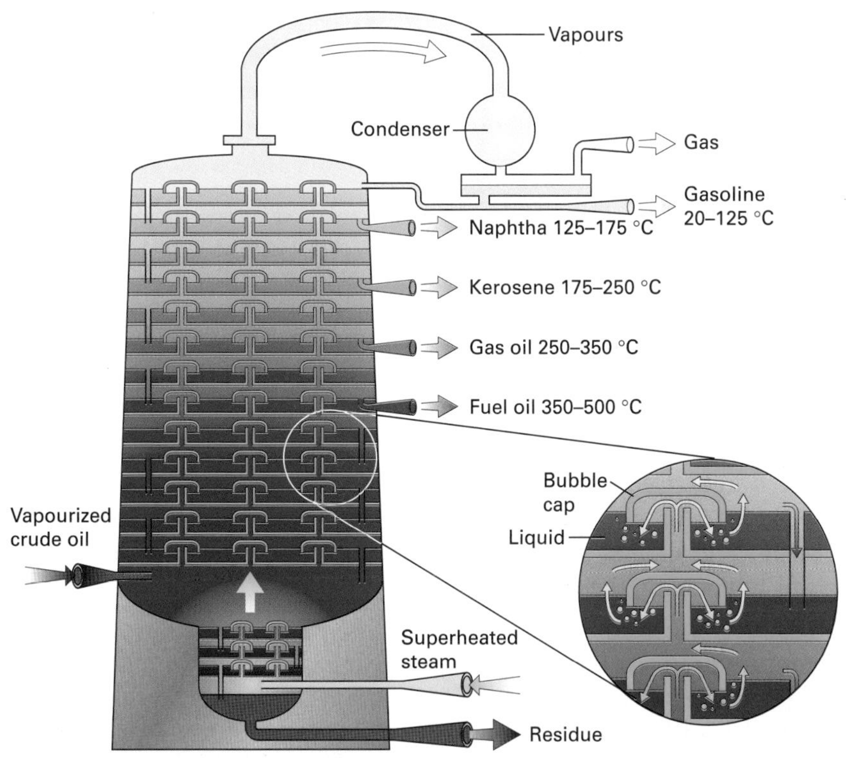

A diagram of a fractionating column from an oil refinery. The bubble caps force the rising vapours to mix with the liquid collected in the trays at each level. This improves the efficiency of the distillation.

Check your understanding

1. What is a fraction?
2. Describe the temperature gradient in a fractionating column.
3. What property of the hydrocarbon molecules in petroleum allows them to be separated by fractional distillation?
4. Describe how different fractions are obtained by fractional distillation.
5. Suggest why the gases from the top of the column are often stored and transported as liquefied petroleum gas, LPG.

fraction	number of carbon atoms per hydrocarbon molecule	approximate range of boiling points (°C)	typical use of fraction
gases	1–4	under 20	camping gas
gasoline (petrol)	4–12	20 to 100	fuel for cars
naphtha	7–14	100 to 150	feedstock for the chemical industry
kerosene (paraffin)	11–15	150 to 250	jet fuel and feedstock for the chemical industry
gas oil (diesel)	15–19	250 to 350	fuel for vehicles and heating
mineral oil	20–30	over 350	lubricating oil
fuel oil	30–40	over 400	fuel for power stations and ships

Typical fractions from petroleum and some of their uses. The fractions from the top are called light fractions *and those from the bottom are called* heavy fractions.

4.05 Cracking

OBJECTIVES

already from GCSE, you know

- that hydrocarbons can be broken down to produce smaller, more useful molecules including alkenes

already from AS Level, you

- understand that different petroleum fractions can be drawn off at different levels in a fractionating column because of the temperature gradient
- can apply IUPAC rules for naming alkenes with up to six carbon atoms

and after this spread you should

- understand that cracking involves the breaking of C—C bonds in alkanes
- know the differences between thermal cracking and catalytic cracking

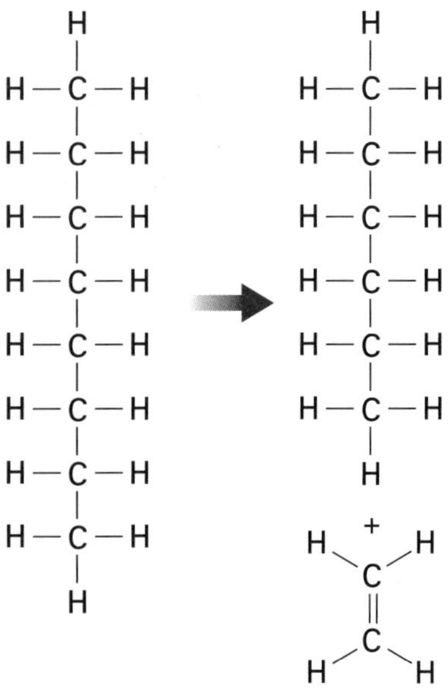

Displayed formulae showing how octane C_8H_{18} can be cracked to form hexane C_6H_{14} and ethene C_2H_4.

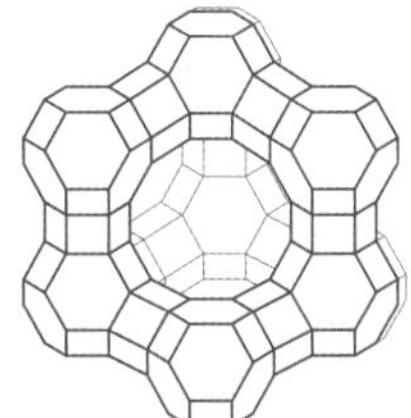

Zeolites have microscopic pores and channels in them.

Petroleum often contains a higher proportion of heavy fractions than is needed. **Cracking** allows these heavy fractions to be used more efficiently.

Overview of cracking

Cracking is a **decomposition reaction**. It involves breaking C—C bonds so that long alkanes are broken down to produce shorter alkanes and alkenes. For example, hexane can be cracked in several ways, including:

$$C_6H_{14} \rightarrow C_4H_{10} + C_2H_4$$

$$C_6H_{14} \rightarrow C_3H_8 + C_3H_6$$

$$C_6H_{14} \rightarrow C_2H_6 + 2C_2H_4$$

Cracking reactions may produce hydrogen, too. For example:

$$C_6H_{14} \rightarrow C_4H_8 + C_2H_4 + H_2$$

Shorter alkanes are more useful as fuels. Alkenes are useful as raw materials for making polymers.

There are two types of cracking, thermal cracking and catalytic cracking.

Thermal cracking

Thermal cracking is the type of cracking you met in your GCSE studies. The bonds in alkanes are relatively strong and a lot of energy is needed to break them.

Thermal cracking happens at high temperatures and pressures. The alkanes are heated to between 450°C and 900°C. Air is kept out and pressures up to 7000 kPa are used. If the alkanes are heated for too long, all their bonds can break. The alkanes will decompose to form carbon and hydrogen. To stop this happening, the alkanes are heated for less than a second.

The products of thermal cracking are different depending on the temperature used. At lower temperatures, the alkane chain tends to break near the middle of the molecule. This produces a higher proportion of medium-sized alkanes and alkenes. At higher temperatures, the alkane chain tends to break near the end of the molecule. This produces a higher proportion of small alkenes such as ethene. The conditions used are adjusted so that the desired products are made. Fractional distillation is used to separate the different products from each other.

Catalytic cracking

Catalytic cracking happens at high temperatures of around 450°C. Unlike thermal cracking, the alkanes are exposed to pressures only slightly above atmospheric pressure. Minerals called **zeolites** are used as catalysts. Zeolites are compounds of aluminium, silicon, and oxygen. They are found naturally but industrial zeolites are manufactured to have precise properties. Zeolites have microscopic pores running through them, which give them a large surface area.

Catalytic cracking is more efficient than thermal cracking. It produces more branched, cyclic, and **aromatic** hydrocarbons than thermal cracking does. These are more useful for making motor fuels than unbranched hydrocarbons.

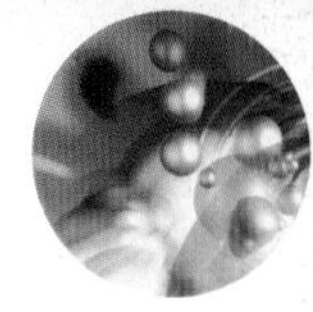

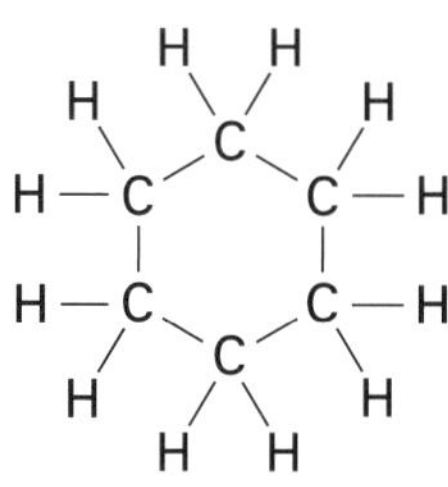

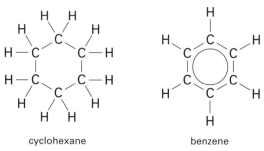

Catalytic cracking produces branched hydrocarbons. It also produces cyclic hydrocarbons such as cyclohexane and aromatic hydrocarbons such as benzene. The circle in the formula of benzene represents delocalized electrons.

Molecular sieves

Zeolites also act as molecular sieves. Unbranched hydrocarbon molecules can enter the pores but branched hydrocarbons cannot. This allows the two types to be separated at the oil refinery. Unbranched hydrocarbons are used as raw materials for the manufacture of detergents.

The detergent industry is a major user of zeolites. Washing powders contain 15–30% zeolites. They remove calcium ions from the water. This softens the water and so improves the efficiency of the detergent. The organic substances also stick to zeolites in the washing water, such as dyes. This stops them from being left on the clothes.

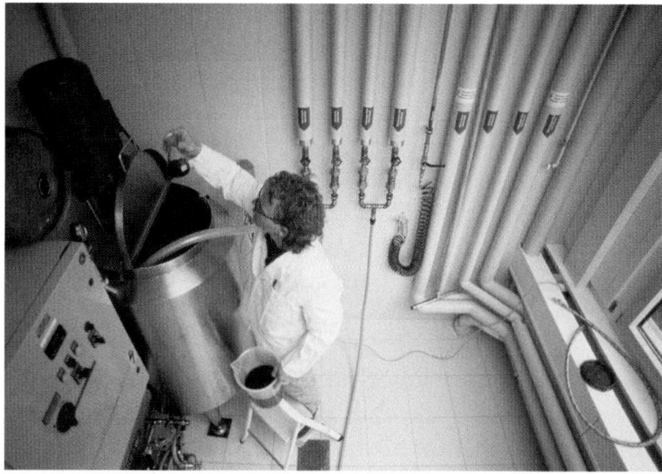

Research is carried out to design and test new washing powders. Standard mixtures of dyes are used to ensure that the tests are fair.

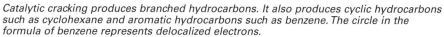

Cracking compared

	thermal cracking	catalytic cracking
temperature	450°C to 900°C	450°C
pressure	high	moderate
catalyst	none	zeolites
products	high proportion of straight-chain alkanes and alkenes useful as raw materials for the chemical industry	high proportion of branched alkanes and alkenes, and cyclic and aromatic hydrocarbons, useful for motor fuels

Thermal cracking and catalytic cracking form different products.

An industrial catalytic cracker or 'cat cracker' at an oil refinery.

Check your understanding

1. a What are the conditions used in thermal cracking?
   b What type of hydrocarbon is produced in high proportions in thermal cracking?
   c Write an equation for the thermal cracking of undecane, $C_{11}H_{24}$, to produce propane and ethene only.
2. a What are the catalysts used in catalytic cracking?
   b What types of hydrocarbon are produced in high proportions in catalytic cracking?
3. Suggest why catalytic cracking is likely to be cheaper than thermal cracking.
4. Write an equation for the thermal cracking of a molecule of dodecane $C_{12}H_{26}$ to produce ethene, propene, and hydrogen gas only.

4.06 A cracking economy

OBJECTIVES

already from GCSE, you know

- that the products from cracking are useful as fuels and raw materials for making ethanol and polymers

already from AS Level, you

- understand that cracking involves the breaking of C—C bonds in alkanes
- know the differences between thermal cracking and catalytic cracking

and after this spread you should

- understand the economic reasons for the cracking of alkanes

Petroleum is a non-renewable resource. There have been oil shortages in the past and there probably will be more in the future.

A spark plug ignites the petrol in a car engine.

Supply and demand

Petroleum often contains more of the heavy fractions than can be sold, and not enough of the lighter, more valuable, fractions. The table shows the typical composition of North Sea oil and the approximate demand for each fraction.

fraction	C atoms per molecule	% supply from distillation	% demand
refinery gas	1–4	2	4
petrol (gasoline)	4–12	12	22
naphtha	7–14	12	5
kerosene	11–15	12	8
gas oil (diesel)	15–19	19	23
mineral oil/fuel oil	20–40	43	38

There is a greater demand for petrol than can be met by fractional distillation alone. There is a greater supply of naphtha than is needed. Cracking allows the conversion of heavy fractions such as naphtha into higher value products. These include ingredients for petrol, and ethene for making ethanol and poly(ethene). It ensures that the petroleum is used more efficiently. This is very important because petroleum is a **non-renewable resource**. It cannot be replaced once it has all been used up.

Making petrol

Petrol manufacturers must mix different hydrocarbons together to produce petrol with the right properties. The petrol must vaporize easily enough to ignite in the engine, but not so easily that excess vapour stops fuel reaching the engine. It must also resist igniting too early in the cylinders of the engine.

Volatility

Short branched alkanes have lower boiling points than long unbranched alkanes. They are more volatile and so vaporize more easily. Petrol contains a mixture of different alkanes to achieve the right amount of volatility. The precise blend of petrol is different in different parts of the world and at different times of year. Winter blends are more volatile than summer blends. A summer blend would not vaporize easily in a cold winter and the engine would be difficult to start. A winter blend would vaporize too easily in the hot summer, causing vapour lock. Liquid fuel would not reach the engine and it would stall.

Classifying oil

Oil is classified according to its density as light, medium, or heavy. The density of *light oils* is less than 0.87 g cm^{-3} and the density of *heavy oils* is more than 0.92 g cm^{3}. The density of a *medium oil* is between these two values. Oil from the North Sea is usually light oil. It has a relatively high proportion of short hydrocarbon molecules. The *API gravity*, devised by the American Petroleum Institute, is used by the oil industry as a measure of the density of oil. Heavy oils with an API of less than 10 sink in water. Light oils from the North Sea have an API of 30 to 40 and float on water.

Octane rating

The petrol should ignite in the engine because of a spark delivered at just the right time. If it ignites before this happens, you hear a characteristic 'knocking' sound. The engine does not perform as well as it should and it can be damaged. Petrol can *auto-ignite* as it is compressed in the cylinder before the spark. The higher the *octane rating* of petrol the less likely it is to auto-ignite and cause knocking. Different hydrocarbons have different octane ratings. They are blended to make petrol with the proper octane rating. This is usually 95 in the UK.

Unbranched hydrocarbons have a greater tendency to auto-ignite than branched, cyclic, and aromatic hydrocarbons. Heptane easily auto-ignites and has a defined octane rating of 0, while 2, 2, 4-trimethylpentane is difficult to auto-ignite and has a defined octane rating of 100. This gives a scale to measure the octane rating of other hydrocarbons.

dodecane octane rating –88

cracking

heptane octane rating 0

+

pent-1-ene octane rating 154

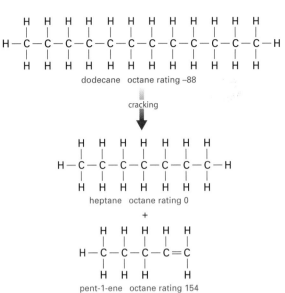

Cracking of dodecane produces hydrocarbons with higher octane ratings.

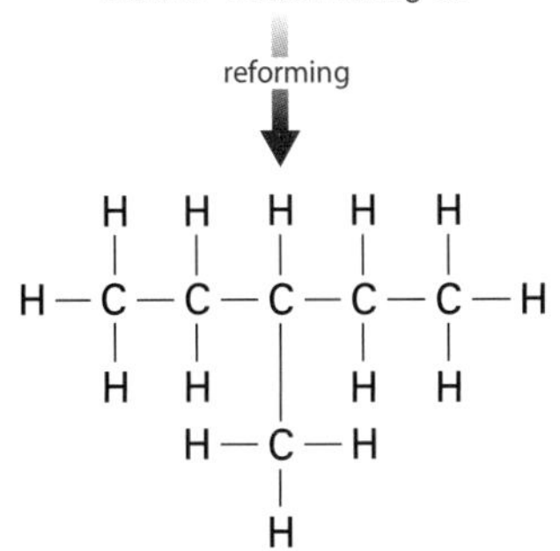

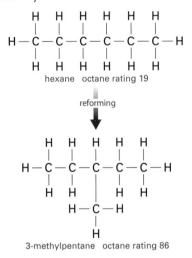

Isomerization *produces branched hydrocarbons with high octane ratings.*

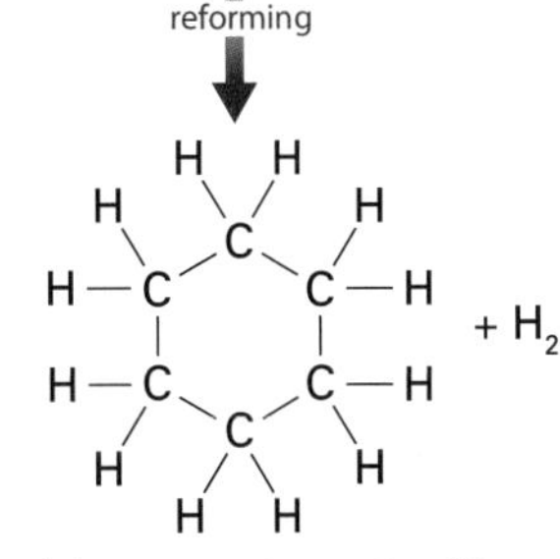

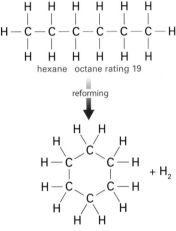

Reforming *produces cyclic hydrocarbons with higher octane ratings. Notice that cyclohexane has the same molecular formula as hexene.*

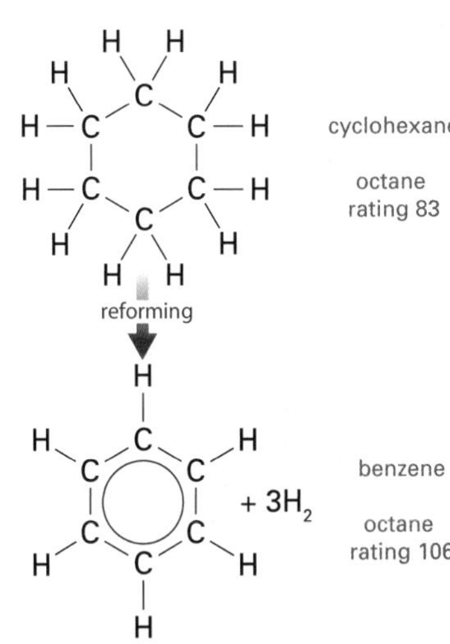

Reforming *also produces aromatic hydrocarbons with high octane ratings.*

Check your understanding

1. Outline the economic reasons for cracking alkanes.
2. Explain why petrol manufacturers produce different blends of petrol during the year.
3. Heptane has an octanerating of 0 and 2, 2, 4-trimethylpentane has an octane rating of 100.
   a. Draw the displayed formula of 2, 2, 4-trimethylpentane.
   b. A mixture of 50% heptane and 50% 2, 2, 4-trimethylpentane has an octane rating of 50. What mixture of the two hydrocarbons will have an octane rating of 95?
4. The products of catalytic cracking include branched hydrocarbons and aromatic hydrocarbons such as benzene. Explain why these are useful ingredients of petrol.

4.07 Alkanes as fuels

OBJECTIVES

already from GCSE, you know

- that alkanes are hydrocarbons
- that carbon dioxide, water vapour, carbon monoxide, and carbon particles may be released when a fuel burns

and after this spread you should

- know that alkanes are used as fuels
- understand that their combustion can be complete or incomplete
- know that the combustion of fossil fuels results in the release of carbon dioxide into the atmosphere
- know that the internal combustion engine produces pollutants that can be removed using catalytic converters

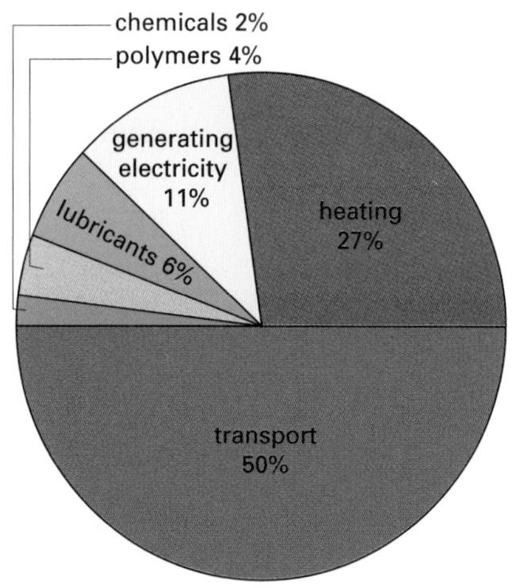

The biggest single of use petroleum products is as fuel for vehicles, ships, and aircraft.

Petroleum products are used to make very many substances from paint, plastics, and polymers to roads, lubricants, and medicines. But the majority of petroleum products are used as fuels for vehicles, heating, and generating electricity. Alkanes and other hydrocarbons react with oxygen in the air when they burn. This combustion can be complete or incomplete.

Complete combustion

Complete combustion happens when there is a good supply of oxygen. The carbon in the alkane is oxidized to carbon dioxide and its hydrogen is oxidized to water. In general:

alkane + oxygen → carbon dioxide + water

Methane is the main gas in natural gas. Here is the equation for the complete combustion of methane.

$$CH_4(g) + 2O_2(g) \rightarrow CO_2(g) + 2H_2O(l)$$

Hexane is a component of petrol. Here is the equation for the complete combustion of hexane.

$$C_6H_{14}(l) + 9\tfrac{1}{2}O_2(g) \rightarrow 6CO_2(g) + 7H_2O(l)$$

Notice that it is acceptable to use a half to balance the equation.

Incomplete combustion

Incomplete combustion happens when there is a limited supply of oxygen. The hydrogen in the alkane is still oxidized to water but the carbon is only partially oxidized to form carbon monoxide instead of carbon dioxide.

Here are equations for the incomplete combustion of methane and hexane.

$$CH_4(g) + 1\tfrac{1}{2}O_2(g) \rightarrow CO(g) + 2H_2O(l)$$

$$C_6H_{14}(l) + 6\tfrac{1}{2}O_2(g) \rightarrow 6CO(g) + 7H_2O(l)$$

Carbon particles may also be released during incomplete combustion, forming soot and smoke.

$$CH_4(g) + O_2(g) \rightarrow C(s) + 2H_2O(l)$$

$$C_6H_{14}(l) + 3\tfrac{1}{2}O_2(g) \rightarrow 6C(s) + 7H_2O(l)$$

NO_x and unburned hydrocarbons

The temperature of the fuel and air mixture in an engine can reach well over 1000°C when the fuel ignites. Under these conditions nitrogen and oxygen in the air can react together to produce various oxides of nitrogen. These are called NO_x. For example, nitrogen monoxide is formed, which easily oxidizes further to form nitrogen dioxide.

$$N_2(g) + O_2(g) \rightarrow 2NO(g)$$

$$2NO(g) + O_2(g) \rightarrow 2NO_2(g)$$

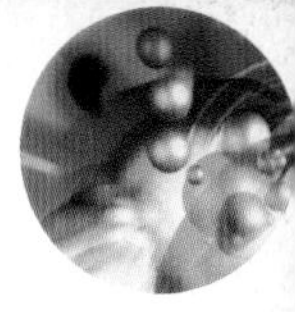

Nitrogen dioxide is a component of acid rain. It reacts with water vapour and oxygen in the clouds to form nitric acid.

$$2NO_2(g) + H_2O(l) + \frac{1}{2}O_2(g) \rightarrow 2HNO_3(aq)$$

Some of the hydrocarbons in the fuel pass through the engine without being oxidized. These are called **unburned hydrocarbons** or volatile organic compounds (VOCs). They can react with NO_x in the presence of sunlight to produce a photochemical smog. This forms an unpleasant haze in the air. It contains many chemicals that can irritate the eyes, throat, and lungs.

Santiago, Chile: photochemical smog can make the symptoms of asthma and bronchitis worse.

Catalytic converters

Catalysts are substances that change the rate of a chemical reaction without being changed by the end of the reaction. A catalytic converter is a device fitted to the exhaust system of cars. It removes pollutants from the exhaust gases. There is a honeycomb in the converter made from ceramic coated with a thin layer of catalyst. The honeycomb allows the exhaust gases to flow through the converter and it provides a large surface area for reactions to happen on. A lambda probe measures the oxygen level in the exhaust gases. Modern *three-way converters* catalyse three reactions, a reduction reaction and two oxidation reactions. Catalytic converters are able to convert more than 90% of the pollutants in exhaust gases into less harmful gases.

Reduction

NO_x is reduced to nitrogen using a platinum-rhodium catalyst. For example:

$$2NO(g) \rightarrow N_2(g) + O_2(g)$$

Oxidation

Carbon monoxide and unburned hydrocarbons are oxidized using a platinum-palladium catalyst.

- carbon monoxide is oxidized to carbon dioxide:

$$CO(g) + \frac{1}{2}O_2(g) \rightarrow CO_2(g)$$

- unburned hydrocarbons are oxidized to carbon dioxide and water vapour. For example:

$$C_9H_{20}(g) + 14O_2(g) \rightarrow 9CO_2(g) + 10H_2O(g)$$

Carbon monoxide

Carbon monoxide is a toxic gas. It binds to haemoglobin in the red blood cells more strongly than oxygen does. This reduces the capacity of the blood to carry oxygen. Carbon monoxide also interferes with the release of oxygen from the red blood cells. The gas is difficult to detect because it is colourless, odourless, and tasteless. Carbon monoxide poisoning causes headaches, unconsciousness, and death. It is important to maintain heating boilers correctly and to allow sufficient ventilation to avoid the release of this gas.

Household carbon monoxide detectors sound an alarm if the air contains excessive amounts of carbon monoxide.

Check your understanding

1. What are the differences between complete combustion and incomplete combustion?
2. Write an equation for the complete combustion of heptane, C_7H_{16}.
3. Write an equation for the incomplete combustion of octane, C_8H_{18}, in which water vapour and one other product are formed.
4. What is NO_x and why is it a problem?
5. Explain with the help of equations how a catalytic converter works.

4.08 Acid rain

OBJECTIVES

already from GCSE, you know

- that sulfur dioxide causes acid rain and carbon dioxide causes global warming

and after this spread you should

- know that combustion of hydrocarbons containing sulfur leads to sulfur dioxide, which causes air pollution
- understand how sulfur dioxide can be removed from flue gases using calcium oxide

Kilauea Volcano in Hawaii emits about 150 tonnes of sulfur dioxide every day. Human activities release around 400 000 tonnes worldwide every day.

These trees have been damaged by acid rain.

Acid rain

Rain is naturally acidic because carbon dioxide from the atmosphere dissolves in it. **Acid rain** is rain that is more acidic than normal. Natural events can cause it. For example, volcanoes throw huge amounts of acidic gases such as sulfur dioxide and hydrogen chloride into the atmosphere when they erupt. But human activity is the major cause of acid rain.

Fossil fuels such as coal, petroleum, and natural gas naturally contain sulfur compounds. Sulfur dioxide is released when these fuels are burned. This dissolves in water vapour to form sulfurous acid:

$SO_2(g) + H_2O(l) \rightleftharpoons H_2SO_3(aq)$ ($\rightleftharpoons$ means that it is a reversible reaction)

The sulfurous acid is then oxidized in a series of reactions to form sulfuric acid, $H_2SO_4(aq)$.

Acid rain causes a lot of damage to the environment. It acidifies lakes, rivers, and soil. This makes it difficult or impossible for living organisms to survive there. The acidic conditions release toxic substances from the soil, poisoning plants. Buildings may be damaged, too, especially those built from limestone.

Flue gas desulfurization

The waste gases from boilers and furnaces are called **flue gases**. **Flue gas desulfurization**, FGD, is a process that removes sulfur dioxide from waste gases. It is particularly useful for coal-fired power stations, where large amounts of sulfur dioxide are produced. The sulfur dioxide is removed using a base such as calcium oxide.

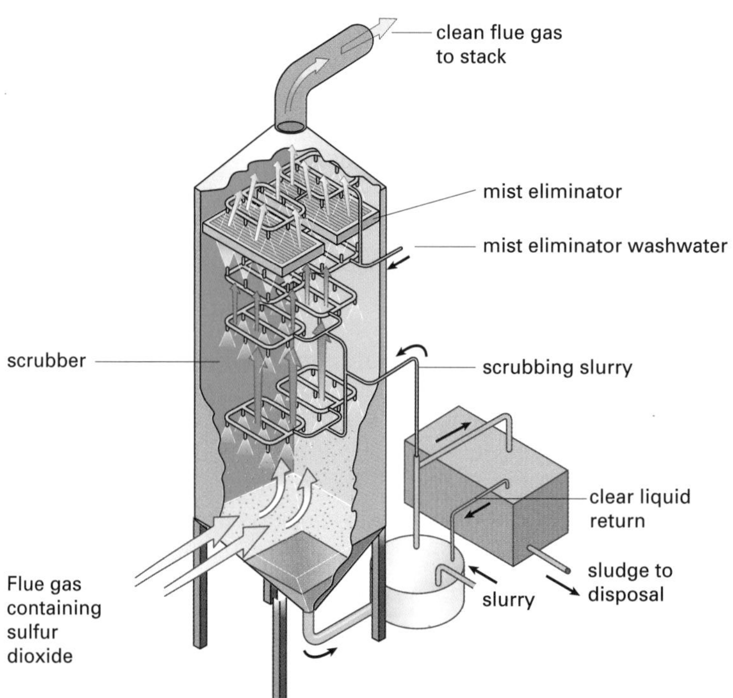

A flue gas desulfurization plant. Notice that the flue gases flow in the opposite direction to the slurry. This makes the process more efficient.

Calcium carbonate

Limestone is nearly pure calcium carbonate. It is plentiful and cheap. Powdered limestone is mixed with water and sprayed through the waste gas. It reacts with the sulfur dioxide to make calcium sulfite, $CaSO_3$:

$$CaCO_3(s) + SO_2(g) \rightarrow CaSO_3(s) + CO_2(g)$$

One drawback is that carbon dioxide is produced as a by-product. This is a greenhouse gas and contributes to global warming.

Calcium oxide

Calcium oxide or lime is produced by heating limestone so that it decomposes. It also reacts with sulfur dioxide in waste gases:

$$CaO(s) + SO_2(g) \rightarrow CaSO_3(s)$$

At first sight this might seem to be a better FGD process, as there is no by-product apart from calcium sulfite. But there is another side to the process. Carbon dioxide is a by-product of making the calcium oxide in the first place:

$$CaCO_3(s) \xrightarrow{\text{heat}} CaO(s) + CO_2(g)$$

Calcium hydroxide

Calcium hydroxide or slaked lime is produced by reacting calcium oxide with water. This reacts with sulfur dioxide in waste gases, too:

$$Ca(OH)_2(aq) + SO_2(g) \rightarrow CaSO_3(s) + H_2O(l)$$

Again this might seem to be a better FGD process, as carbon dioxide is not a by-product. But you will recall that the production of calcium oxide to make the calcium hydroxide releases carbon dioxide, so overall this gas is still produced.

Gypsum

Calcium sulfite is almost insoluble in water and is difficult to dispose of. Fortunately it can be oxidized in the presence of water to make hydrated calcium sulfate. This is *gypsum*, used to make plasterboard for lining interior walls and ceilings.

$$CaSO_3(s) + \tfrac{1}{2}O_2(g) + 2H_2O(l) \rightarrow CaSO_4.2H_2O(s)$$

Check your understanding

1. With the help of an equation, explain how sulfur dioxide emissions cause acid rain.
2. State two environmental effects of acid rain.
3. Describe how calcium oxide is used in flue gas desulfurization.
4. Suggest why international treaties are needed to combat acid rain, rather than just individual countries working alone.
5. Discuss the economic advantages and disadvantages of limiting emissions of sulfur dioxide.

Science @ Work — Treaties

Sulfur dioxide emissions are falling because of FGD and other technologies. The *Convention on Long-range Transboundary Air Pollution* was signed in 1979 by thirty countries. It was aimed at controlling emissions of sulfur to reduce the effects of acid rain. The 1999 *Gothenburg Protocol* is a later addition to the Convention. It sets country-by-country targets for reducing emissions of sulfur dioxide and other pollutants.

Science @ Work — Energy considerations

Thermal energy is needed to decompose limestone to make calcium oxide. Electrical energy drives the machinery in the FGD plant. Fossil fuels are the world's main source of energy for heating and generating electricity. So carbon dioxide is released in both cases. The extra energy costs are passed on to the consumer in higher electricity prices.

It is impractical to fit an FGD plant to a car. The sulfur is removed from the fuel at the oil refinery to make "low sulphur" diesel and petrol. The sulfur can be used in the manufacture of sulfuric acid.

4.09 Global warming

OBJECTIVES

already from GCSE, you know

- that emissions of carbon dioxide from burning fossil fuels are linked to global warming

and after this spread you should be able to

- understand the link between absorption of infrared radiation and global warming
- know that carbon dioxide, methane, and water vapour are greenhouse gases and may contribute to global warming
- know that combustion of fossil fuels, including the alkanes, releases carbon dioxide into the atmosphere

Infrared radiation

Infrared radiation is part of the electromagnetic spectrum. This includes visible light, ultraviolet light, and radio waves. You cannot see infrared radiation but you do feel it as heat.

The greenhouse effect

When electromagnetic radiation from the Sun reaches the Earth, some of it is reflected back into space. The rest is absorbed and heats the Earth up. As a result of this, infrared radiation is radiated from the Earth towards space.

Certain gases in the atmosphere, called **greenhouse gases**, are particularly good at absorbing infrared radiation. Water vapour, carbon dioxide, and methane molecules are like this. They absorb infrared radiation and then re-emit it in a random direction. Some infrared radiation eventually escapes into space after being randomly absorbed and re-emitted by many molecules in the atmosphere. The rest warms the atmosphere up. This is called the **greenhouse effect**.

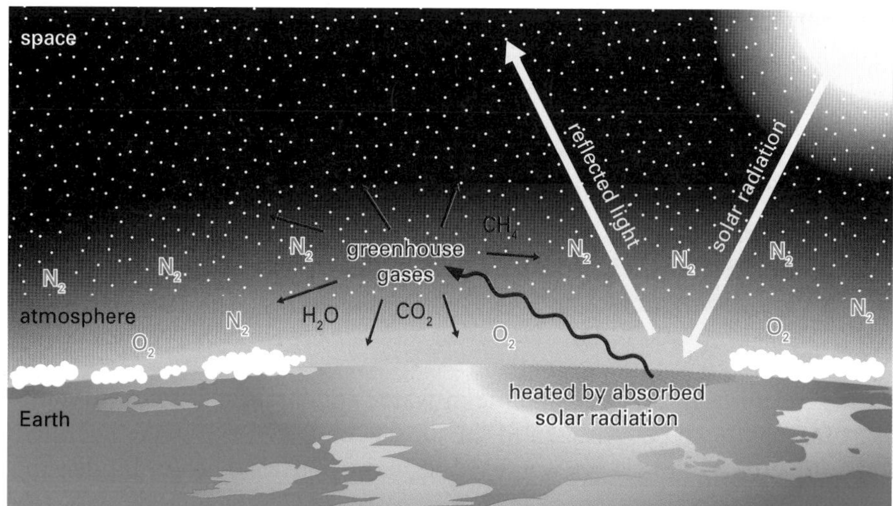

The Earth would be about 33°C colder without the greenhouse effect.

Radiative forcing

Some greenhouse gases are better than others at warming the atmosphere. *Radiative forcing* is a measure of the ability of a gas to warm the atmosphere compared to what would happen if it was not there.

gas	approximate relative radiative forcing
water vapour	1
methane	7
carbon dioxide	24

Carbon dioxide is a particularly effective greenhouse gas.

An enhanced greenhouse effect

The combustion of fossil fuels releases carbon dioxide and water vapour to the atmosphere. Methane is released as result of using natural gas and petroleum, and through mining coal. Farming also produces methane. Rice paddy fields, cattle, and decomposing plant and animal material are significant sources of methane. Chemical reactions in the upper atmosphere eventually oxidize methane to carbon dioxide and water vapour.

The release of additional greenhouse gases like these by human activities has the potential to cause an enhanced greenhouse effect. In turn this can increase the temperature of the atmosphere. It can cause **global warming** and lead to climate change.

Climate change as a result of global warming has brought increased rainfall to the UK. This causes rivers such as the Ouse at York to overflow, damaging farmland and buildings.

Check your understanding

1. Name three greenhouse gases.
2. Explain why the combustion of fossil fuels is thought to contribute to global warming.
3. Outline some reasons why has it taken many years for scientific opinion to reach a consensus on climate change.

A scientific consensus

It has taken many years for scientists to reach a consensus of scientific opinion on global warming and climate change. The *Intergovernmental Panel on Climate Change* has considered the scientific evidence. It concludes that the rise in average temperatures since the middle of the last century is very likely to be due to increased greenhouse gases, caused by human activities. They further conclude that continued emissions of greenhouse gases will cause further warming, and that the effects on the climate are very likely to be larger than the changes seen in the last century.

Why does the panel say 'very likely', since this introduces the idea of uncertainty? The Earth and its atmosphere form a very large and complex system. This makes it difficult to collect and interpret evidence in a way that cannot be disputed. There is no way to carry out a controlled fair test in the way the scientists would do normally. In a sense, the human race is carrying out a huge experiment on its home planet. Scientists can make predictions based on computer models and historical data, but the actual outcome of the experiment will only be revealed as time goes by.

Unit 2: *Chemistry in Action* introduces more chemical principles and covers their applications. Extending work from GCSE, you find out more about energy changes in reactions, rates of reaction, and chemical equilibria. The properties of the elements in groups 2 and 7 are investigated, together with the practical uses of these elements and some of their compounds. Environmental issues are explored, such as sustainable development, extraction of metals, carbon neutral fuels, and damage to the Earth's ozone layer. The organic chemistry from Unit 1 is developed to include alkenes, haloalkanes, alcohols, and modern analytical techniques.

AQA Approved Specification (July 2007)

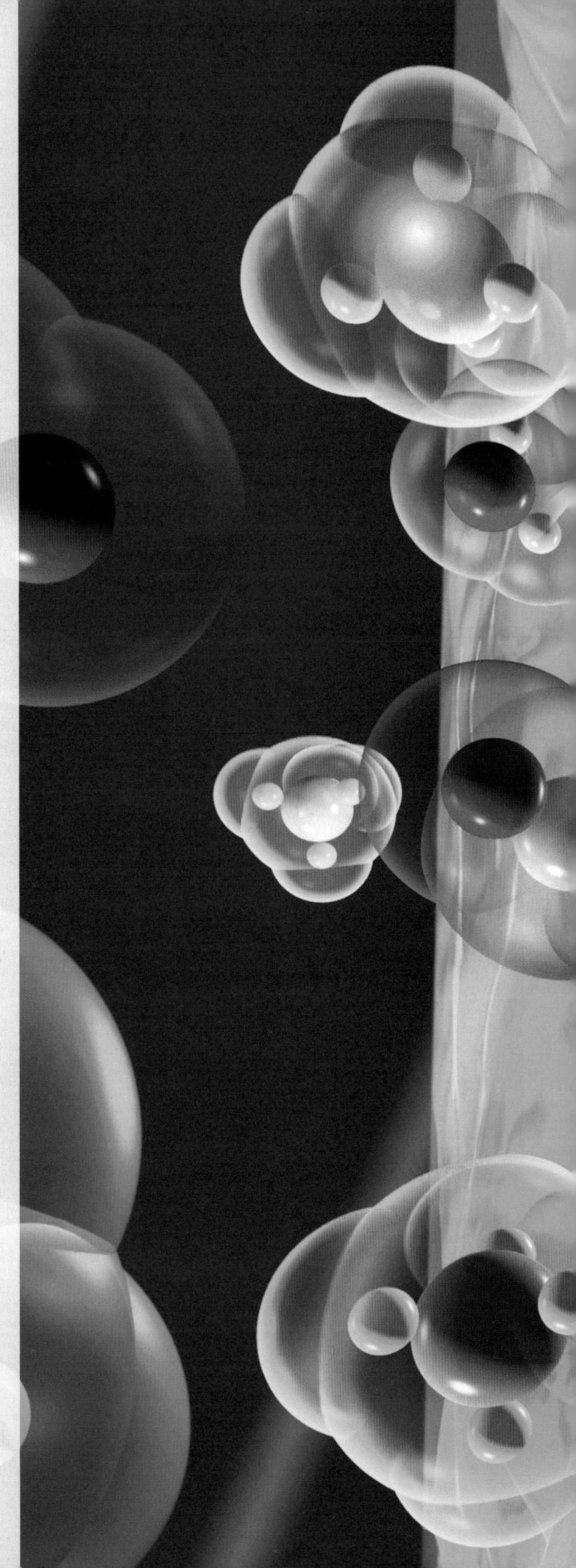

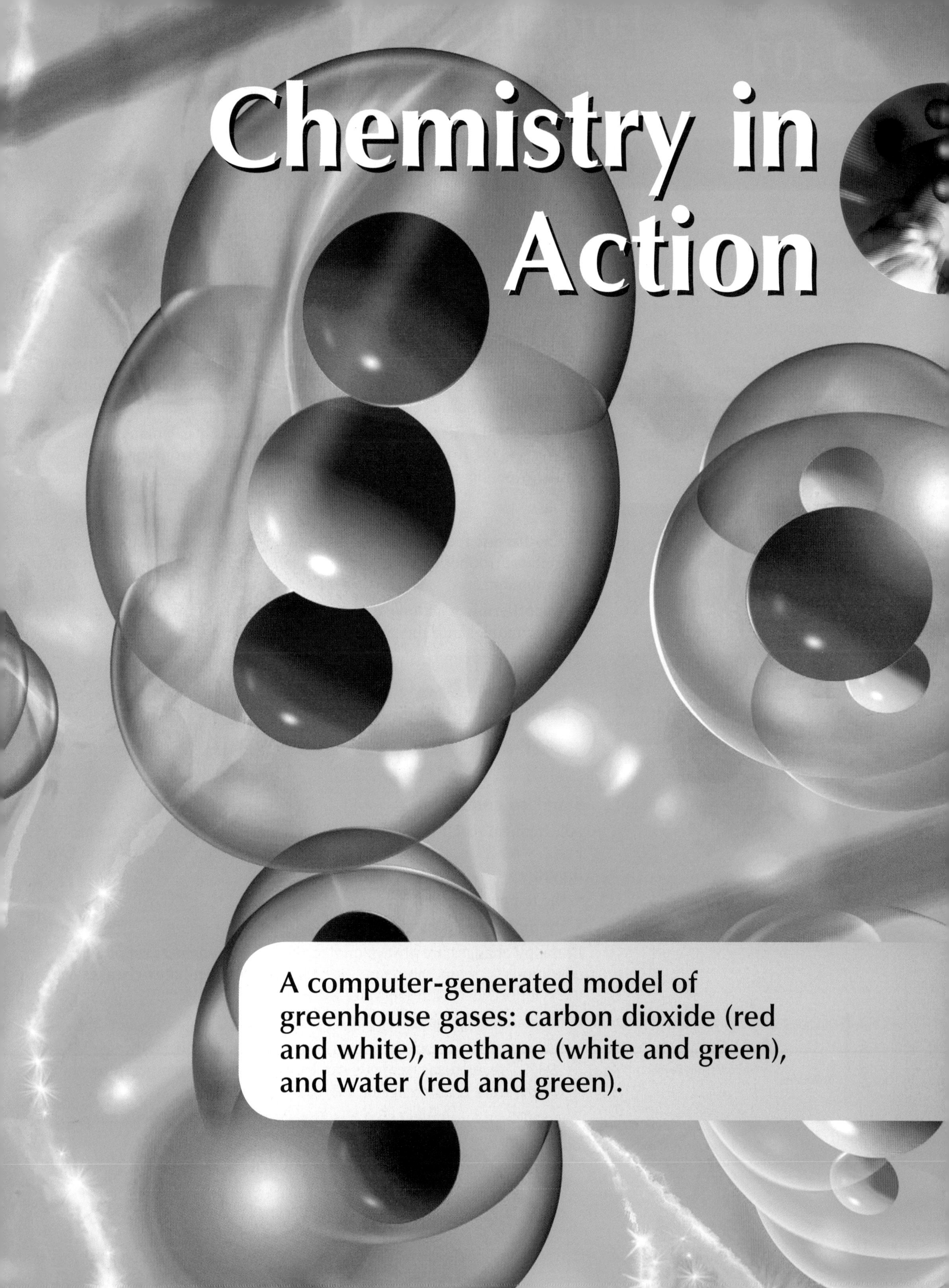

Chemistry in Action

A computer-generated model of greenhouse gases: carbon dioxide (red and white), methane (white and green), and water (red and green).

5.01 Enthalpy changes and standard conditions

OBJECTIVES

already from GCSE, you know

- that some reactions release energy and some absorb energy

and after this spread you should be able to

- define enthalpy
- explain the difference between exothermic and endothermic reactions
- state standard conditions
- define standard enthalpy of combustion and standard enthalpy of formation

The combustion of rocket fuel is a highly exothermic process.

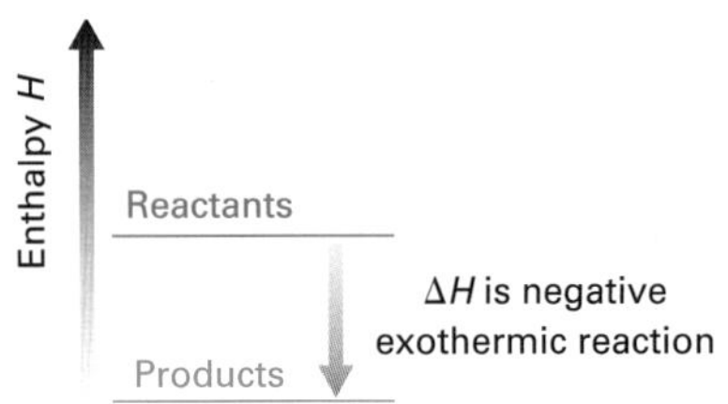

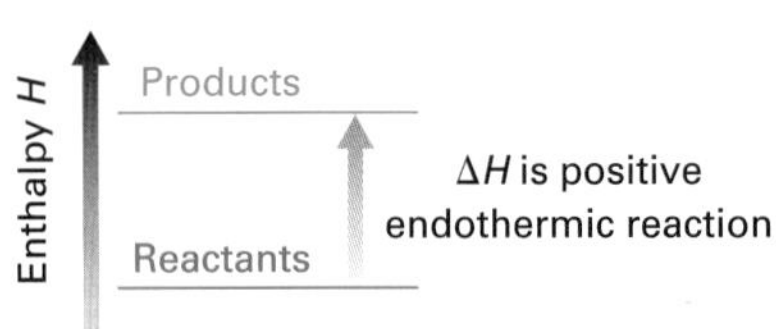

For an exothermic reaction, the enthalpy change, ΔH, is negative. For an endothermic reaction, the enthalpy change, ΔH, is positive.

Chemical reactions involve

- the breaking of bonds between atoms
- the making of new bonds between atoms

These processes involve energy exchanges between the reacting system and its surroundings.

Enthalpy

Enthalpy is the term used to describe the heat (energy) content of a system. It has the symbol H and is measured in kilojoules. It is only really possible to measure enthalpy changes (delta H, written as ΔH). This can be calculated from measured changes in temperature. Enthalpy change is defined as 'the heat energy change measured under conditions of constant pressure'.

In a chemical reaction the enthalpy change is easily calculated by subtracting the (total) enthalpy of the reactants (H_r) from the (total) enthalpy of the products (H_p).

$$\Delta H = H_p - H_r$$

- **Exothermic** reactions *release* heat energy (heat flows from the system to the surroundings). ΔH is always negative, so the value always has a minus sign.
- **Endothermic** reactions *absorb* heat energy (heat flows from the surroundings into the system). ΔH is always positive.

These changes can be shown on enthalpy change diagrams.

Standard conditions

100 kPa and a stated temperature (normally 298 K).

Standard conditions

All chemical reactions have enthalpy changes associated with them. For any reaction, the value of the enthalpy change is dependent on two things:

- the temperature and pressure
- the amount of substance used

Enthalpy changes are always quoted relative to a *standard* set of conditions of temperature and pressure. *Standard* enthalpy changes use the symbol $\Delta H^{\ominus}$. This signifies that

- the standard pressure is 100 kPa
- the substances are in their standard states (this is normally at 298 K, *but*, as measurements can be made at other temperatures the temperature is always given e.g. $\Delta H^{\ominus}$ (298 K))

The value (in kilojoules) of ΔH is always given *per mole of substance*.

Standard enthalpy change of formation $\Delta H_f^{\ominus}$

This is the enthalpy change when one mole of a compound is formed from its elements in their standard states.For example, when producing magnesium oxide the standard enthalpy of formation is

$$Mg(s) + \tfrac{1}{2}O_2(g) \rightarrow MgO(s)$$

$$\Delta H_f^{\ominus}\ (298K) = -602\,kJ\,mol^{-1}$$

(Note: this is for one mole of magnesium oxide; all state symbols are given in addition to the temperature at which the enthalpy change is measured; the formation of one mole of magnesium oxide releases 602 kJ of energy into the surroundings – had two moles of magnesium oxide been formed then the amount of energy released would have been 1204 kJ.)

Standard enthalpy change of combustion $\Delta H_c^{\ominus}$

This is the enthalpy change when 1 mole of a substance is completely burned in oxygen under standard conditions (100 kPa and 298 K).

For example the standard enthalpy of combustion of ethane is:

$$C_2H_6(g) + \tfrac{7}{2}O_2(g) \rightarrow 2CO_2(g) + 3H_2O(l)\ \Delta H_c^{\ominus}\ (298\,K) = -1560\,kJ\,mol^{-1}$$

(Note: this is for one mole of ethane; all state symbols are quoted; when completely burnt in oxygen, one mole of ethane releases 1560 kJ of energy into the surroundings.)

Other enthalpy changes

Many other types of enthalpy change can be measured and the term 'standard enthalpy change of reaction' can be applied to many other reactions. At AS level you will be concerned with formation and combustion reactions.

Check your understanding

1. Enthalpy changes are always quoted under standard conditions. Explain the reasons for this.
2. Write equations for reactions involving the following enthalpy changes:
   a Enthalpy of formation of methane, CH_4
   b Enthalpy of formation of ammonia, NH_3
   c Enthalpy of combustion of butane, C_4H_{10}
3. The standard enthalpy of combustion of methane at 298 K is $-890\,kJ\,mol^{-1}$ State the amount of energy released when the following quantities of methane are completely combusted in oxygen under standard conditions:
   (a) 1 mol (b) 5 mol (c) 30 g (d) 250 g
4. The formation of 32 g of sulfur dioxide under standard conditions releases 197.5 kJ of energy. Calculate the standard enthalpy of formation of sulfur dioxide.

Enthalpy of formation $\Delta H_f^{\ominus}$

The enthalpy change when one mole of a compound is formed from its elements in their standard states.

Enthalpy of formation of elements

By definition the standard enthalpy of formation of an element in its standard state is zero.

Enthalpy of combustion $\Delta H_c^{\ominus}$

The enthalpy change when one mole of a substance is completely burned in oxygen under standard conditions.

Worked example

Octane (C_8H_{18}) is a key component of petrol. Chemical engineers need to be able to calculate how much energy is released per gram of octane used.

$\Delta H_c^{\ominus}\ C_8H_{18}(l) = -5512\,kJ\,mol^{-1}$

- 1 mol of octane releases 5512 kJ of energy
- 1 mol of octane has a mass of 114 g.

So 1 g of octane releases 5512/114 = 48.3 kJ of energy

5.02 Calculating enthalpy of reaction

OBJECTIVES

already from GCSE, you know

- that some reactions release energy and some absorb energy

and after this spread you should be able to

- define specific heat capacity
- calculate the heat energy released/absorbed in a reaction
- calculate enthalpy of reaction

Simple calorimetry

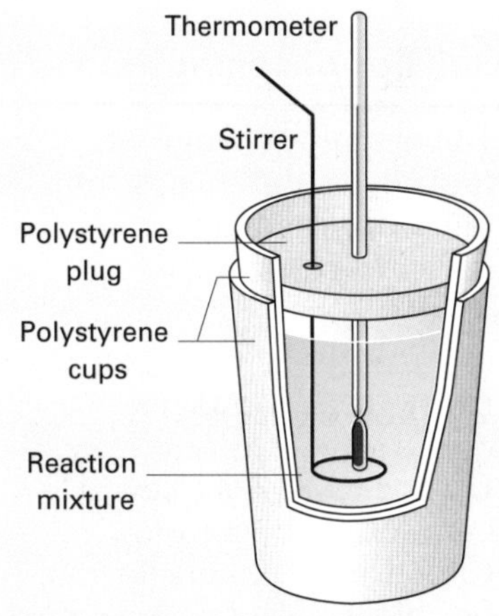

This simple apparatus can be adapted to measure the heat change for many different reactions.

Two expanded polystyrene cups are placed one inside the other. A wire stirrer is used to ensure thorough mixing. If the mass and specific heat capacity of the reaction mixture are sufficiently greater than those of the thermometer and stirrer, it can be assumed that all the heat exchanged in the reaction is used to alter the temperature of the water.

The value of enthalpy changes accompanying chemical reactions can be determined by the use of a **calorimeter**. The purpose of a calorimeter is to insulate the reaction system from its surroundings. The reaction is carried out in the calorimeter and the change in temperature of the calorimeter is measured. This enables the enthalpy change for the reaction to be determined.

Calculating energy changes

There are several types of enthalpy change that can be measured using a calorimeter. These include enthalpy changes of

- dissolving
- neutralization reactions between acids and bases
- formation
- combustion

The heat energy change for any reaction can be calculated using the relationship below:

heat = mass of substance × specific heat capacity × temperature change

$q = mc\Delta T$

Specific heat capacity (c) is the heat energy required to raise the temperature of 1g of a substance by 1K. For pure water the specific heat capacity is 4.18 $JK^{-1}g^{-1}$.

Enthalpy of solution

A known mass of solid is totally dissolved in a large excess of water whose mass is known. The change in temperature of the water is measured.

Worked example

In an experiment using a simple calorimeter, 8.00g of ammonium nitrate (NH_4NO_3) was dissolved in 50.0g of water. The temperature fell by 10.1°C. Calculate the enthalpy change for this process. (The specific heat capacity of water is 4.18 $JK^{-1}g^{-1}$. The heat capacity of the container can be ignored.)

Step 1 *Calculate the heat change from the change in temperature using* $q = mc\Delta T$

q = mass x specific heat capacity of water × change in temperature

$q = 58 \times 4.18 \times 10.1 = 2448\,J$

Step 2 *Calculate the moles of ammonium nitrate dissolved*

$$\text{moles} = \frac{\text{mass}}{M_r} = \frac{8}{80} = 0.10$$

Step 3 *Calculate the enthalpy change per mole of ammonium nitrate (ΔH).*

0.10mol of ammonium nitrate absorbed 2.448kJ

1mol of ammonium nitrate would absorb $\frac{2.448}{0.10} = 24.48$ kJ

The reaction is endothermic, so the enthalpy change is positive

$$\Delta H = +24.48\,kJ\,mol^{-1}$$

Enthalpy of combustion

A known mass of a substance is combusted. The heat energy released in the reaction is absorbed in a known volume of water and the temperature change of the water is measured and recorded.

Worked example

An experiment was carried out to determine a value for the enthalpy of combustion of liquid hexane using the apparatus shown. Burning 1.75 g of hexane caused the temperature of 250 g of water to rise by 78°C. Use this information to calculate a value for the enthalpy of combustion of hexane. (The specific heat capacity of water is 4.18 J K^{-1}g^{-1}. The heat capacity of the container can be ignored.)

Step 1 *Calculate the heat change from the change in temperature using* $q = mc\Delta T$

$q = 250 \times 4.18 \times 78 = 81510\,\text{J} = 81.51\,\text{kJ}$

Step 2 *Calculate the moles of hexane combusted*

$$\text{moles} = \frac{\text{mass}}{M_r}$$

moles = 1.75 / 86 = 0.020

Step 3 *Calculate the enthalpy change per mole of hexane (ΔH)*

0.02 mol of hexane released 81.51 kJ

$$1 \text{ mole of hexane would release } \frac{81.51}{0.020} = 4075.5\,\text{kJ}$$

The reaction is exothermic, so the enthalpy change is negative.

$\Delta H = -4075.5\,\text{kJ mol}^{-1}$

The quoted enthalpy combustion of hexane is –4194 kJ mol^{-1}. The less exothermic value calculated here is due to heat loss from the calorimeter used in this experiment.

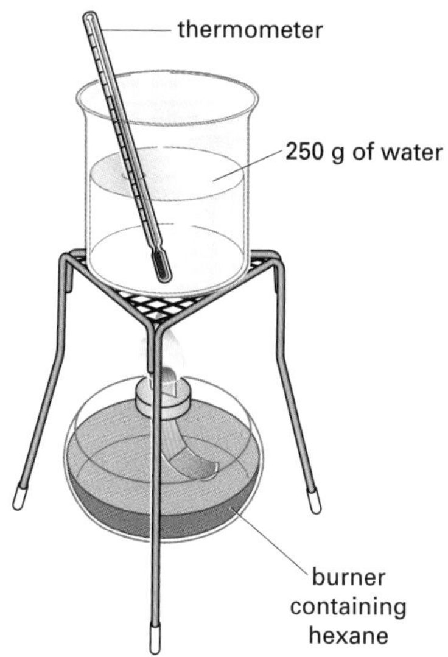

The fuel is burnt in a simple spirit burner and the heat energy released is used to heat a known mass of water.

Cooling curves

For reactions carried out in a simple calorimeter, allowance can be made for heat losses by plotting a graph of temperature against time. If it is assumed that heat loss in the experiment is constant then an accurate figure for the maximum temperature change is obtained by extrapolation of the graph.

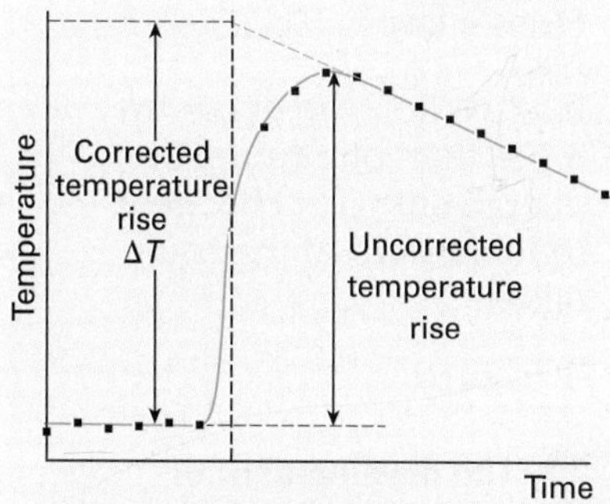

The initial temperature line and the final temperature curve are extrapolated to enable the temperature change to be accurately determined at the point of mixing.

Check your understanding

1. The specific heat capacity of water is 4.18 J K^{-1}g^{-1}. Use this data to calculate the heat energy needed to raise the temperature of a typical mug of water (containing 350 cm^3) of water from 5°C to 100°C. Compare this value with that needed to raise the temperature of a whole kettle of water (containing 1700 cm^3 water) by the same amount. How could this information be used to explain why we should not always boil a full kettle of water?
2. When 1.50 g of ethanol was burnt and the heat energy evolved used to heat 500 g of water, the temperature of the water rose by 18°C.
   - **a** Calculate the enthalpy of combustion of ethanol.
   - **b** The value you have calculated is lower than the listed value. Give two possible reasons for this.
3. The enthalpy of solution of magnesium bromide is –186 kJ mol^{-1}. Assuming no heat is lost to the surroundings, calculate the temperature change when 4 g of magnesium bromide is dissolved in 400 cm^3 of water.

5.03 Energy cycle diagrams

OBJECTIVES

after this spread you should be able to

- recall Hess's law
- construct simple energy cycles
- carry out calculations using enthalpy of formation

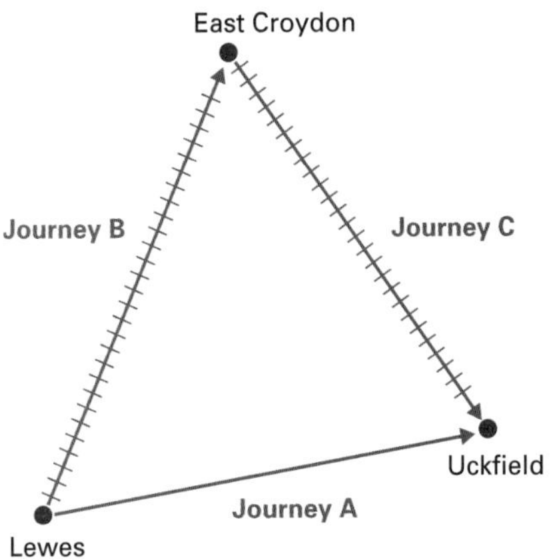

A journey that cannot be made directly can be made using two (or more) different journeys. The same principle can be applied to reactions.

Hess's law

If a reaction can occur by more than one route, the overall enthalpy change is independent of the route taken.

The development of scientific ideas

Germain Hess used his knowledge and understanding to pose scientific questions and develop a theory which became universally accepted.

Chemical reactions involve a change in enthalpy. The value of this enthalpy change is dependent on the bonds that are broken and made during the reaction.

The law of conservation of energy

The conservation of energy is a driving principle behind much chemistry. It states in simple terms that 'energy cannot be created or destroyed, only changed from one form to another'. You use this principle in terms of atoms when you write balanced equations (in this instance the law relates to conservation of matter rather than conservation of energy). The law can be applied to enthalpy changes in order to help you determine the enthalpy change for reactions that you cannot carry out under standard conditions.

An impossible journey?

A chemical reaction can be thought of as a journey between reactants and products. There may be many ways of carrying out the journey but the overall enthalpy change for the journey is constant. In the journey shown in the diagram, the passenger needs to travel from Lewes to Uckfield but there is not track between the stations. However, the journey is possible if the passenger catches two trains: one to East Croydon and then a second to Uckfield. This can be described as an indirect route. The net effect is the same as if the passenger had travelled directly from Lewes to Uckfield.

enthalpy change A = enthalpy change B + enthalpy change C

Hess's Law

Germain Hess was a Russian chemist who in 1840 developed a thermochemistry version of the law of conservation of energy that is known as **Hess's law**. This states that 'If a reaction can occur by more than one route, the overall enthalpy change is independent of the route taken.' You can use this law to construct an energy cycle in which enthalpy changes for reactions are connected together.

Calculating the standard enthalpy of reaction using standard enthalpy of formation data

The equation represents the complete combustion of methane.

$$CH_4(g) + 2O_2(g) \rightarrow CO_2(g) + 2H_2O(l)$$

You can determine the enthalpy change for this reaction by using the enthalpy of formation data for each of the reactants and products.

$CH_4(g)$	$C(s) + 2H_2(g) \rightarrow CH_4(g)$	$-74.9\,kJ\,mol^{-1}$
$\Delta H_f^{\ominus}$ $CO_2(g)$	$C(s) + O_2(g) \rightarrow CO_2(g)$	$-394\,kJ\,mol^{-1}$
$\Delta H_f^{\ominus}$ $H_2O(l)$	$H_2(g) + \frac{1}{2}O_2(g) \rightarrow H_2O(l)$	$-286\,kJ\,mol^{-1}$

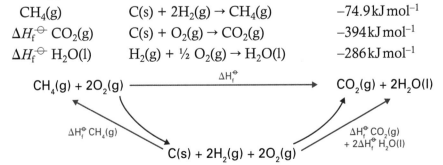

$CH_4(g) + 2O_2(g)$ $\xrightarrow{\Delta H_f^{\ominus}}$ $CO_2(g) + 2H_2O(l)$

$\Delta H_f^{\ominus}$ $CH_4(g)$

$C(s) + 2H_2(g) + 2O_2(g)$

$\Delta H_f^{\ominus}$ $CO_2(g)$ + $2\Delta H_f^{\ominus}$ $H_2O(l)$

$\Delta H_r^{\ominus} = -[\Delta H_f^{\ominus}\ CH_4(g)] + [\Delta H_f^{\ominus}\ CO_2(g) + 2\Delta H_f^{\ominus}\ H_2O(l)]$

$\Delta H_r^{\ominus} = -[-74.9] + [-394 + 2(-286)] = -[-74.9] + [-966]$

$\Delta H_r^{\ominus} = -891.1\ kJ\ mol^{-1}$

Note that we subtract the enthalpy change of formation of methane. This is because we move against the arrow to move round the cycle. Using square brackets for calculating the overall value for each enthalpy change ensures that the signs of all enthalpy changes are correct.

This calculation can be simplified using the relationship:

$\Delta H_r^{\ominus} = \Sigma\Delta H_f^{\ominus}$ (products) – $\Sigma\Delta H_f^{\ominus}$ (reactants)

The Greek symbol Σ (sigma) means 'the sum of'. This relationship is obtained from the energy cycle; it is essential that you are able to use either method.

Worked example

For the reaction of lead(II) oxide with carbon monoxide it is possible to calculate the enthalpy change from enthalpy of formation data.

$PbO(s) + CO(g) \rightarrow Pb(s) + CO_2(g)$

$\Delta H_f^{\ominus}$ PbO(s) –219 kJ mol⁻¹

$\Delta H_f^{\ominus}$ CO(g) –111 kJ mol⁻¹

$\Delta H_f^{\ominus}$ CO_2(g) –394 kJ mol⁻¹

$\Delta H_r^{\ominus} = \Sigma\Delta H_f^{\ominus}$ (products) - $\Sigma\Delta H_f^{\ominus}$ (reactants)

$\Delta H_r^{\ominus} = [\Delta H_f^{\ominus}\ CO_2(g)] - [\Delta H_f^{\ominus}\ PbO(s) + \Delta H_f^{\ominus}\ CO(g)]$

$\Delta H_r^{\ominus} = [-394] - [-219 + (-111)] = [-394] - [-330]$

$\Delta H_r^{\ominus} = -64\ kJ\ mol^{-1}$

Check your understanding

1. Construct an energy cycle and use it to calculate the standard enthalpy change when ammonia gas reacts with hydrogen chloride gas to form ammonium chloride using the standard enthalpy of formation data given.

   $\Delta H_f^{\ominus}$ NH_3(g) –46.2 kJ mol^{-1}

   $\Delta H_f^{\ominus}$ HCl(g) –92.3 kJ mol^{-1}

   $\Delta H_f^{\ominus}$ NH_4Cl(s) –315 kJ mol^{-1}

2. Ethanol is being used as an alternative fuel to petrol. A knowledge of the standard enthalpy of combustion of ethanol is useful to chemical engineers carrying out testing. Use the data below to construct an energy cycle and use it to determine the standard enthalpy of combustion of ethanol.

   $\Delta H_f^{\ominus}$ C_2H_5OH(l) –278 kJ mol^{-1}

   $\Delta H_f^{\ominus}$ CO_2(g) –394 kJ mol^{-1}

   $\Delta H_f^{\ominus}$ H_2O(g) –286 kJ mol^{-1}

5.04 Calculating enthalpy changes

OBJECTIVES

after this spread you should be able to

- construct energy cycles
- use energy cycles to calculate enthalpy changes from enthalpy of formation and enthalpy of combustion data

Standard enthalpy change of combustion $\Delta H_c^{\ominus}$

The enthalpy change when 1 mole of a substance is completely burned in oxygen under standard conditions (100 kPa and 298 K).

You can calculate standard enthalpy changes for reactions using many different kinds of standard enthalpy data. Simple calorimetry experiments will give you standard enthalpies of combustion, and you can apply Hess's law to construct energy cycles from these data.

Using enthalpy of combustion data

The equation represents the formation of ethane gas.

$\Delta H_f^{\ominus}$ $C_2H_6(g)$ $2C(s) + 3H_2(g) \rightarrow C_2H_6(g)$

You can determine the enthalpy change for this reaction by using the enthalpy of combustion data for each of the reactants and products.

$\Delta H_c^{\ominus}$ C(s)	$C(s) + O_2(g) \rightarrow CO_2(g)$	$-394\,kJ\,mol^{-1}$
$\Delta H_c^{\ominus}$ $H_2(g)$	$H_2(g) + \frac{1}{2}O_2(g) \rightarrow H_2O(l)$	$-286\,kJ\,mol^{-1}$
$\Delta H_c^{\ominus}$ $C_2H_6(g)$	$C_2H_6(g) + \frac{7}{2}O_2(g) \rightarrow 2CO_2(g) + 3H_2O(l)$	$-1560\,kJ\,mol^{-1}$

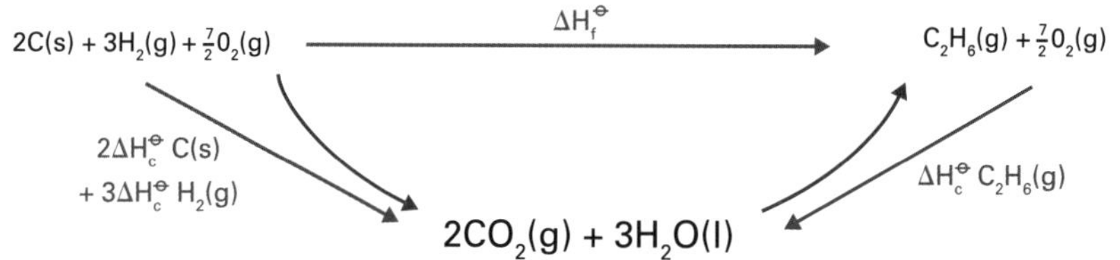

$$\Delta H_f^{\ominus}\, C_2H_6(g) = [2\Delta H_c^{\ominus} C(s) + 3\Delta H_c^{\ominus}\, H_2(g)] - [\Delta H_c^{\ominus}\, C_2H_6(g)]$$

$$\Delta H_f^{\ominus}\, C_2H_6(g) = [2(-394) + 3(-286)] - [-1560] = [-1646] - [-1560]$$
$$= -86\,kJ\,mol^{-1}$$

Note that you subtract the enthalpy change of combustion for ethane. This is because you are moving against the arrow to move round the cycle. Using of square brackets for calculating the overall value for each enthalpy change ensures that the signs of all enthalpy changes are correct.

Worked example

Calculate the enthalpy change for the reaction below using the data given.

$\Delta H_c^{\ominus}$ $C_2H_5OH(l)$ $C_2H_5OH(l) + 3O_2(g) \rightarrow 2CO_2(g) + 3H_2O(l)$

Data:

$\Delta H_c^{\ominus}$ C(s)	$C(s) + O_2(g) \rightarrow CO_2(g)$	$-394\,kJ\,mol^{-1}$
$\Delta H_c^{\ominus}$ $H_2(g)$	$H_2(g) + \frac{1}{2}O_2(g) \rightarrow H_2O(l)$	$-286\,kJ\,mol^{-1}$
$\Delta H_f^{\ominus}$ $C_2H_5OH(l)$	$2C(s) + 3H_2(g) + \frac{1}{2}O_2(g) \rightarrow C_2H_5OH(l)$	$-278\,kJ\,mol^{-1}$

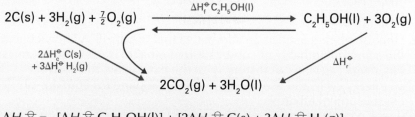

2C(s) + 3H₂(g) + 7/2 O₂(g) — ΔH_f C₂H₅OH(l) → C₂H₅OH(l) + 3O₂(g)
2ΔH_c C(s) + 3ΔH_c H₂(g)
ΔH_r
2CO₂(g) + 3H₂O(l)

$$\Delta H_r^{\ominus} = -[\Delta H_f^{\ominus}\, C_2H_5OH(l)] + [2\Delta H_c^{\ominus}\, C(s) + 3\Delta H_c^{\ominus}\, H_2(g)]$$
$$\Delta H_r^{\ominus} = -[-278] + [2(-394) + 3(-286)] = -[-278] + [-1646]$$
$$\Delta H_r^{\ominus} = -1368\,kJ\,mol^{-1}$$

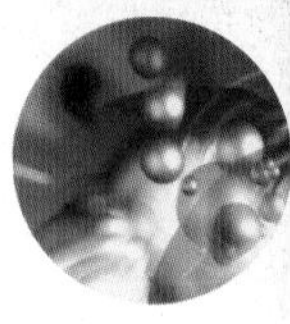

Combustion of alcohol is a very exothermic reaction.

Check your understanding

1. Using the data provided construct an energy cycle and use it to calculate the standard enthalpy change for the reaction shown:
   $6C(s) + 6H_2(g) + 3O_2(g) \rightarrow C_6H_{12}O_6(s)$
   Data:
   $\Delta H_c^\ominus$ C(s) –394 kJ mol^{-1}
   $\Delta H_c^\ominus$ $H_2(g)$ –286 kJ mol^{-1}
   $\Delta H_c^\ominus$ $C_6H_{12}O_6(s)$ –1264 kJ mol^{-1}
2. Calculate the standard enthalpy of formation of nitromethane, $CH_3NO_2(l)$from the standard enthalpies of combustion provided:
   Data:
   $\Delta H_c^\ominus$ C(s) –394 kJ mol^{-1}
   $\Delta H_c^\ominus$ $H_2(g)$ –286 kJ mol^{-1}
   $\Delta H_c^\ominus$ $N_2(g)$ +34 kJ mol^{-1}
   $\Delta H_c^\ominus$ $CH_3NO_2(l)$ –709 kJ mol^{-1}
3. Use the following data to calculate the standard enthalpy of formation of propane:
   $\Delta H_c^\ominus$ C(s) –394 kJ mol^{-1}
   $\Delta H_c^\ominus$ $H_2(g)$ –286 kJ mol^{-1}
   $\Delta H_c^\ominus$ $C_3H_8(g)$ –2220 kJ mol^{-1}

5.05 Bond enthalpies

OBJECTIVES

after this spread you should be able to

- define mean bond enthalpy
- calculate mean bond enthalpy from bond enthalpy data
- calculate enthalpy of reaction from mean bond enthalpy data

You can use standard enthalpy of combustion data with Hess's law to determine the enthalpy change for a reaction. It is also possible to calculate standard enthalpy changes by considering the enthalpy associated with each individual covalent bond. This enthalpy is referred to as '**bond enthalpy**'.

Bond enthalpy

Bond enthalpy is the standard enthalpy change associated with breaking one mole of bonds in a gaseous substance into individual gaseous atoms.

A–A(g) → A(g) + A(g)

These values are always positive because *breaking bonds requires energy.* The value for making a bond is the same but has the opposite sign.

For example, the standard bond enthalpy for H–Br is 366 kJ mol^{-1}. This means that 366 kJ of energy are needed to break one mole of gaseous H–Br molecules into individual H atoms and Br atoms. If one mole of gaseous H–Br molecules are formed then 366 kJ of energy will be released.

Mean bond enthalpy

Enthalpy (or energy) needed to break (or dissociate) a bond averaged over different molecules.

Mean bond enthalpy data (kJ mol^{-1})

H–H	436	H–C	413
C–C	348	H–O	463
C=O	743	C–O	360
O–O	146	O=O	498

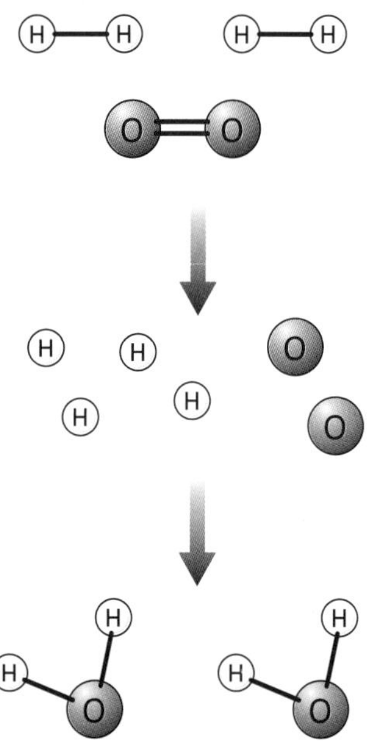

Bonds are broken in hydrogen and oxygen, and new bonds are formed in water.

Mean bond enthalpies

Bond enthalpy values are dependent on the attraction between the nuclei of the atoms involved in the bond and the electrons shared between them. The enthalpy values will therefore change slightly depending on the environment of the particular atom. For example, the C–H bond enthalpy value in methane will be different from the C–H bond enthalpy value in ethanol. In fact each of the C–H bond enthalpies in methane will be slightly different.

The equation represents the atomization of methane:

$CH_4(g) \rightarrow C(g) + 4H(g) \qquad \Delta H = 1662\,kJ$

In this process four moles of C–H bonds are broken so the **mean bond enthalpy** is 1662/4 = 415.5 kJ mol^{-1}.

The data quoted for enthalpy calculations is always a mean bond enthalpy.

Making and breaking bonds

When a chemical reaction takes place energy is put into the system to break the covalent bonds (bond breaking is endothermic) and released by the system when bonds are formed (bond making is exothermic). The overall enthalpy change for a reaction is the difference between these two values. In the reaction of hydrogen and oxygen to make water two H–H and one O=O bond are broken then four H–O bonds are made.

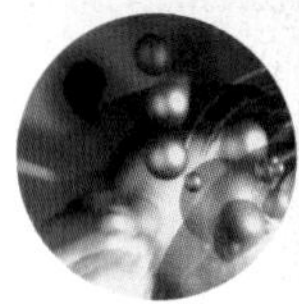

Calculating enthalpy changes

The standard enthalpy change for a reaction is calculated by

- calculating the energy needed to break all the bonds in the reactant molecules
- calculating the energy released when all the bonds in the product molecules are made
- subtracting the energy released in making bonds from the energy needed to break bonds

Example

For the formation of water from hydrogen and oxygen the standard bond enthalpies are:

Oxygen	O=O	+496 kJ mol^{-1}
Hydrogen	H–H	+436 kJ mol^{-1}
Water	O–H	+463 kJ mol^{-1}

The overall equation for the formation of water in the gas phase is

$H_2(g) + \frac{1}{2}O_2(g) \rightarrow H_2O(g)$

Note that this is for the formation of *one mole* of water.

energy needed to break bonds		energy released by making bonds	
H–H	436	2 O–H	2 × 463
½ O=O	496 / 2		
Total	684	**Total**	926

$\Delta H_r^{\ominus} = \Sigma\Delta H_r^{\ominus}$ (bonds broken) – $\Sigma\Delta H_r^{\ominus}$ (bonds made)

$\Delta H_r^{\ominus} = 684 - 926 = -242$ kJ mol^{-1}

Check your understanding

1. Use the mean bond enthalpy data given on page 130 to calculate the standard enthalpy changes below:
   - **a** enthalpy of combustion of ethane
   - **b** enthalpy of combustion of ethanol
   - **c** enthalpy of combustion of butane
2. Calculate the mean bond enthalpy for the O–H bond in water using the two enthalpy changes given below:

   $H–O–H(g) \rightarrow H(g) + O–H(g)$ $\quad \Delta H_r^{\ominus} = 502$ kJ mol^{-1}

   $O–H(g) \rightarrow O(g) + H(g)$ $\quad \Delta H_r^{\ominus} = 427$ kJ mol^{-1}
3. 1-bromobutane, 1-chlorobutane, and 1-iodobutane all react with aqueous sodium hydroxide solution to form butan-1-ol. These reactions occur at different rates. Explain this difference in rate using the bond enthalpy data given:

C–Cl	338 kJ mol^{-1}
C–Br	276 kJ mol^{-1}
C–I	238 kJ mol^{-1}

5.06 Measuring an enthalpy change

OBJECTIVES

already from AS you can

- define mean bond enthalpy
- calculate mean bond enthalpy from bond enthalpy data
- calculate enthalpy of reaction from mean bond enthalpy data

after this spread you should be able to

- understand how to measure enthalpy changes

You can determine the enthalpy change for a reaction in the laboratory by carrying out a simple calorimetry experiment. In this practical skills assessment you would measure the enthalpy change for the reaction of anhydrous copper(II) sulfate with water to produce hydrated crystals. This reaction cannot be carried out directly so you carry out two experiments and apply Hess's law to your results in order to determine the enthalpy change for the unknown reaction. You are assessed on how accurately you carry out your practical work and the precision with which you record all your measurements.

The hydration of anhydrous copper sulphate

$$CuSO_4(s) + 5H_2O(l) \xrightarrow{\Delta H_r} CuSO_4.5H_2O(s)$$

anhydrous copper sulfate → hydrated copper sulfate

You carry out two experiments. In the first you dissolve a known mass of anhydrous copper sulfate in a known excess of water. In the second you disolve a known mass of hydrated copper sulfate in a known excess of water. You use a simple calorimeter for both experiments and measure the temperature change in each case as accurately as you can. In both cases a solution of aqueous copper sulfate is formed. From the temperature changes you can calculate the enthalpy change for the reaction.

Anhydrous copper sulphate ΔH_1

$CuSO_4(s) + \text{excess } H_2O(l) \rightarrow CuSO_4(aq)$

Hydrated copper sulphate ΔH_2

$CuSO_4.5H_2O(s) + \text{excess } H_2O(l) \rightarrow CuSO_4(aq)$

Procedure

1. Use a measuring cylinder to measure a volume of water and pour the water into a simple calorimeter.
2. Record the volume of the water.
3. Use a thermometer to read the temperature of the water and record it.
4. Obtain a known mass of anhydrous copper sulfate, using a 2dp balance.
5. Pour the anhydrous copper sulfate into the calorimeter and stir the solution until all the solid has dissolved.
6. Record the maximum or minimum temperature reached by the solution.
7. Repeat steps **1** to **6** using hydrated copper sulfate instead of anhydrous copper sulfate.

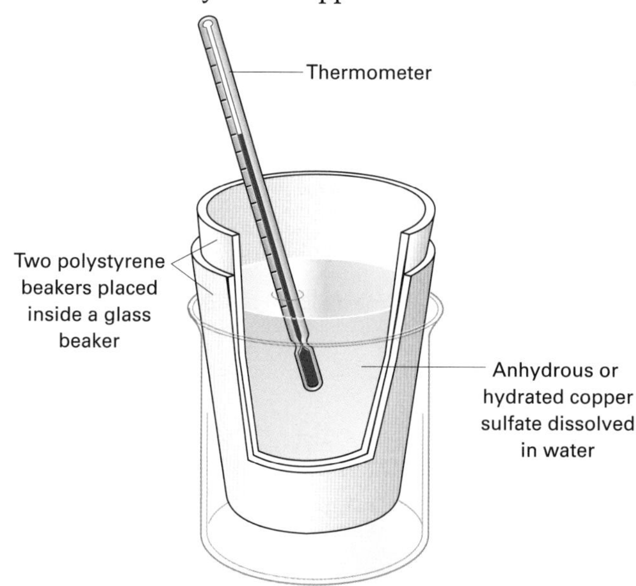

The actual apparatus used may look slightly different from this.

Points to consider:

- How many decimal places should your mass reading be recorded to?
- You will need to choose the type of thermometer that you use for this experiment. Examine those available to you, look carefully at the divisions on them and consider to what degree of precision you can obtain results.
- Why is the solution stirred during the dissolving process?

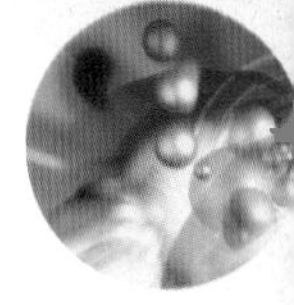

Choosing a thermometer

A 0–100°C thermometer has divisions of 1°C.
A 0–50°C thermometer has divisions of 0.5°C

Safety

Before starting work think about the hazards presented by each of the substances that you would use and the precautions that you would need to take. Constructing a table for completion before you start will help you think clearly about this.

substance	hazard	precaution

Analysing the results

From the results of each of these experiments you can calculate the overall enthalpy change for both dissolving reactions.

- Calculate the heat change using $q = mc\Delta T$.
- Calculate the number of moles of solid dissolved in the water
- Convert the heat change into ΔH in kJ mol^{-1}.
- Ensure that your answer has the correct sign

Construct an energy cycle that links the enthalpy changes from these two experiments. Use this cycle to calculate ΔH_r. Remember that the reaction which you want to find out about is:

$$CuSO_4(s) \;+\; 5H_2O(l) \;\rightarrow\; CuSO_4.5H_2O(s) \quad \Delta H_r$$

Evaluating the results

It is important to consider the validity of any results that you obtain in the laboratory from a practical procedure. Do not dwell on any mistakes you have made, instead focus on the procedure you have carried out and the equipment you have used.

You should consider:

- any anomalous results obtained and possible reasons for them (these are results in your experimental data that do not match the general trend of the data)
- any errors connected to the procedure you have carried out (for example: how many times was the experiment repeated, was there any heat loss in the procedure, were any colour changes obscured)
- the measurement errors for each piece of equipment you have used (for example: a 0–100°C thermometer has 1°C divisions so can be read accurate to 0.5°C. If this was used to measure a temperature change of 2°C there would be a measurement error of ±25% in the temperature reading.)

Identify the most significant of these errors and suggest improvements that you could make in order to minimize this.

6.01 Collision theory

OBJECTIVES

already from GCSE, you know

- that particles must collide in order to react
- how to interpret information about the control of rates of reaction in chemical synthesis
- how to interpret results from experiments that investigate rates of reactions

and after this spread you should be able to

- recall the definition of reaction rate
- understand collision theory
- recall the definition of activation energy

Chemists can find out many different pieces of information about particular reactions. The most useful are the enthalpy changes associated with reactions and information about how fast a reaction will occur. Controlling reaction rates and understanding the processes by which reactants change into products are essential skills for a chemist.

Reaction rate

You will be familiar with many everyday uses of the term 'rate'. For example the rate of movement of a car is referred to as speed: the distance travelled by the car in relation to the time taken for the journey. The rate of a chemical reaction is measured as the amount of product made or reactant used up in a certain time.

$$\text{rate of reaction} = \frac{\text{increase in concentration}}{\text{time}}$$

Collision theory

When a chemical reaction takes place the reactant particles must collide. The theory which explains the reactions that take place as the result of collisions is called **collision theory**. Imagine a gaseous substance in which the particles are in constant random motion. The particles are continuously colliding with each other and with the walls of their container. Not all of the collisions between the gaseous particles result in a reaction.

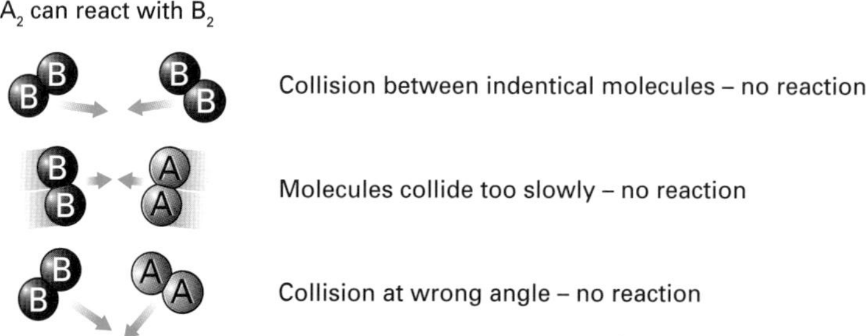

A_2 can react with B_2
B B B B
Collision between indentical molecules – no reaction
B B A A
Molecules collide too slowly – no reaction
B B A A
Collision at wrong angle – no reaction

The diagram shows that in order for a reaction occur

- the correct particles must collide
- the collision must be of the correct energy

In order to maximize the energy of the collision the particles must be moving quickly and collide head on. Reaction conditions can be altered to maximize the probability of a collision occurring or to increase the energy with which particles collide. The factors that affect the rates of chemical reactions are

- the temperature of the reactants
- the concentration of the reactants and products
- the surface area of reactants
- the presence of any catalysts

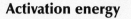

Activation energy

The minimum energy with which particles must collide in order for a reaction to occur is referred to as the 'activation energy'. If the particles collide with energy less than the activation energy then the particles will simply bounce off each other and no reaction will occur. If the particles collide with energy equal to or greater than the activation energy the collision is described as successful and a reaction will take place.

Activation energy and enthalpy changes

The diagram shows the relationship between the activation energy, E_a, of a reaction and the enthalpy change of the reaction.

Monitoring reaction rate

For the reaction of calcium carbonate with hydrochloric acid the rate can be monitored by collecting the carbon dioxide gas released.

$CaCO_3(s) + 2HCl(aq) \rightarrow CaCl_2(aq) + H_2O(l) + CO_2(g)$

The graph obtained will be a curve until the reaction has finished. The gradient or slope at any point on the rate curve gives the rate of the reaction at that time.

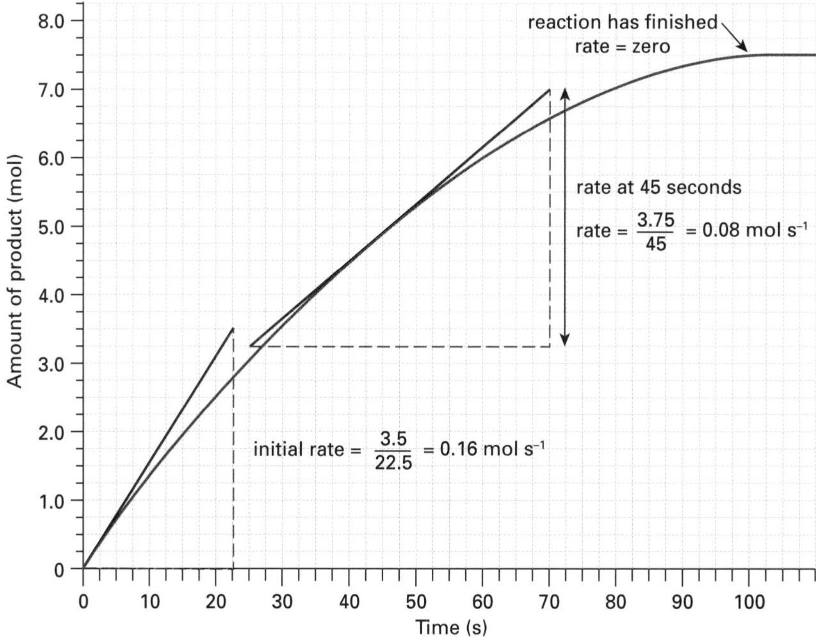

Activation energy

The *minimum energy* with which particles must collide in order for a reaction to occur.

(a)

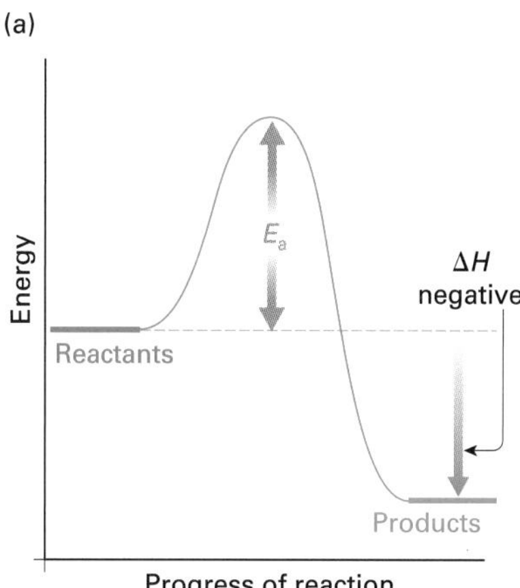

(b)

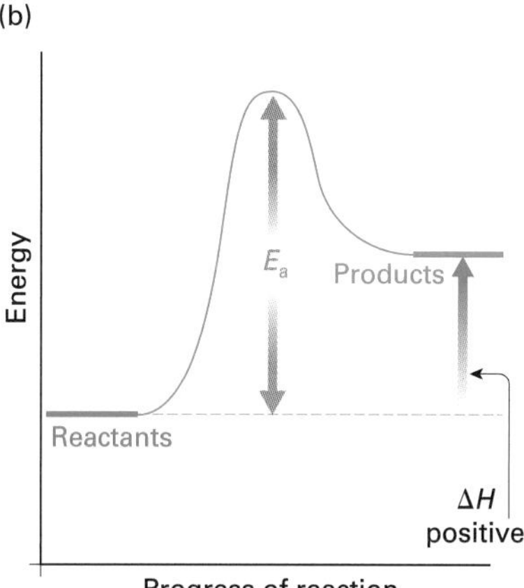

The reaction profiles for (a) an exothermic reaction and (b) an endothermic reaction. E_a is the activation energy barrier that reactants must overcome before they can change into products. ΔH indicates the overall enthalpy change for the reaction.

Check your understanding

1. Sketch an enthalpy profile diagram for the combustion of methane and use it to explain why a room full of gas does not explode until a source of ignition is provided.
2. For each of the factors that affect the rate of a chemical reaction explain how they maximize the chance of a collision or the energy of the collisions.
3. Describe with sketch diagrams as appropriate the methods that could be used to follow the rate of reaction for sodium carbonate with sulfuric acid.

6.02 The effect of temperature on reaction rate

OBJECTIVES

already from GCSE, you know

- that increasing temperature increases the rate of a reaction

and after this spread you should be able to

- sketch a Maxwell–Boltzmann distribution curve
- explain the effect of temperature on the rate of a reaction using the Maxwell–Boltzmann distribution curve

Scientific ideas

Scientific progress is made when scientists contribute to the development of new ideas, materials and theories. James Maxwell proposed a theory which was then used by Ludwig Boltzmann twelve years later.

Temperature is used as a measure of the amount of energy of the particles in a substance. For example if a sample of gas is heated up to a higher temperature this tells us that the average amount of energy each particle possesses has increased.

You know that whether a reaction takes place or not is dependent on

- the energy of the colliding particles
- the number of collisions

Maxwell–Boltzmann distribution of molecular energies

In a sample of gas at a given temperature, the molecules do not all have the same energy. They are all moving at different speeds. At any instant some of the particles have a very low energy, a small proportion have a very high energy while the majority of the particles have an intermediate energy.

In 1859 the Scottish physicist James Clerk Maxwell calculated this distribution of energies in a sample of gas. His ideas were applied by the Austrian physicist Ludwig Edward Boltzmann in 1871. The resulting graph showing the distribution of molecular energies is known as the **Maxwell–Boltzmann distribution**.

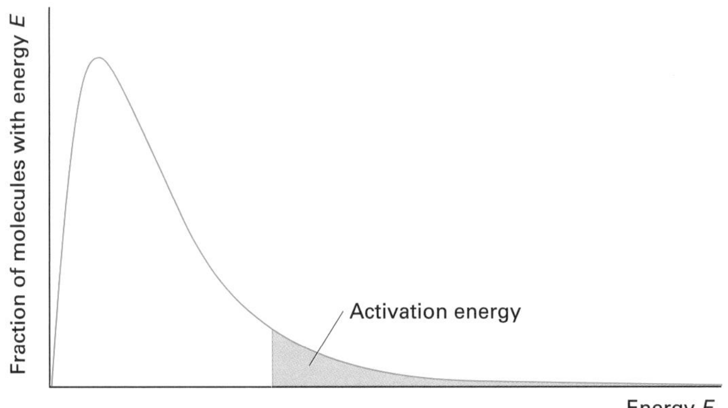

Only molecules in the shaded area collide with enough energy to react.

Note the following features of the curve:

- Only very small fractions of the molecules have extremely high or extremely low energies.
- The curve is not symmetrical; the average energy is to the right of the peak of the curve.
- The curve passes through the origin.
- The curve does not touch the x (horizontal) axis on the right hand side. It is an asymptote, a line or curve which approaches another but never touches it.

The shaded area under the right of the curve shows the proportion of molecules that possess the activation energy (the minimum energy with which particles must collide in order to react). Only molecules in this portion of the curve are able to react. Each reaction has its own activation energy. At a given temperature a reaction with higher activation energy will be slower than one with lower activation energy.

Effect of temperature

The shape of the Maxwell-Boltzmann distribution changes as temperature is altered. As the temperature increases the energy distribution moves to the right and the height of the peak decreases. The total area under the curve is constant as this represents the total number of particles.

For a small increase in temperature the shape of the graph remains broadly the same but note that the area of the shaded portion has increased, so more molecules have energy greater than or equal to the activation energy

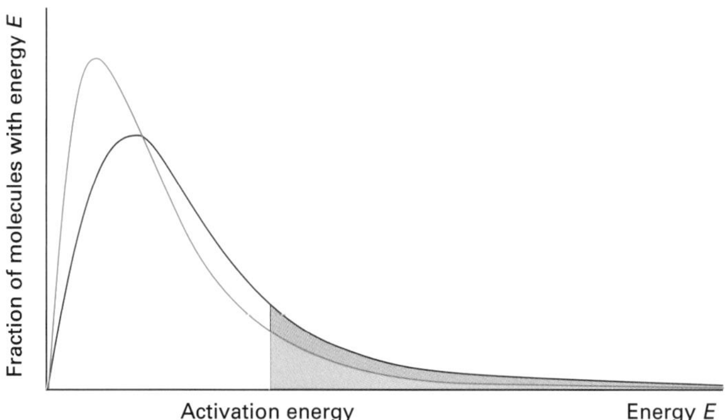

For a large increase in temperature the shape of the graph alters more dramatically.

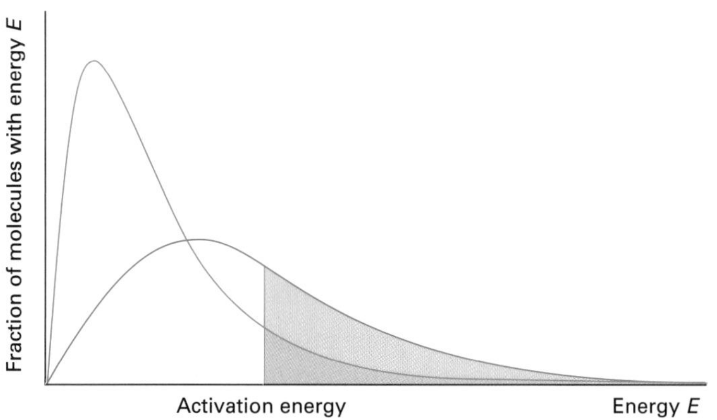

Note that altering the temperature does not have any effect on the value of the activation energy; this is constant for a given reaction. Increasing the temperature does not influence this value it increases the energy with which particles collide so that more of these collisions possess the activation energy.

Check your understanding

1. Sketch two Maxwell–Boltzmann curves and use them to explain why food does not go off so quickly in a refrigerator.
2. Use collision theory to explain why increasing the temperature of a reaction mixture increases the rate of the reaction.
3. Explain why altering the temperature of a reaction has no impact on the value of the activation energy.

The fire at Buncefield oil depot

This fire was caused by leaking fuel. The very fast reaction that began the fire only took a second. It was very fast because it had a low activation energy.

6.03 Investigating rates of reaction

In this practical skills assessment you would carry out the reaction of sodium thiosulfate with hydrochloric acid. The rate of this chemical reaction can be altered by altering the temperature of the reagents. This effect can investigated by carrying out the same reaction at different temperatures while controlling all other factors that may affect reaction rate. You will be assessed on choosing appropriate quantities of each reagent, working accurately and safely, and obtaining final results within the expected range.

The reaction of sodium thiosulfate with acid

Hydrochloric acid reacts readily with sodium thiosulfate forming a yellow **precipitate** of sulfur. The rate of the reaction can be measured by timing how long it takes for the precipitate to be formed.

When the colourless solutions are mixed, a yellow precipitate of sulfur is formed.

$$Na_2S_2O_3(aq) + 2HCl(aq) \rightarrow SO_2(g) + S(s) + H_2O(l) + 2NaCl(aq)$$

Since reaction rate is the change in concentration of either reactants or products relative to the change in time the rate can be determined by calculating 1/time.

$$\text{rate of reaction} = \frac{\text{increase in concentration}}{\text{time}}$$

$$\text{rate of formation of sulfur} = \frac{1}{\text{time taken for sulfur to form}}$$

Planning the experiment

In planning your experiment you need to think about several key areas.

- What equipment will you use so that you can make precise measurements?
- You will have to measure the time taken for the sulfur precipitate to form with a stopwatch. What quantities of sodium thiosulfate solution and hydrochloric acid should you use to give a suitable time? A preliminary experiment at room temperature may help you to decide.
- How will you follow the formation of the precipitate so that your results are consistent?
- Over what range of temperatures will you make measurements? How many measurements will you make?
- How will you heat the sodium thiosulfate and hydrochloric acid solutions? How will you keep the temperature of the mixture constant while the sulfur precipitate is forming?
- What are the hazards presented by the reactants and products? What safety precautions will you take?

Safety

Before starting work consider the hazards presented by each of the substances that you are going to use and the precautions that you will need to take. Constructing and completing a table before you start will help you think clearly about this.

substance	hazard	precaution

Analysis of the results

If you record your results in a table like this you can easily use them to find how the reaction rate changes with temperature.

temperature(°C)	time (s)	rate (s^{-1})

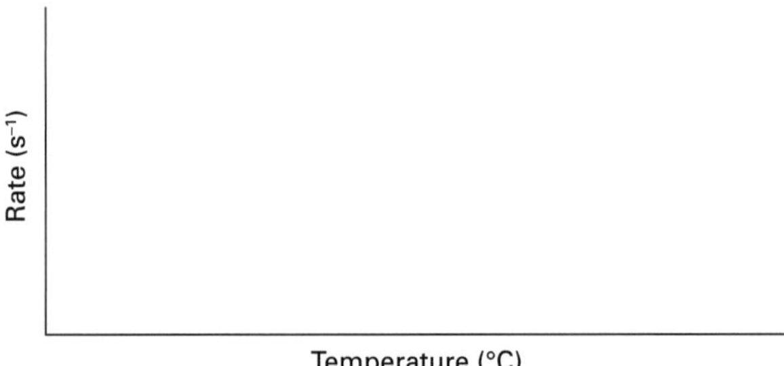

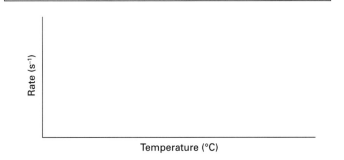

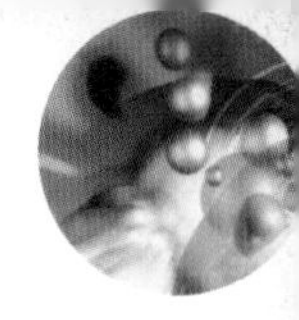

Plotting a graph of rate against temperature allows you to determine the effect of temperature on the reaction rate.

Drawing a best fit line through the data will show if the rate is proportional to the temperature of the sodium thiosulfate. A straight line relationship indicates direct proportionality.

You can draw conclusions from this graph and then explain them using the Maxwell–Boltzmann distribution curve.

Sodium thiosulphate is a practical example - as 'fixer' in photography.

Evaluating the results

You should evaluate your experimental data in order to confirm the validity of your conclusions. Focus on the procedure you have carried out and the equipment you have used. Do not comment on mistakes you have made.

You should consider

- any anomalous results obtained and possible reasons for them (these are results in your experimental data that do not match the general trend of the data)
- any errors connected to the procedure you have carried out (for example: how many times was the experiment repeated, was there any heat loss in the procedure, were any colour changes obscured)
- the measurement errors for each piece of equipment you have used (think about the apparatus you have used to measure the rate of the reaction)

Identify the most significant of these errors and suggest improvements which could be made in order to minimize this error.

6.04 Catalysis

OBJECTIVES

already from GCSE, you know

- that catalysts speed up chemical reactions while not being used up in the chemical changes

and after this spread you should be able to

- define the term catalyst
- draw a reaction profile showing the impact of a catalyst
- draw a Maxwell–Boltzmann distribution and use it to explain how a catalyst increases reaction rate.

Catalyst

A substance which alters the rate of a chemical reaction but remains unchanged at the end of the reaction.

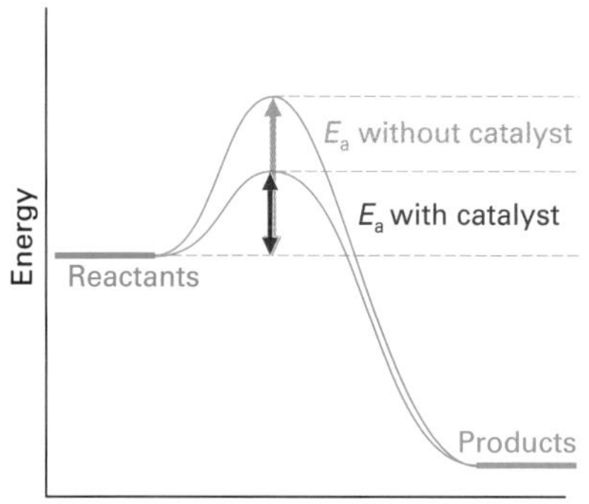

The reaction profiles for an uncatalysed reaction (upper curve) and the same reaction with a catalyst present (lower curve).

Many of the reactions essential for our everyday life would not occur without the presence of **catalysts**. They can be recovered at the end of the reaction and used many times over. Industrially, catalysts are used in the manufacture of ammonia, sulfuric acid, margarines, plastics, and fertilizers. Without the presence of enzymes (biological catalysts) our bodies would not function, we would not have biological washing powder, nor have bread to eat or alcohol to drink.

Catalysts

A catalyst can be described as a substance that alters the rate of a chemical reaction and remains unchanged at the end of the reaction. The catalyst acts by providing an alternative route of lower activation energy. This reduction in activation energy enables many more of the collisions between reactants to achieve this minimum requirement for a reaction to take place. Therefore the rate of the reaction increases.

Catalysts and the Maxwell–Boltzmann distribution

The addition of a catalyst to a reaction has no effect on the energies of the reactant molecules or on the total number of molecules in the reaction system. It therefore has no effect on the shape or size of the Maxwell–Boltzmann distribution. The activation energy line is marked further to the left hand side of the curve to show the reduction in its value. This has a significant effect on the number of particles that are found in the shaded area of the curve and can therefore take part in the reaction.

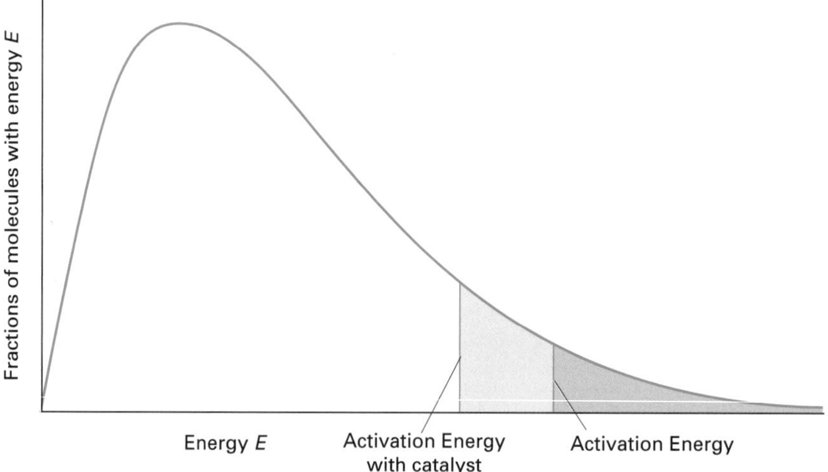

Adding a catalyst to a reaction lowers the activation energy. More molecules can collide with enough energy to react.

Heterogeneous and homogeneous catalysts

There are two important classes of catalysts: heterogeneous and homogeneous.

A **heterogeneous catalyst** is in a different phase from the reactants. For example in the hardening of vegetable oils for the production of margarine a nickel catalyst is used to reduce the activation energy for the reaction of two gases: hydrogen and an alkene. The reactant molecules

attach themselves to the nickel surface breaking the C=C in the alkene as they attach. The reaction then takes place on the surface and the alkane molecules formed detach from the surface. You will learn about the catalytic converter, an important use of heterogeneous catalysts, when you study the combustion of fossil fuels.

A **homogeneous catalyst** is in the same phase as the reactants. Chlorine radicals act as a homogeneous catalyst in the upper atmosphere, and have a devastating effect on the sequence of reactions which take place constantly making and destroying ozone. One chlorine radical can catalyse as many as one hundred thousand reactions. The series of reactions that take place is complex and you will study them in more detail elsewhere. Some of them are given here:

$Cl\bullet + O_3 \rightarrow ClO\bullet + O_2$

$ClO\bullet + O_3 \rightarrow Cl\bullet + 2O_2$

The chlorine radical is represented by $Cl\bullet$. It destroys ozone in the first reaction and is regenerated in the second.

Catalysis and the ozone layer

Scientists have worked together to understand the origins of the hole in the ozone layer. The evidence provided by the scientists has been used to inform decision making.

Catalytic converters in the exhaust systems of modern cars show all the main features of a heterogeneous catalyst. The catalyst consists of about 2g of finely divided platinum/rhodium, on a rigid ceramic support. The primary effect is to catalyse the conversion of the pollutants carbon monoxide and nitrogen monoxide to carbon dioxide and nitrogen.

$2CO(g) + 2NO(g) \rightarrow 2CO_2(g) + N_2(g)$

Note that leaded petrol will rapidly poison a catalytic converter.

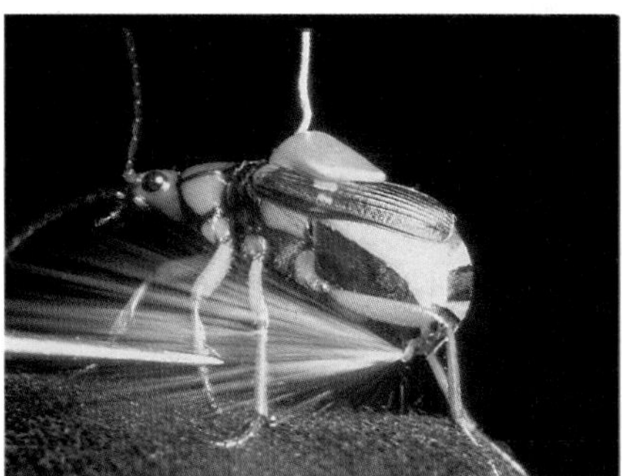

The bombardier beetle stores hydrogen peroxide, water, and noxious substances in an abdominal sac. When threatened, it injects a catalyst into this mixture. The almost instantaneous exothermic decomposition of hydrogen peroxide generates steam, which ejects the contents of the sac as a hot and highly offensive spray.

Check your understanding

1. Explain the meaning of the following terms:
   a activation energy
   b catalyst
   c Maxwell–Boltzmann distribution
   d homogeneous
   e heterogeneous
2. Carry out some research to find the names of the catalysts used for each of the following industrial processes:
   a manufacture of ammonia
   b manufacture of nitric acid
   c manufacture of sulfuric acid
3. Find out about the use of enzymes as biological catalysts. Make brief notes on their importance with some examples of their uses.

6.05 The effect of concentration on reaction rate

OBJECTIVES

already from GCSE, you know

- that reaction rate varies with the concentrations of solutions
- some simple collision theory

and after this spread you should be able to

- describe the effect of concentration on rate using collision theory
- interpret graphs to explain the effect of concentration on rate

Zinc reacts more rapidly with hydrochloric acid when the acid is concentrated.
$Zn(s) + 2HCl(aq) \rightarrow ZnCl_2(aq) + H_2(g)$

The rate of a chemical reaction is dependent on the reactant particles colliding with the activation energy for the reaction. Increasing the concentration of reactants will in the majority of cases increase the rate of the reaction. There are some instances where this is not the case. (These reactions will be considered in A2.)

Collision theory

If a reaction is to take place between two or more reactant particles, the particles must collide. Increasing the concentration of one or more of the reactants increases the probability of the reactant particles colliding. It therefore usually also increases the rate of the reaction, because the rate of collisions with the activation energy increases. The diagram represents reactants in solution at low and higher concentrations.

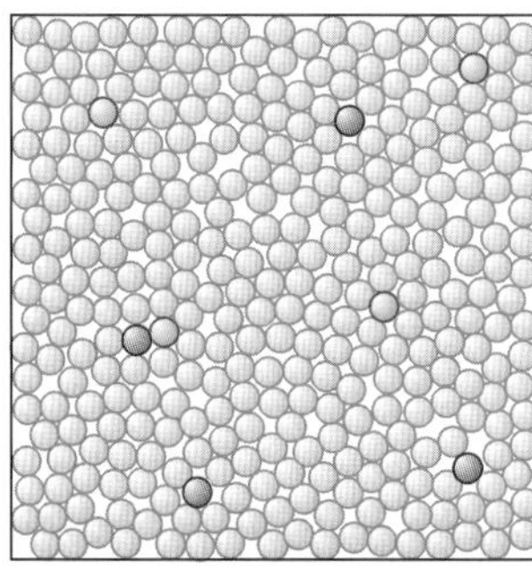

Low concentration

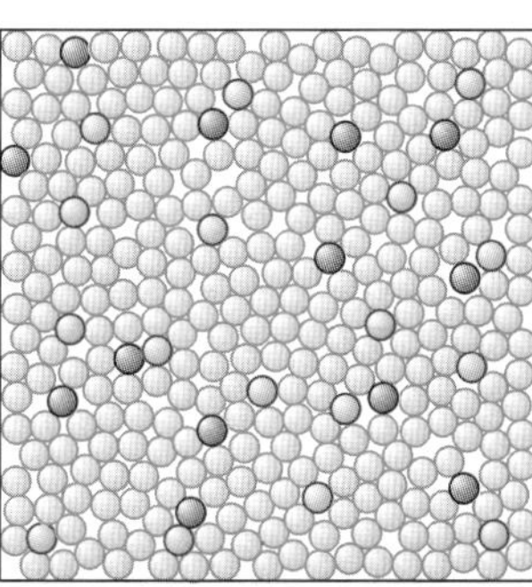

High concentration

Reactions which start with the decomposition of one reactant also increase in rate when the concentration of the reactant is increased. In this case, rate is not dependent on two particles colliding but is dependent on the number of reactant particles which possess the activation energy. Increasing concentration will increase the number of particles which possess the activation energy and hence increases rate.

Interpreting graphical data

The decomposition of hydrogen peroxide occurs very slowly at room temperature. In the presence of a catalyst it occurs more rapidly. The rate can easily be measured by collecting the oxygen that is formed.

$$2H_2O_2(aq) \rightarrow 2H_2O(l) + O_2(g)$$

If the decomposition of 50 cm^3 of 1.0 mol dm^{-3} hydrogen peroxide is carried out at 298 K the blue line is obtained on the graph.

The decomposition of 25 cm^3 of 2.0 mol dm^{-3} hydrogen peroxide at 298 K gives line a. The volume of hydrogen peroxide has been decreased and the concentration doubled. This means that the same number of moles of hydrogen peroxide take part in the reaction so the same maximum volume of oxygen will be formed. It is, however, formed twice as fast because the chance of hydrogen peroxide molecules decomposing has doubled.

The decomposition of 100 cm^3 of 0.5 mol dm^{-3} hydrogen peroxide at 298 K gives line b (concentration halved so rate halved, volume doubled so same number of moles).

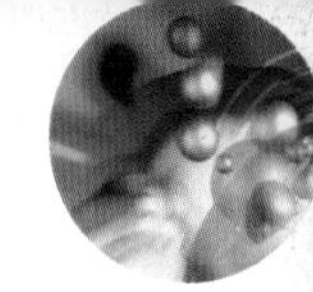

The decomposition of $100\,cm^3$ of $2.0\,mol\,dm^{-3}$ hydrogen peroxide at 298 K gives line c (concentration doubled so rate doubled, volume doubled so four times as many moles)

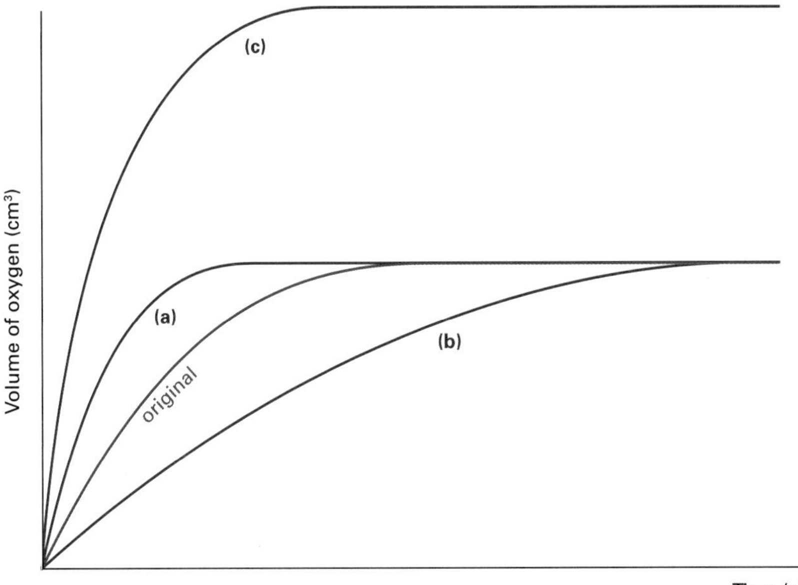

The graphs represent the catalytic decomposition of hydrogen peroxide with different concentrations and volumes.

Check your understanding

1. Describe what is meant by the term 'successful collision'. How does increasing the concentration of one of the reactants affect the chance of a successful collision?
2. The graph below was obtained for the reaction of $25\,cm^3$ of $1\,mol\,dm^{-3}$ hydrochloric acid with an excess of magnesium. Make a copy of this graph and sketch onto it the curves that would be obtained for
   a $50\,cm^3$ of $0.5\,mol\,dm^{-3}$ hydrochloric acid with excess magnesium
   b $50\,cm^3$ of $1\,mol\,dm^{-3}$ hydrochloric acid with excess magnesium
   c $25\,cm^3$ of $2\,mol\,dm^{-3}$ hydrochloric acid with excess magnesium

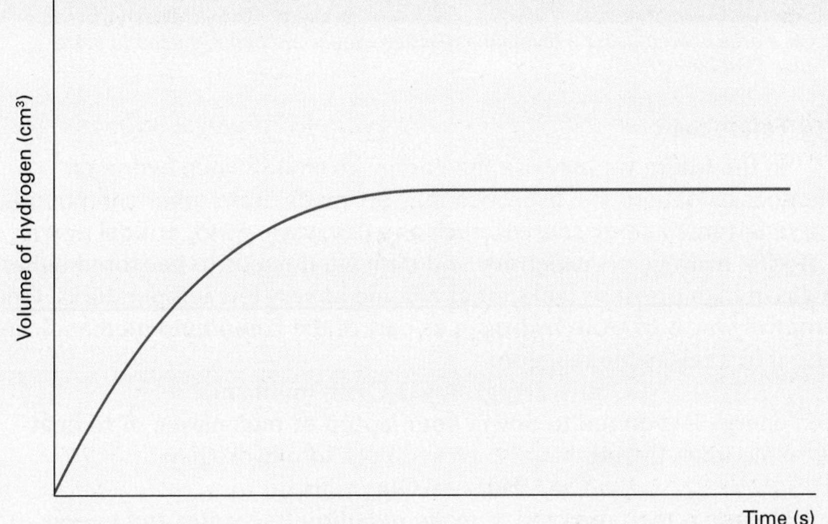

7.01 Reversible reactions and dynamic equilibria

OBJECTIVES

after this spread you should be able to

- recall what a reversible reaction is
- recall what a dynamic equilibrium is
- give examples of reversible reactions

Going both ways

The symbol '$\rightleftharpoons$' is used to show that a reaction is reversible.

A reversible reaction: water drips onto anhydrous copper(II) sulfate which turns blue.

Photosynthesis is an example of a reaction that may be reversed.

carbon dioxide + water + energy $\rightleftharpoons$ sugar + oxygen

The reverse is respiration:

sugar + oxygen $\rightleftharpoons$ carbon dioxide + water + energy

Reversible reactions

Many reactions around you are reversible. They can go backwards *and* forwards. For a chemical reaction it means that the reactants will react to make the products, but also the products will react to make the reactants. The result is that it is difficult to fully change the reactants into products! This is important in industrial processes because it would be better if the reactants reacted completely.

Everybody's favourite: copper sulfate

You may have heated blue hydrated copper(II) sulfate crystals to form white copper(II) sulfate powder and water:

$$CuSO_4.5H_2O(s) \overset{heat}{\rightleftharpoons} CuSO_4(s) + 5H_2O(l)$$

Blue crystals + heat white powder + water

When water is added to the white powder then the reaction is reversed, and so the blue solid is made:

$$CuSO_4(s) + 5H_2O(l) \rightleftharpoons CuSO_4.5H_2O(s)$$

This car stores energy using a ***reversible reaction*** *where methanol is used to store hydrogen (see below)*

Future fuel?

In the future we may use the energy given out when hydrogen undergoes oxidation. The hydrogen may be 'made' from other compounds using renewable energy sources, such as solar, wave, wind, or tidal power. But storing hydrogen is expensive and difficult. It needs to be stored either as a gas in high pressure tanks, or as a liquid at very low temperatures. One alternative way is to store hydrogen as part of the compound methanol, and re-use it in a reversible reaction:

$$CO(g) + 2H_2(g) \rightleftharpoons CH_3OH(l) \text{ (methanol)}$$

When energy is required to power your laptop or mp3 player, or to heat your home, then the methanol is turned back into hydrogen:

$$CH_3OH(l) \rightleftharpoons CO(g) + 2H_2(g)$$

The hydrogen is then oxidized to make pollution free water and energy.

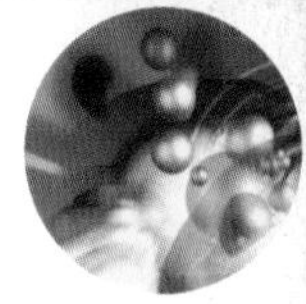

Dynamic equilibrium

In a reversible reaction, when the concentration of reactants and products are constant, then the reaction is at **equilibrium**. At equilibrium, the reactants are still making the products, and the products are still reacting to make the reactants, so the reaction is called a **dynamic equilibrium**. The word 'dynamic' means 'active'.

The speed at which reactants react is called the **forward rate**, and the speed at which the products react is called the **backward rate**. It is important to remember that in a dynamic equilibrium:

- the forward rate equals the backward rate

At equilibrium the concentrations of the reactants and products may be constant, but that does not mean that the reactant concentration equals the product concentration. Really there is a fixed ratio between the reactants and products.

Methanol fuel tanks

Suppose you had a fuel tank storing methanol ready to make hydrogen. In the tank is a mixture of methanol, hydrogen, and carbon monoxide. The methanol is always breaking down to make carbon monoxide and hydrogen, which are always reacting to make methanol. All three chemicals are in dynamic equilibrium.

Methanol powers this laptop using a reversible reaction.

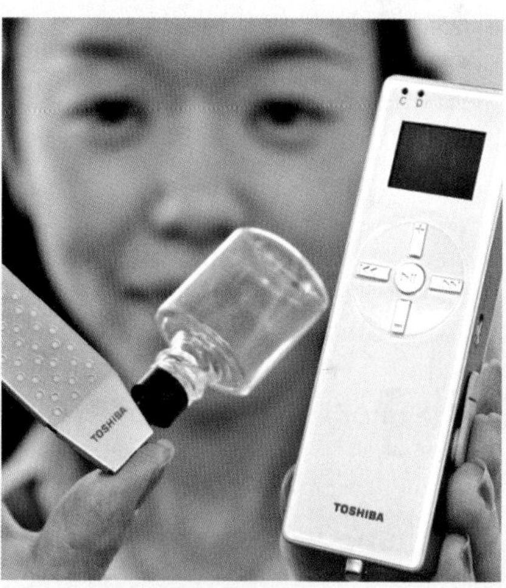

This mp3 player is powered by methanol which is turned into hydrogen by a reversible reaction.

At equilibrium the concentration of the chemicals do not change until the conditions are changed. *What happens when the conditions are changed is described in the next section.*

Check your understanding

1. What does the sign '$\rightleftharpoons$' mean?
2. Which of these are the same when a reaction is at equilibrium?
   a the amount of product and reactant
   b the concentration of reactant and product
   c the forward rate and the backward rate
3. Ethanol is made by reacting ethene with steam:
   $CH_2CH_2(g) + H_2O(g) \rightleftharpoons CH_3CH_2OH(g)$
   a Not all the ethene will react. Why is that?
   b What do you think is done with the left-over ethene?
   c What is meant by 'dynamic equilibrium'?

7.02 Le Chatelier's Principle 1

OBJECTIVES

after this spread you should know

- what Le Chatelier's Principle states
- how an equilibrium shifts when the amount of reactants or products is changed
- how an equilibrium shifts when the concentrations of the reactants are changed

Stop or resist?

Notice that the equilibrium resists the change. It lessens the change. It does not completely stop any change.

Just a shift to the left?

When an equilibrium shifts to make more reactants, and so less products, then you can say that

- the equilibrium shifts to the left.

Or a jump to the right?

When an equilibrium shifts to make more products, and so less reactants, then you can say the that

- the equilibrium shifts to the right.

Large industrial plants use reactions that are in chemical equilibrium. As a product is made, then it is removed to allow more product to form.

The world tries to resist you: Le Chatelier's Principle

Have you noticed that many things that are in balance will resist you changing them? Chemical equilibrium is like that. When a reaction is at equilibrium and you try to change the amounts of the chemicals, then the **position of equilibrium** will shift to resist the change.

This is described in **Le Chatelier's Principle** which says:

- When the conditions of an equilibrium are changed then the position of equilibrium will shift to minimize the change.

What happens if you change the amount of chemicals?

Look again at the reversible reaction that stores hydrogen fuel as methanol:

$$CO(g) + 2H_2(g) \rightleftharpoons CH_3OH(l) \text{ (methanol)}$$

What will happen as the hydrogen is taken away to make energy? The answer is that more hydrogen is made because some methanol reacts. This is because

- At equilibrium the amounts of all three chemicals remain constant.
- When some hydrogen is removed the position of equilibrium will shift to resist the change.
- More hydrogen is made because some methanol reacts to make carbon monoxide and hydrogen.

If this equilibrium mixture is in a car fuel tank, then the energy from the car braking could be used to make electricity, which is used to make hydrogen that is pumped into the fuel tank. What will happen then? The position of equilibrium will shift to the right, to make more methanol, and so use up most of the hydrogen.

Effect of changing a concentration

The position of equilibrium will change if the concentration of one reactants or products is changed. Here is a reaction that is used to make the common alcohol, ethanol;

$$\text{ethene} + \text{steam} \rightleftharpoons \text{ethanol vapour}$$

$$CH_2CH_2(g) + H_2O(g) \rightleftharpoons CH_3CH_2OH(g)$$

Think of the mixture being at equilibrium.

If the concentration of ethene is increased then

- the position of equilibrium will shift to resist the change,
- the position of equilibrium will shift to the right,
- more ethanol will be made

If the concentration of water is decreased then

- the position of equilibrium will shift to resist the change
- the position of equilibrium will shift to the left
- less ethanol will be made

If the concentration of ethanol is decreased then

- the position of equilibrium will shift to resist the change
- the position of equilibrium will shift to the right
- more ethanol will be made

When a chemical is added to a mixture at equilibrium, the position of equilibrium shifts to resist the increase.

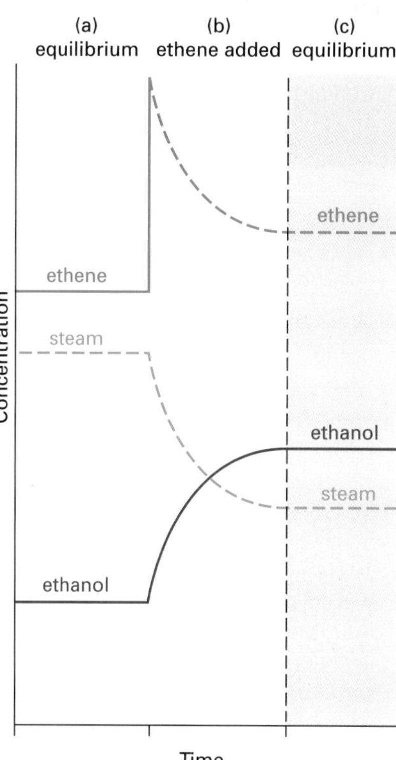

This graph shows the change in concentration of ethene, steam, and ethanol over time. (a) At the start the mixture is at equilibrium. (b) After a time more ethene is added. (c) Later the concentrations level off when the mixture is again at equilibrium.

Fuel on the go?

Imagine the future when fuel cells could be used to make electricity for your laptop, or mobile phone. The hydrogen and methanol are in equilibrium, so as the hydrogen is used then the methanol reacts to make more. Also when recharging in a wall socket, the electricity makes more hydrogen, and so more methanol is made.

Check your understanding

1. Esters are used to add fragrances to your food and cosmetics. Here is how one ester, ethyl ethanoate, is made:
   ethanol + ethanoic acid ⇌ ethyl ethanoate + water
   $CH_3CH_2OH(l) + CH_3COOH(l) \rightleftharpoons CH_3COOCH_2CH_3\ (l) + H_2O(l)$
   a If the mixture is at equilibrium, what would happen if
      i the concentration of ethanol is increased
      ii the mixture is diluted by adding water
      iii the ester is removed
      iv the ethanoic acid concentration is decreased, by neutralizing the acid with alkali?
   b What would you do to make sure all the ethanol was used up to make as much ester as possible?
2. The toxic gas chlorine is used to kill harmful micro-organisms in drinking water. In fact the micro-organisms are killed by chlorate(I) ions, ClO^-, that come from chloric(I) acid, HClO. They are made like this:
   $$Cl_2(g) + H_2O(l) \rightleftharpoons HCl(aq) + HClO(aq)$$
   Notice that hydrochloric acid, HCl, is on the right.
   a What would happen to the concentration of HClO if more hydrochloric acid was added?
   b If sodium hydroxide was added to neutralize the acid, what would happen to the concentration of ClO^- (chlorate(I)) ions?
   c What would be the best way to remove the smell of chlorine gas?
   d What would happen if acid was accidentally mixed with a bleach containing HClO? Why would that be dangerous?

7.03 Le Chatelier's Principle 2

OBJECTIVES

after this spread you should know

- how an equilibrium involving gases will shift if the pressure is changed
- how an equilibrium will shift if the temperature is changed
- that catalysts do not change the position of an equilibrium

More = more

Two moles of gas molecules would produce more pressure than one mole of gas molecules in the same volume, and at the same temperature.

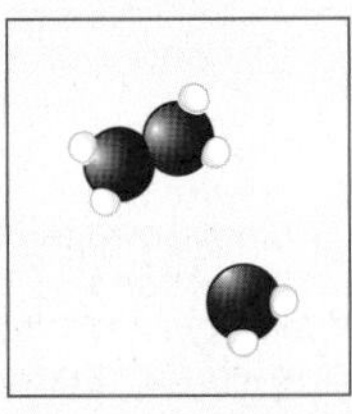
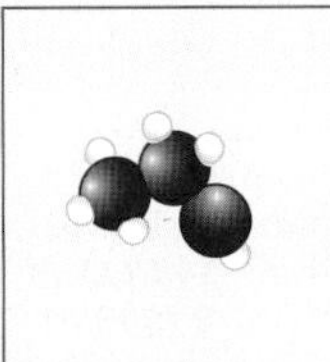

The box on the left contains two gas molecules which exert more pressure than the one gas molecule in the box on the right.

Exothermic?

If you are told that a reaction is exothermic, and so produces heat, then

- It is the forward reaction that is exothermic.
- The enthalpy change of this forward reaction will have a negative sign.
- The backward reaction will be endothermic.

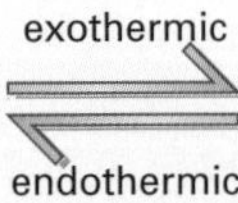

When the pressure is on

Which way will the position of equilibrium shift when you change the *pressure*? The equilibrium will shift to resist the change. Here is an example;

ethene + water as steam ⇌ ethanol vapour

$$CH_2CH_2(g) + H_2O(g) \rightleftharpoons CH_3CH_2OH(g)$$

If this mixture reaction is at equilibrium, and the pressure is increased, then which way would the position of equilibrium shift?

- There are two gas molecules on the left and one gas molecule on the right.
- Two gas molecules would produce more pressure than one gas molecule in the same box.
- To reduce the pressure, the number of gas molecules must get smaller.
- In this example, an increase in pressure would shift the position of equilibrium to the right, to produce more ethanol, and fewer ethene and water molecules.

It is important to remember that;

- When the pressure is increased the position of equilibrium will shift in the direction of the side with the fewer gas molecules.

When the heat is on

Which way does the position of equilibrium shift when you change the temperature?

The position of equilibrium will shift to resist the change.

The reaction $CH_2CH_2(g) + H_2O(g) \rightleftharpoons CH_3CH_2OH(g)$ is exothermic.

If this reaction mixture was at equilibrium, what would happen if the temperature was increased?

- The reaction is labelled as an exothermic reaction, which means the forward reaction is exothermic.
- The backward reaction is endothermic.
- If the temperature is increased then the position of equilibrium will shift to resist the change and so cool the mixture,
- The endothermic reaction is favoured, which is the backward reaction,
- In this example, the position of equilibrium shifts to the left.

It is important to remember that

- If the temperature *increases* then the position of equilibrium will shift in the *endothermic* direction.

Cool it!

For the same reaction, what would happen if the temperature was decreased? The equilibrium would resist the change by shifting in the exothermic direction, and so make the mixture heat up.

It is important to remember that

- If the temperature *decreases* then the position of equilibrium will shift in the *exothermic* direction.

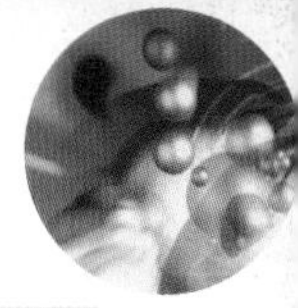

Hydrogen from coal

This reaction uses coal and steam to make hydrogen:

$C(s) + 2H_2O(g) \rightleftharpoons CO_2(g) + 2H_2(g)$. This reaction is endothermic.

When heated up, will more hydrogen or less hydrogen be made?

- The forward reaction is endothermic.
- If the temperature is increased the position of equilibrium will shift to resist the change and so lower the temperature.
- The endothermic reaction is favoured, which, in this example, is the forward reaction.
- The position of equilibrium shifts to the right, and so more hydrogen is produced.

"Cats have no effect on Le Chat"

Note that catalysts do not affect the position of an equilibrium. Although the reaction rate does increase, the amount of product is unchanged. It is important to remember that in an equilibrium:

- Catalysts have no effect on the positon of equilibrium.

Endothermic?

If you are told that a reaction is endothermic, and so takes in heat, then

- It is the forward reaction that is endothermic.
- The enthalpy change of this forward reaction will have a positive sign.
- The backward reaction will be exothermic.

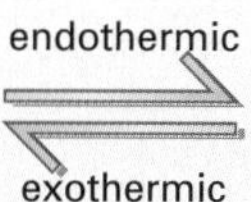

In the future coal and steam may be combined to make hydrogen to power the future economy. Also made is the greenhouse gas, carbon dioxide, which would have to be stored underground.

Check your understanding

1. This reaction shows the dissociation of a pale coloured gas, dinitrogen tetraoxide:
   $N_2O_4(g) \rightleftharpoons 2NO_2(g)$ This reaction is endothermic.
   The nitrogen dioxide, NO_2, is a dark brown colour. State any colour change and explain each answer. If a mixture of the gases are at equilibrium then,
   - **a** What would happen if the temperature is increased?
   - **b** What would happen if the temperature is decreased?
   - **c** What would happen if the pressure is increased?
   - **d** What would happen if the pressure is decreased?
   - **e** What would happen is a platinum catalyst is added?
2. The reaction of hydrogen and iodine produces hydrogen iodide:
   $H_2(g) + I_2(g) \rightleftharpoons 2HI(g)$
   State what would happen, if anything, if the pressure is increased. Explain your answer.
3. $CH_3CH_2OH(l) + CH_3COOH(l) \rightleftharpoons CH_3COOCH_2CH_3(l) + H_2O(l)$
   ethanol + ethanoic acid ⇌ ethyl ethanoate + water
   The enthalpy change for this reaction is zero.
   If the reactants and products are at equilibrium what would happen, if anything, if the temperature is increased? Explain your answer.

7.04 Le Chatelier's Principle and compromise

OBJECTIVES

after this spread you should know

- the conditions required to make ethanol from ethene and water
- the conditions required to make methanol from carbon monoxide and hydrogen
- that rate and equilibrium are both important when choosing the optimum conditions for a reaction
- that sometimes a compromise temperature and pressure may be used in practice.

Methanol, our future car fuel? What will be the cost to the environment?

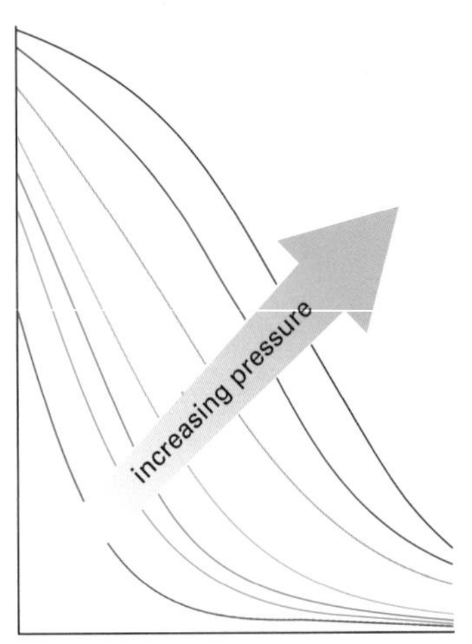

A generalized trend graph of yield of methanol against temperature for varying pressure. Lower temperatures produce higher yields, but at lower rates. Higher temperatures may produce faster rates, but with lower yields. A compromise moderate temperature is required.

Making methanol is a compromise

Methanol may be the fuel of the future. It is made by this reaction:

$CO(g) + 2H_2(g) \rightleftharpoons CH_3OH(l)\ \Delta H = -90\,kJ\,mol^{-1}$.

What are the optimum conditions that will produce the most methanol per day?

The effect of pressure

- The *higher* the pressure then the faster the rate.
- Also the *higher* the pressure the greater the yield, as the position of equilibrium will shift to the right because there are fewer molecules on that side.
- So a *high* pressure will give both a high rate and a high yield.
- To use a very high pressure would be very expensive.

The effect of temperature

- The higher the temperature, the higher the rate.
- A higher temperature decreases the yield, as the position of equilibrium shifts to the left, in the endothermic direction.
- A lower temperature increases the yield, as the position of equilibrium shifts to the right, in the exothermic direction.
- But a lower temperature decreases the rate.
- So a *compromise moderate temperature* is chosen that will give a high enough yield while ensuring a high enough rate.

The effect of a catalyst

- Adding a catalyst will increase the rate of both forward and reverse reactions by the same ratio.
- Adding a catalyst will not change the yield.
- So a catalyst is used.

The actual conditions required for the optimum production of methanol are

- 300°C temperature,
- 300 atmospheres pressure (30 MPa),
- and a catalyst made of zinc and chromium(III) oxide, Zn and Cr_2O_3.

In industry, ethanol is made by reacting ethene with steam:
$CH_2CH_2(g) + H_2O(g) \rightleftharpoons CH_3CH_2OH(g)\ \Delta H = -287\,kJ\,mol^{-1}$
The reasons for these optimum conditions are similar to those stated for methanol production.
The optimum conditions required:

- **300°C temperature**
- **90 atmospheres pressure**
- **and a phosphoric acid, H_3PO_4, catalyst**

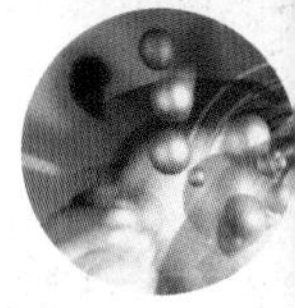

Is this a natural way of producing hydrogen?

This reaction uses natural gas, which is methane, and steam to make hydrogen and carbon monoxide, which is called synthesis gas;

$CH_4(g) + H_2O(g) \rightleftharpoons CO(g) + 3H_2(g)$. $\Delta H = +206 \text{ kJ mol}^{-1}$.

What are the optimum conditions that will produce the most hydrogen per day from methane?

How does *temperature* influence the rate and yield?

- The higher the temperature, the higher the rate,
- The higher the temperature the higher the yield, because the position of equilibrium shifts to the right, in the endothermic direction,
- So high temperature will give both a high rate and high yield.
- The higher temperature will have an economic cost as fuel must be used.

How does *pressure* influence the rate and yield?

- The higher the pressure then the faster the rate.
- The higher the pressure then the lower the yield, as the position of equilibrium will shift to the left, because there are fewer gas molecules on that side.
- So a compromise moderate pressure is chosen that will give a high enough yield while ensuring a high enough rate.

How does adding a *catalyst* influence the rate and yield?

- Adding a catalyst will increase the rate.
- Adding a catalyst will not change the yield.
- So a catalyst is used.

The actual conditions required for the optimum production of hydrogen and carbon monoxide are

- 900°C temperature,
- 30 atmospheres pressure (3 MPa),
- and a nickel catalyst.

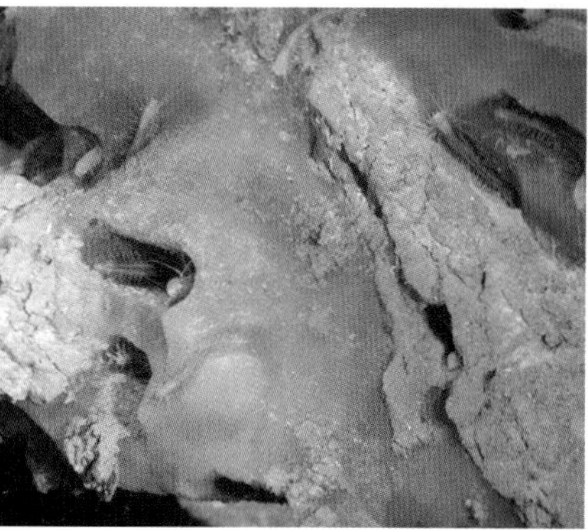

Huge amounts of methane are found crystallized with water in the deep ocean. Could this be a possible future source of the fuel methane? Here animals are living off the methane.

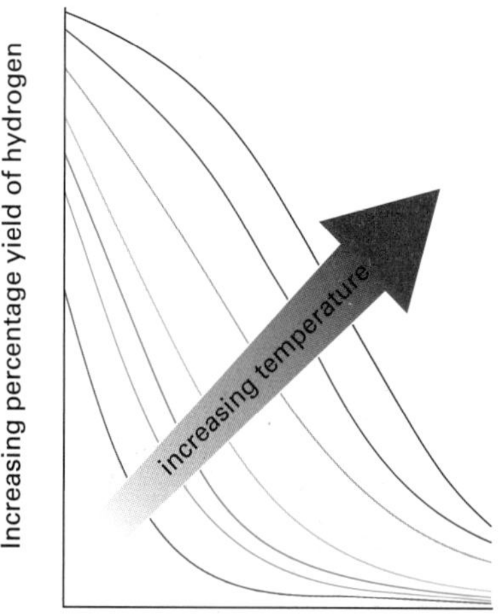

A generalized trend graph of yield of hydrogen against pressure for varying temperature. Lower pressures produce higher yields, but at lower rates. Higher pressures may produce faster rates, but with lower yield. A compromise moderate pressure is required.

Check your understanding

1. The Haber Process is used to produce ammonia:
   $N_2(g) + 3H_2(g) \rightleftharpoons 2NH_3(g)$ $\Delta H = -92.4 \text{ kJ mol}^{-1}$.
   a State the effect of increasing temperature on the reaction rate and the yield. Why is the moderate temperature of 450°C used?
   b State the effect of increasing pressure on the reaction rate and the yield. Why is a high pressure of 200 atmospheres used?
   c Why is an iron catalyst used?
2. Ammonia is often then converted into nitrogen oxide which is used to make nitric acid:
   $4NH_3(g) + 5O_2(g) \rightleftharpoons 4NO(g) + 6H_2O(g)$ $\Delta H = -902 \text{ kJ mol}^{-1}$.
   a High pressure would increase the reaction rate. Why is a low pressure used?
   b Suggest why a high temperature of 850°C is used.
   c The catalyst is a mixture of platinum and rhodium. Why is it used as a woven wire mesh rather than a sheet?

7.05 Carbon neutral fuels

OBJECTIVES

after this spread you should

- know what is meant by a carbon neutral fuel
- be able to give examples of fuels that are or are not carbon neutral

What are carbon neutral fuels?

Fossil fuels release carbon dioxide into the atmosphere when they burn. This contributes to global warming because carbon dioxide is a greenhouse gas. Carbon neutral fuels do not contribute to global warming. The phrase **carbon neutral** refers to an activity that has no net annual carbon (greenhouse gas) emissions to the atmosphere. The production and use of carbon neutral fuels does not cause an overall release of carbon dioxide, taken over a year.

That means there is no 'overall' production of carbon dioxide. It may be produced, but it is reabsorbed in another stage of the process.

1 Ethanol from ethene?

- The ethene comes from crude oil.
- When the fuel is combusted then the carbon dioxide produced is not reabsorbed.
- All the carbon dioxide is released into the atmosphere.

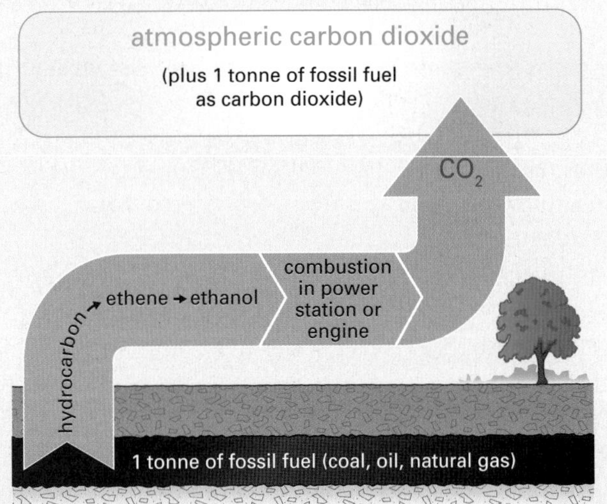

So ethanol from ethene is NOT carbon neutral.

2 Growing corn for ethanol

- Uses huge amounts of fossil fuel.
- Little carbon dioxide is absorbed.
- Most carbon dioxide is released into the atmosphere.

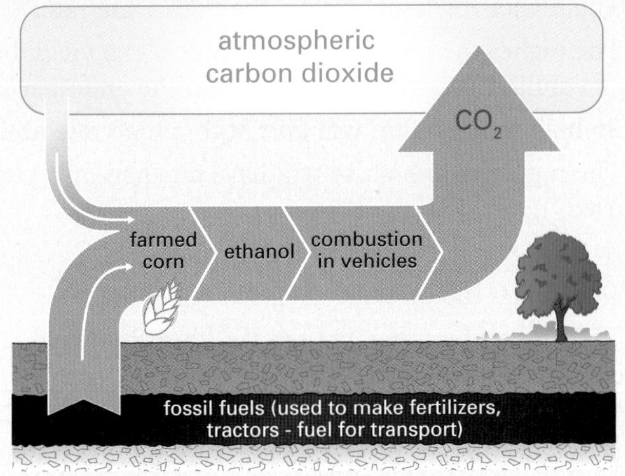

So ethanol from corn is NOT carbon neutral.

3 Ethanol fuel from sugar cane?

- The sugar cane comes from plants, which absorb carbon dioxide.
- The process to make ethanol uses the waste sugar cane to fuel the process.
- It requires huge amounts of land.
- Sugar cane production could harm the environment.

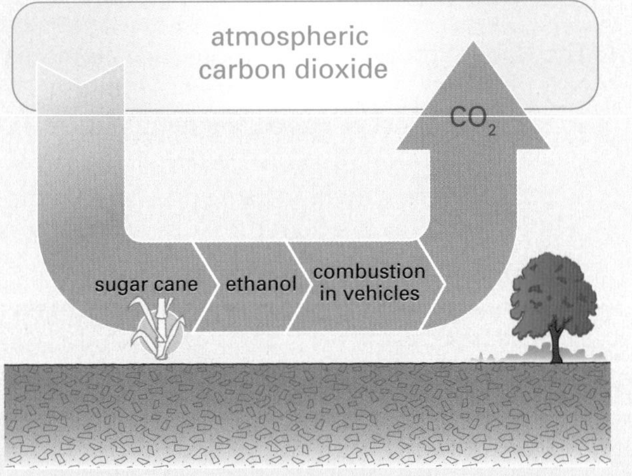

So ethanol from sugar cane IS carbon neutral.

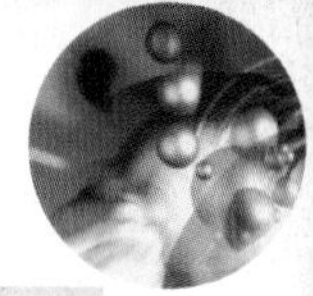

4 Coal as a fuel if the carbon dioxide made is stored underground?

- Power stations could burn fossil fuels to produce energy.
- The carbon dioxide produced could be collected and stored underground in old coal mines, or oil wells.
- Some scientists are concerned that the carbon dioxide could escape.

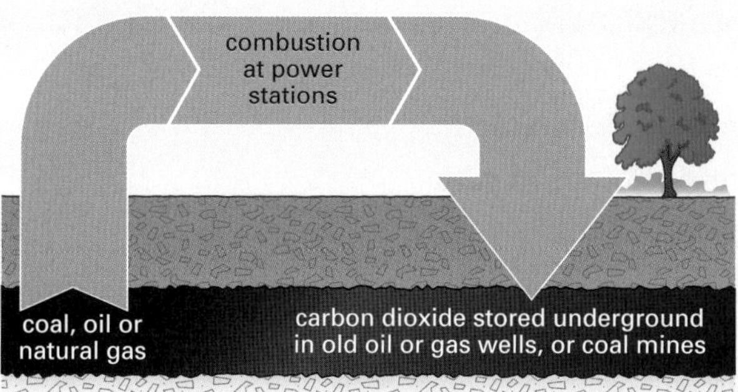

This IS a carbon neutral way of using fossil fuels.

5 Ethanol fuel from hydrogen using nuclear energy?

- Nuclear energy is used to make heat, which is used to make steam, which is used to drive turbines which make electricity which could make hydrogen.
- The hydrogen could be made into methanol fuel, or used directly.
- Building a nuclear plant produces large amounts of carbon dioxide because concrete is used.
- When the plants are shut then cleaning up the site would require energy and more concrete. This would produce more carbon dioxide.
- The amount of carbon dioxide produced would be less than from a coal-fired power station, for the same energy.
- None of the carbon dioxide produced would be recycled.
- There are also environmental worries about nuclear energy.

So nuclear energy is NOT FULLY carbon neutral.

6 Methanol fuel from hydrogen?

- The production of methanol from carbon monoxide and hydrogen does not have to release any carbon dioxide.
- The hydrogen could be made from renewable sources such as wave, wind, tidal, and solar power.
- Also the carbon dioxide produced from combustion would be recycled, or the original carbon monoxide could be made from plant waste.

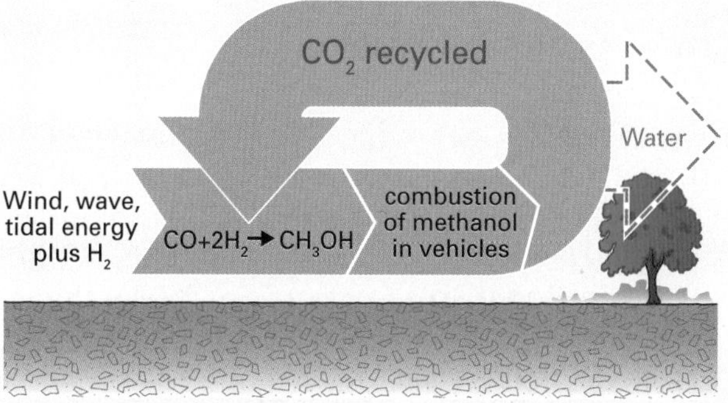

If the above processes are used then methanol fuel would be carbon neutral. Note that hydrogen made from crude oil or natural gas (methane) would not be carbon neutral.

7 Ethanol fuel from intensively farmed grain?

- The grain comes from plants, which absorb carbon dioxide.
- Farming requires crude oil to fuel the tractors, transport the grain, and to make the fertilizer.
- Ethanol produced directly from crude oil uses more than it saves.
- Much more carbon dioxide is produced than is absorbed by photosynthesis.
- So ethanol from grain is not carbon neutral.

Check your understanding

1. What does 'carbon neutral' mean?
2. How could you decrease the amount of carbon dioxide you produce?
3. Which of these are carbon neutral?
   a ethanol made from grain
   b ethanol made from sugar cane
   c methanol made from hydrogen

8.01 OIL-RIG

OBJECTIVES

already from GCSE, you

- know that some reactions involve oxidation and reduction
- are familiar with dot and cross diagrams

and after this spread you should be able to

- explain what is meant by oxidation and reduction
- explain what is meant by an oxidizing agent and a reducing agent

To remember these definitions you could learn;

OIL Oxidation Is Loss (of electrons)

RIG Reduction Is Gain (of electrons)

Redox reactions are all around and in you

You are full of **redox** reactions. The redox reaction respiration releases the energy you need to live, and the food you eat ultimately comes from the redox reaction photosynthesis. Around you, mobile and laptop batteries work using redox reactions, as do fuel cells, bleaches, old fashioned wet photography, and metal corrosion, such as rusting iron.

So what are redox reactions?

The redox comes from two words: *Red*uction and *Ox*idation.

In a reaction, **oxidation** is the *loss* of electrons from an atom or ion, and **reduction** is the *gain* of electrons by an atom or ion. In a redox reaction there is the gain of electrons by one chemical, reduction, and the loss of electrons by another chemical, oxidation, so both are found together.

You must remember

- oxidation is the process of electron loss
- reduction is the process of electron gain

An example of redox

When the metal sodium reacts with the green gas chlorine then the white solid sodium chloride is made:

$$2Na(s) + Cl_2(g) \rightarrow 2NaCl(s)$$

The sodium has been oxidized and the chlorine has been reduced.

Oxidation of sodium

Sodium metal is made of sodium atoms. In a reaction, each sodium atom *loses* an electron to a chlorine atom, so this is called *oxidation*.

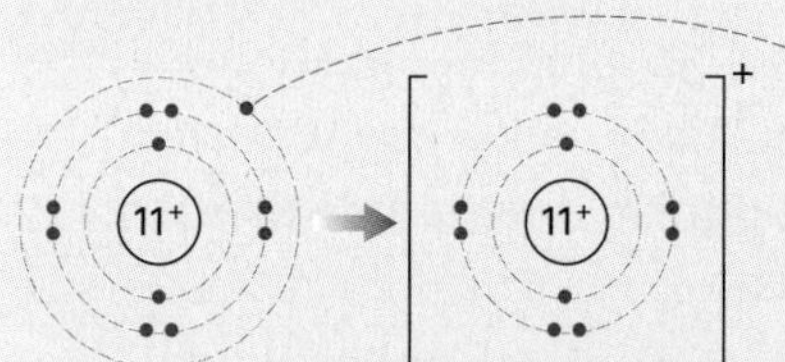

Each sodium atom, Na, has the electronic configuration 2,8,1. It reacts to make a sodium ion, Na^+, with an electronic configuration 2,8. Each sodium atom loses an electron so oxidation has occurred.

Reduction of chlorine

The chlorine gas is made of chlorine atoms. When the chlorine reacts then each chlorine atom gains one electron to become a chloride ion, Cl^-. This is reduction because each chlorine atom gains an electron.

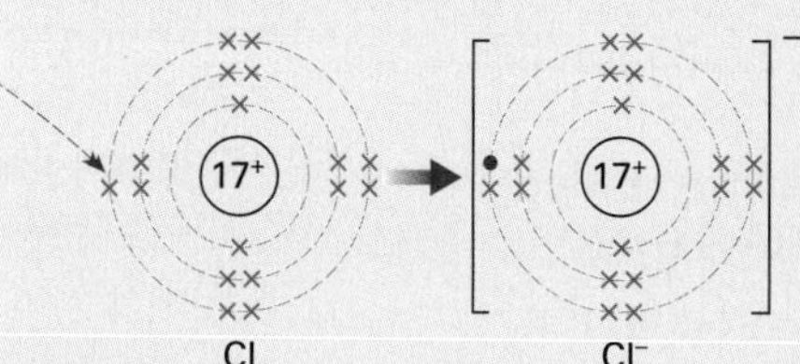

Each chlorine atom, Cl, has the electron configuration of 2,8,7. it reacts to make a chloride ion, Cl^-, with an electronic structure of 2,8,8. Each chlorine atom gains an electron so reduction has occured.

Ancient reduction

You may wonder why the word reduction is used to mean a gain of electrons. Thousands of years ago it was noticed that the metal made by smelting had less mass than the original ore. The ore was made smaller, reduced. The loss of oxygen made the mass less. More recently it was realised that the metal atoms were gaining electrons.

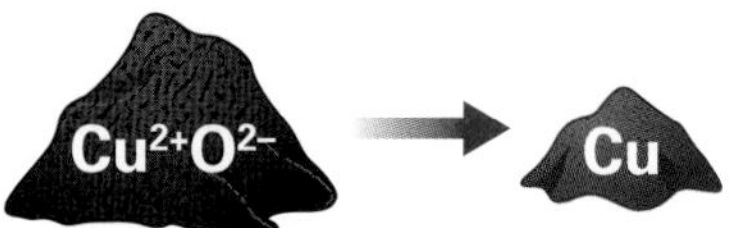

Copper ore is reduced to copper metal. The copper metal has less mass than the copper ore. The copper ions have gained electrons. Here reduction means 'make smaller' and 'gain of electrons'.

Oxidizing agents and reducing agents

Oxidizing agents oxidize other chemicals, so oxidizing agents are themselves reduced, so oxidizing agents gain electrons. Reducing agents reduce other chemicals, so they lose electrons.

You must remember that

- **oxidizing agents** are electron acceptors
- **reducing agents** are electron donors

Why 'oxidation'?

Oxidation used to mean just 'gain of oxygen', but it was realized, more importantly, that the other chemical was losing electrons to the oxygen. So it was decided that 'oxidation' should have the broader meaning 'loss of electrons'.

White hot fire

To allow trains to travel at high speeds the rails must be welded together so there are no gaps. This must be done in isolated places so molten iron is made using the Thermit reaction:

$Fe_2O_3(s) + 2Al(s) \rightarrow Al_2O_3(s) + 2Fe(l)$

Here the iron(III) oxide, Fe_2O_3, is reduced to iron, Fe, so

- the iron(III) oxide is reduced,
- the iron(III) oxide is the oxidizing agent

The aluminium powder, Al, is oxidized to aluminium oxide, Al_2O_3, so

- the aluminium is oxidized
- the aluminium is the reducing agent

The Thermit reaction is a dramatic way to weld iron rails together. Here the iron(III) oxide is the oxidizing agent, and the aluminium is the reducing agent.

Check your understanding

1. Define: (a) oxidation, (b) reduction, (c) oxidizing agents, (d) reducing agents.
2. For these reactions, state which substance is being oxidized:
   a $Fe_2O_3 + 3CO \rightarrow 2Fe + 3CO_2$
   b $Mg + PbO \rightarrow MgO + Pb$
   c $2KI + Cl_2 \rightarrow 2KCl + I_2$
3. For these reactions, state which substance is the oxidizing agent:
   a $Fe_2O_3 + 3CO \rightarrow 2Fe + 3CO_2$
   b $Mg + PbO \rightarrow MgO + Pb$
   c $2KI + Cl_2 \rightarrow 2KCl + I_2$
4. For this reaction, state which change is oxidation and which change is reduction: $Mg + Cl_2 \rightarrow MgCl_2$

The green gas chlorine oxidizes the colourless aqueous bromide ions to make red-brown bromine. This is a redox reaction involving two non-metals.

A green gas to a red-brown liquid

When green chlorine gas reacts with aqueous potassium bromide then the bromide ions are reduced to the red-brown solution of bromine:

$Cl_2(g)$ +	$2KBr(aq)$ →	$Br_2(aq)$ +	$2KCl(aq)$
green chlorine gas	colourless potassium bromide solution	red-brown solution of bromine	colourless potassium chloride solution

- In the KBr the bromide ions, Br^-, have lost electrons,
- This change from bromide ions, Br^-, to bromine atoms, Br, is oxidation.
- The bromide ions are electron donors, so the bromide ion is the reducing agent.
- You could also say that the potassium bromide is the reducing agent because it contains the bromide ions.
- The chlorine atoms in the Cl_2 molecules, have changed into chloride ions.
- The chlorine atoms have gained electrons.
- This change from chlorine atoms, Cl, to chloride ions, Cl^-, is reduction.
- The chlorine atoms are the electron acceptors.
- The chlorine is the oxidizing agent.

8.02 Half-equations

OBJECTIVES

after this spread you should know

- what is meant by a 'half-equation'
- how to label a half-equation as either oxidation or reduction
- how to write simple half-equations

A **half-equation** shows
- the gain OR loss of electrons by one chemical.

Why use half-equations?

Half-equations help you see what is happening to one chemical without any other distracting chemicals.

Copper metal reacts with silver ions, aqueous silver nitrate to produce silver. The half-equations help you to understand the reaction.

- $Ag^+(aq) + e^- \rightarrow Ag(s)$
- $Cu(s) \rightarrow Cu^{2+}(aq) + 2e^-$
- *This time the nitrate ions are the spectator ions.*

Which ions are spectators?

Spectator ions are those which;
- Are not oxidized or reduced,
- Do not change state (e.g. from solid to aqueous solution)

Ions that you must include are;
- Those which are oxidized or reduced,
- Those that change state.

What are half-equations?

Half equations show the gain or loss of electrons by one chemical.

For example, this is the full equation for when sodium reacts with chlorine:

$$2Na(s) + Cl_2(g) \rightarrow 2NaCl(s)$$

Each sodium atom is losing an electron to a chlorine atom, so you could write this half equation:

$$Na \rightarrow Na^+ + e^-$$

(You use e^- to stand for an electron)

Each chlorine atom in a chlorine molecule gains an electron, so you could write this half equation:

$$Cl_2 + 2e^- \rightarrow 2Cl^-$$

(You need to show two chloride ions because each chlorine molecule, Cl_2, contains two chlorine atoms.)

The way sodium reacts does not depend on the other reactant. For example, if sodium reacted with bromine instead of chlorine the equation would be:

$$2Na(s) + Br_2(g) \rightarrow 2NaBr(s)$$

Sodium is still gaining electrons in the same way;

$$Na \rightarrow Na^+ + e^-$$

and the half equation for bromine becomes:

$$Br_2 + 2e^- \rightarrow 2Br^-$$

Half-equations and redox

Half-equations involve electron gain or loss, so they always are either oxidation *or* reduction.

In this reaction an electron is lost, so this is oxidation:

$$Na \rightarrow Na^+ + e^-$$

This time electrons are gained, so this is reduction;

$$Br_2 + 2e^- \rightarrow 2Br^-$$

Why 'half'?

In a reaction, when one chemical loses electrons then another must gain them. Half-equations only show half the story, either the gain or the loss of electrons.

Some ions are spectators

Ions that take no part in the reaction are called **spectator ions**. When zinc is put into copper sulfate solution then the full equation is:

$$Zn(s) + CuSO_4(aq) \rightarrow Cu(s) + ZnSO_4(aq)$$

Each zinc atom is oxidized:

$$Zn(s) \rightarrow Zn^{2+}(aq) + 2e^-$$

Each copper ion is reduced:

$$Cu^{2+}(aq) + 2e^- \rightarrow Cu(s)$$

Notice that the sulfate ions, SO_4^{2-}, do not appear in the equations. They do not change during the reaction. They are dissolved in the water, so are

just floating around. As they are said to only 'watch' the reaction they are called 'spectator ions'.

A piece of grey zinc metal dipped into a blue copper(II) sulfate solution. The grey metal is plated with brown copper metal. The colourless zinc ions replace the blue copper(II) ions.

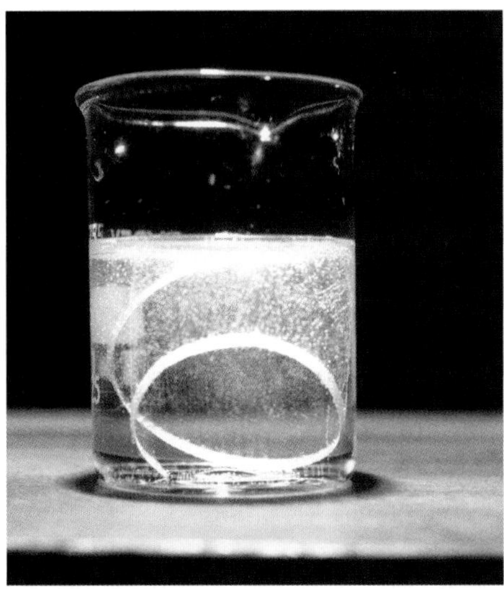

Magnesium ribbon put into dilute sulfuric acid. The bubbles are hydrogen gas.

Check your understanding

1. State whether these reactions are oxidation of reduction;
   - **a** $Ca^{2+}(aq) + 2e^- \rightarrow Ca(s)$
   - **b** $H_2(g) \rightarrow 2H^+(aq) + 2e^-$
   - **c** $Fe^{2+} \rightarrow Fe^{3+} + e^-$
   - **d** $2O_2(g) + 4e^- \rightarrow 2O^{2-}(aq)$
   - **e** $Cl_2(g) + 2H_2O(l) \rightarrow 2HClO(aq) + 2H^+(aq) + 2e^-$
2. Write half-equations for these changes:
   - **a** an iodide ion changes into iodine
   - **b** a fluorine atom changes into a fluoride ion
   - **c** a potassium atom changes into a potassium ion
   - **d** a barium atom changes into a barium ion
   - **e** an aluminium ion is changed into an aluminium atom

Metal and acid

What happens when you mix metal and acid? As an example, we will look at the reaction between magnesium and dilute sulfuric acid which makes explosive hydrogen gas:

$$Mg(s) + H_2SO_4(aq) \rightarrow MgSO_4(aq) + H_2(g)$$

It is easier to see what is happening by looking at the half-equations.

Each magnesium atom is losing two electrons to make a magnesium ion:

$Mg \rightarrow Mg^{2+} + 2e^-$
(this is oxidation)

The hydrogen ions have gained the electrons and then joined to make covalent molecules:

$2H^+(aq) + 2e^- \rightarrow H_2(g)$
(This is reduction)

Notice that again the sulfate ions are spectator ions. This shows that it does not matter which acid is used in a reaction between a metal and an acid, because the reaction is really between the metal and the hydrogen ions. You can combine the two half-equations to give an ionic equation.

You will learn how to combine half-equations later but for now you can put the two half-equations together:

$$Mg + 2H^+(aq) + 2e^- \rightarrow Mg^{2+} + 2e^- + H_2(g)$$

As there are the same number of electrons on each side we can cancel them to make:

$$Mg + 2H^+(aq) \rightarrow Mg^{2+} + H_2(g)$$

So this equation shows what happens when magnesium reacts with any acid.

8.03 Oxidation states and their calculation 1

OBJECTIVES

after this spread you should be able to

- recall what is meant by oxidation state
- calculate the oxidation state of an element in a compound

Using oxidation states

Here is an example of a simple covalent compound, hydrogen chloride

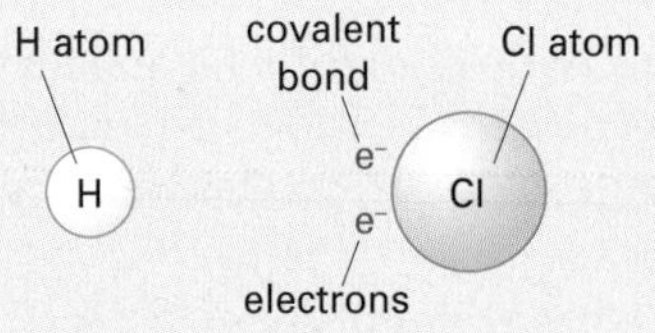

- The two electrons in the covalent bond in HCl are closer to the Cl than the H so the Cl has control of both electrons.
- Cl has control of the electron from the hydrogen atom.
- The Cl is said to have an oxidation state of –1 because it has gained control of an electron.
- The H has lost control of its electron so it has a +1 oxidation state.

Although oxidation states are similar to ionic charges they are written differently.

- The *oxidation state* for magnesium in MgO is written as +2,
- The *charge* on a magnesium ion is written as 2+.

What is an oxidation state?

An **oxidation state** is

- it is the number of electrons needed to be gained or lost to make a neutral atom

Using oxidation states is a way of working out how oxidized or reduced something is. It is similar to the charge on ions, except that it is also used for covalent compounds.

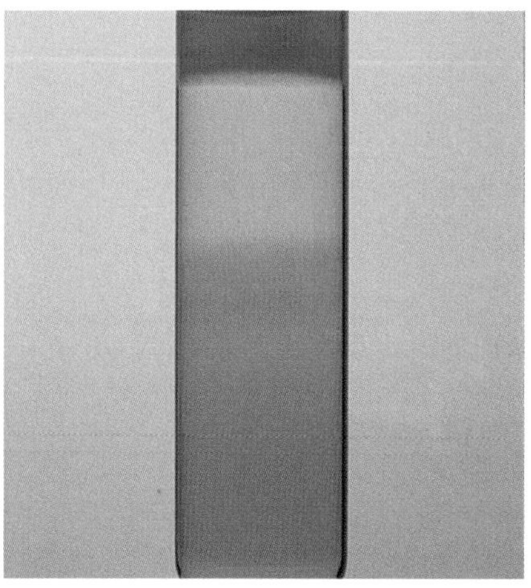

Different oxidation states may have different colours. In the test tube are all the oxidation states of vanadium from pale yellow +5 to violet +2 at the bottom. Two oxidation states of manganese produced the colours at the top.

How work out an oxidation state

Use these rules to calculate oxidation state:

- Elements always have an oxidation state of zero.
- In a compound, the sum of the oxidation numbers equals zero.
- In an ion, the sum of the oxidation numbers equals the charge.

In a compound

- Group 1 atoms always have a +1 oxidation state, e.g. Na is +1 in NaCl.
- Group 2 atoms always have a +2 oxidation state, e.g. Mg is +2 in $MgCl_2$.
- Group 3 atoms always have a +3 oxidation state, e.g. Al is +3 in $AlCl_3$.
- Fluorine always has a –1 oxidation state, e.g. F is –1 in KF.
- Oxygen has a –2 oxidation state, unless it is in a peroxide compound, such as H_2O_2, when O is –1, or with fluorine (as F is more electronegative than O); e.g. O is –2 in MgO, but is –1 in Na_2O_2, and +2 in OF_2
- Chlorine has a –1 oxidation state, unless it is with F or O (as they are more electronegative than Cl), e.g. Cl is –1 in NaCl, but +1 in Cl_2O, and +3 in ClF_3.
- Hydrogen is +1 except in metal hydrides where it has an oxidation state of –1, e.g. H is +1 in HCl, +1 in H_2O, but –1 in NaH.

Here are some examples of common compounds with all the oxidation numbers:

sodium chloride (common salt), NaCl, Na = +1, Cl = –1

sodium carbonate (washing soda), Na_2CO_3, Na = +1, C = +4, O = –2

calcium fluoride (fluorspar), CaF_2, Ca = +2, F = –1

calcium hydroxide (lime water), $Ca(OH)_2$, Ca = +2, O = –2, H = +1

potassium nitrate (saltpetre), KNO_3, K = +1, N = +5, O = –2

iron(III) oxide (haematite), Fe_2O_3, Fe = +3, O = –2

copper(II) sulfate, $CuSO_4$, Cu = +2, S = +6, O = –2

Old and new names

Compounds used to be named differently. At one time each writer would have their own names for compounds. It was very confusing, so internationally chemists agreed standard names. Later it was thought that the words used were difficult or confusing, so internationally it was agreed to use numbers.

Here are some examples;

- KNO_3 was called common nitre or saltpetre, then potassium nitrate, but now is called potassium nitrate(V), because the N has an oxidation state of +5.
- KNO_2 was called potassium nitrite, but now is called potassium nitrate(III), because the N has an oxidation state of +3.
- As further examples here are some chlorine compounds;

formula	old name	new name
KClO	potassium hypochlorite	potassium chlorate(I)
$KClO_2$	potassium chlorite	potassium chlorate(III)
$KClO_3$	potassium chlorate	potassium chlorate(V)
$KClO_4$	potassium perchlorate	potassium chlorate(VII)

Some of the old names are still used. For example, $KMnO_4$ should be called potassium manganate(VII), but is was known as potassium permanganate.

Writing oxidation states

An oxidation state is

- for one atom only,
- written vertically above the symbol

The charge is always written at the top right.

$\overset{+2}{Mg}$ (oxidation state) Mg^{2+} (charge)

Writing the oxidation state in names

You are likely to have met oxidation states already in some names. Roman numerals are used.

CuO is called copper(II) oxide because the copper ions have a 2+ charge, so the oxidation state is +2.

Another example is iron(III) chloride, which contains Fe^{3+} ions (oxidation state +3)and chloride ions, Cl^-.

Check your understanding

1. Define oxidation state.
2. Work out the oxidation states of the underlined element
   (a) $\underline{N}O_2$, (b) $\underline{N}H_3$, (c) $\underline{N}O$, (d) $\underline{S}O_2$, (e) $H_2\underline{S}$, (f) $Mg\underline{S}O_4$, (g) $Ca\underline{Cl}_2$, (h) $H\underline{N}O_3$, (i) $H_2\underline{O}$.
3. Write the formula for these compounds
   (a) lead(II) oxide, (b) silver(I) nitrate, (c) lead(IV) oxide, (d) sulfur(VI) oxide.
4. Work out the oxidation states of the underlined element
   (a) $Na_2\underline{O}_2$, (b) $Mg\underline{H}_2$, (c) $\underline{Na}$, (d) $\underline{Cl}_2$, (e) $K\underline{Cl}O_4$, (f) $H\underline{N}O_3$, (g) $K_2\underline{Cr}O_4$, (h) $H_2\underline{S}O_4$, (i) $Mg\underline{Br}_2$.

8.04 Oxidation states and their calculation 2

OBJECTIVES

after this spread you should be able to

- calculate more difficult oxidation states
- use oxidation states to write half-equations

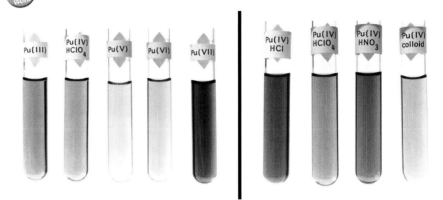

The colours of the radioactive element plutonium.

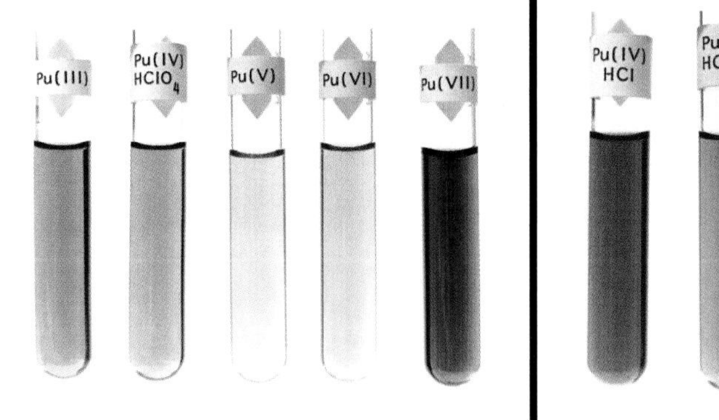

These are the colours of the oxidation states of the radioactive element plutonium which vary from Pu(III) to Pu(VII). Plutonium could be used in nuclear power stations or to make nuclear bombs. Understanding the oxidation states of plutonium will help to clear up the waste from the Cold War.

How to work out oxidation states

You need to be able to calculate oxidation states in various situations. These worked examples will help you when you meet more difficult questions.

Step 1 Write down the formula.

Step 2 For the oxidation states known, write the oxidation states *above* the symbol. Remember an oxidation state is for *one* atom.

Step 3 For the oxidation states known, write the sum of the oxidation states *below* the symbol.

Step 4 Work out the oxidation state of the unknown. For a compound, the sum of the oxidation states must equal zero. For an ion, the sum of the oxidation states must equal the charge.

If there is more than one atom of the element, then its number is the sum of the oxidation states.

Show the steps in calculating the oxidation states.

Worked example 1

What is the oxidation state of sulfur in Na_2SO_3?

Step 1 $Na_2\ S\ O_3$

Step 2 $\overset{+1}{Na_2}\ S\ \overset{-2}{O_3}$

Step 3 $\overset{+1}{Na_2}\ S\ \overset{-2}{O_3}$ — below: $+2\ \square\ -6 = 0$

Step 4 $\overset{+1}{Na_2}\ S\ \overset{-2}{O_3}$ — below: $+2\ \boxed{+4}\ -6 = 0$ The oxidation state of S is +4

Worked example 2

What is the oxidation state of sulfur in sodium thoisulfate, $Na_2S_2O_3$?

Step 1 $Na_2\ S_2\ O_3$

Step 2 $\overset{+1}{Na_2}\ S_2\ \overset{-2}{O_3}$

Step 3 $\overset{+1}{Na_2}\ S_2\ \overset{-2}{O_3}$ — below: $+2\ \square\ -6 = 0$

Step 4 $\overset{+1}{Na_2}\ S_2\ \overset{-2}{O_3}$ — below: $+2\ \boxed{+4}\ -6 = 0$ $\boxed{\frac{+4}{2} = +2}$

The oxidation state of S_2 is +2

Worked example 3

What is the oxidation state of chlorine in the chlorate ion, ClO_3^-.

Step 1 $Cl\ O_3^-$

Step 2 $Cl\ \overset{-2}{O_3^-}$

Step 3 $Cl\ \overset{-2}{O_3^-}$ — below: $\square\ -6 = -1$

Step 4 $Cl\ \overset{-2}{O_3^-}$ — below: $\boxed{+5}\ -6 = -1$ The oxidation state of Cl is +5

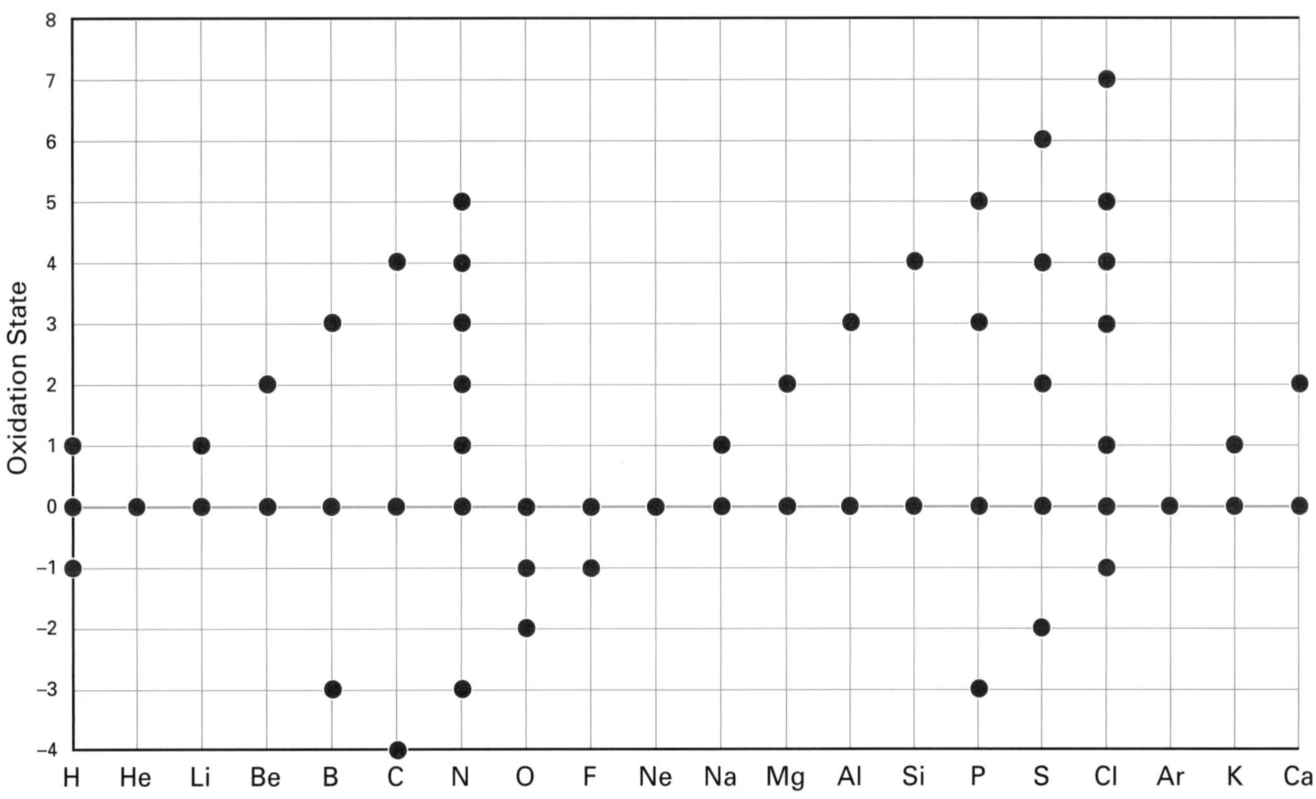

Here are the more common oxidation states of the first twenty elements.

Check your understanding

1. Work out the oxidation states of the underlined element:
   - a $\underline{N}_2O_4$
   - b $\underline{S}O_3$
   - c $\underline{S}_2Cl_2$
   - d $\underline{P}_4$
   - e $\underline{Cl}_2O_7$
   - f $Na_2\underline{O}_2$
   - g $K_2Cr_2\underline{O}_7$
   - h $\underline{P}Cl_5$
   - i $LiAl\underline{H}_4$
2. Work out the oxidation states of the underlined element in these ions:
   - a $\underline{S}O_4^{2-}$
   - b $\underline{N}O_3^-$
   - c $\underline{C}O_3^{2-}$
   - d $\underline{P}Cl_6^-$
   - e $\underline{P}Cl_4^+$
   - f $\underline{O}H^-$
   - g $\underline{S}_2O_7^{2-}$
   - h $\underline{N}O_2^-$
   - i $\underline{N}H_4^+$

8.05 How to combine half-equations

OBJECTIVES

after this spread you should be able to

- combine two half-equations
- use oxidation numbers to balance half-equations

The Daniell cell

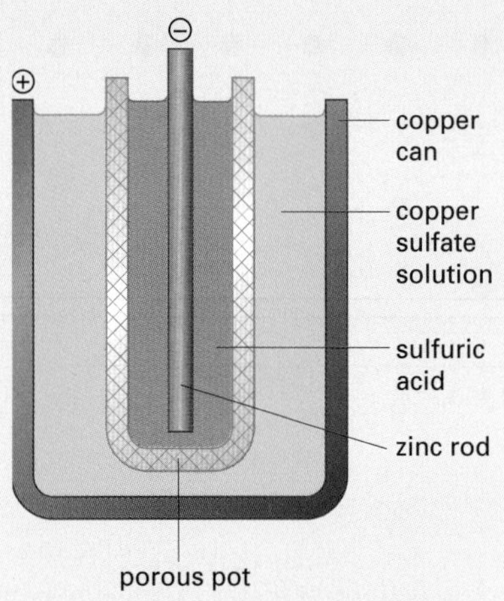

This is the Daniell Cell from 1836, an early source of electricity that used the oxidation of copper and the reduction of zinc ions to produce electricity. This cell could make electricity continuously, but it was heavy and used liquids. The liquid would start to steam and boil if the cell was used for long periods.

The electrode half-equations are:

$$Zn \rightarrow Zn^{2+} + 2e^-$$
$$Cu^{2+} + 2e^- \rightarrow Cu$$

so the overall reaction is:

$$Zn + Cu^{2+} \rightarrow Zn^{2+} + Cu$$

Combining half-equations

When you are given two half-equations, sometimes you will need to join them to make one full equation. The aim here is make sure that the same number of electrons that are donated by one half-equation are accepted by the other.

How to combine two half-equations

The easiest way to combine two half-equations is to work in steps. These worked examples will help you when you meet more difficult questions. Refer to the examples while reading these steps;

Step 1 Write out the two half-equations.

Step 2 Note the number of electrons each half-equation gains or loses. So that both equations involve the same number of electrons, you may have to multiply up one or both equations.

Step 3 Multiply up the reactants and products.

Step 4 Write all the reactants together, and the products together.

Step 5 There should be the same number of electrons on both sides of this equation. Cancel them. What is left is the full balanced equation.

Worked example 1

Balance these two half equations:

$$(A)\ Fe^{3+} + e^- \rightarrow Fe^{2+}$$
$$(B)\ Zn \rightarrow Zn^{2+} + 2e^-$$

Step 1 We are given the equations.

Step 2 Equation (A) is gaining 1 electron. Equation (B) is losing 2 electrons. We need to multiply equation (A) by 2.

Step 3 Here all parts of equation (A) were multiplied by 2:

$$(A)\ 2Fe^{3+} + 2e^- \rightarrow 2Fe^{2+}$$
$$(B)\ Zn \rightarrow Zn^{2+} + 2e^-$$

Step 4 The equations combined:

$$2\,Fe^{3+} + 2e^- + Zn \rightarrow 2Fe^{2+} + Zn^{2+} + 2e^-$$

Step 5 So, with the electrons cancelled, the final full equation is:

$$2\,Fe^{3+} + Zn \rightarrow 2Fe^{2+} + Zn^{2+}$$

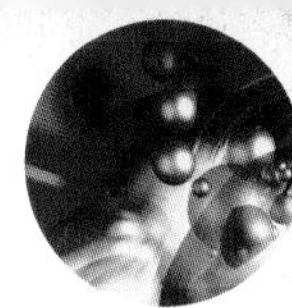

Worked example 2

Aluminium metal may be oxidized to make aluminium ions by bleach, which contains chlorate(I) ions, ClO^-. Using the given half-equations, write a full equation for the reaction:

$$\text{(C) } Al \rightarrow Al^{3+} + 3e^-$$

$$\text{(D) } ClO^- + 2H^+ + 2e^- \rightarrow Cl^- + H_2O$$

Step 1 We are given the equations.

Step 2 Equation (C) is losing 3 electrons. Equation (D) is gaining 2 electrons. The number divisible by both 3 and 2 is 6. We need to multiply equation (C) by 2, and equation (D) by 3.

Step 3 Here all parts of equation (C) are multiplied by 2, and equation (D) by 3.

$$\text{(C) } 2Al \rightarrow 2Al^{3+} + 6e^-$$
$$\text{(D) } 3ClO^- + 6H^+ + 6e^- \rightarrow 3Cl^- + 3H_2O$$

Step 4 The equations combined:

$$2Al + 3ClO^- + 6H^+ + 6e^- \rightarrow 2Al^{3+} + 6e^- + 3Cl^- + 3H_2O$$

Step 5 So, with the electrons cancelled, the final full equation is:

$$2Al + 3ClO^- + 6H^+ \rightarrow 2Al^{3+} + 3Cl^- + 3H_2O$$

The cell in the photograph uses zinc and silver oxide to store the energy. The half-equations are:

$$Zn(s) + 2OH^-(aq) \rightarrow Zn(OH)_2(s) + 2e^-$$

$$Ag_2O(s) + H_2O(l) + 2e^- \rightarrow 2Ag(s) + 2OH^-(aq)$$

The overall discharge equation is:

$$Zn(s) + Ag_2O(s) + H_2O(l) \rightarrow Zn(OH)_2(s) + 2Ag(s)$$

The cell has a high energy density, but is very expensive.

This is the latest rechargeable electrical cell, as used by the European Space Agency on the International Space Station.

Check your understanding

1. Combine these two half-equations into one balanced ionic equation:

$$Na \rightarrow Na^+ + e^-$$

$$O_2 + 4e^- \rightarrow 2O^{2-}$$

2. When potassium chlorate(V) reacts with iron(II) sulfate then iron(III) ions are made. Combine these two half-equations to form a full balanced equation:

$$ClO_3^- + 6H^+ + 6e^- \rightarrow Cl^- + 3H_2O$$

$$Fe^{2+} \rightarrow Fe^{3+} + e^-$$

9.01 Group 7: boiling points

OBJECTIVES

already from GCSE, you know

- what a covalent bond is
- that there are intermolecular forces between molecules
- the names and symbols of the halogens

and after this spread you should know

- that the halogen elements are in group 7 of the periodic table
- the colours of the halogens
- the trend in the boiling points of the halogens and be able to explain it
- the trend in the electronegativity of the halogens

The salt makers

The group 7 elements are called the '**halogens**' because they make salts such as the common salt, sodium chloride, that you put on your food. 'Halo' is Greek for 'salt' and 'gen' means 'make', as in the word 'generate'. This makes the 'halo-gens' the 'salt-makers'. They are in many soluble salts, are all commonly found in sea salt, and we need them to live.

Colourful elements

The halogens appear very different from each other. They have different colours and may be solids, liquids, or gases.

element	symbol	formula	colour	standard State
fluorine	F	F_2	pale yellow	gas
chlorine	Cl	Cl_2	green-yellow	gas
bromine	Br	Br_2	red-brown	liquid
iodine	I	I_2	black	solid

You may have seen some of these halogens. They all are dangerous.

The many colours of iodine

You may see pure iodine as a black solid, but you can make it change colour.

- Heat iodine and make hot iodine vapour which is a stunning purple colour.
- Iodine does not dissolve in water but it does make a brown solution in aqueous potassium iodide.
- Dissolve iodine in a non-polar solvent such as hexane, and you will see the strong purple colour again.
- Have you ever tested for starch using iodine? If you have, then you will know that iodine turns blue-black with starch.

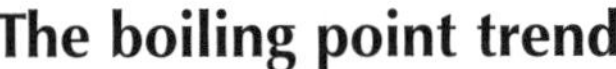

The boiling point trend

You can see from the photographs that the halogens at the top of the periodic table are gases, bromine in the middle is a liquid, and at the bottom is solid iodine. So as you look down group 7 the halogen boiling point increases. We can explain this.

The boiling point trend explained

Down group 7, the halogen boiling point increases because

- All the halogen molecules are two-atom covalent molecules.
- There are only weak van der Waals' forces between the molecules.
- Down the group the atoms and so the molecules get larger.
- The strength of the 'van der Waals' forces between the molecules increase, so the molecules are more strongly attracted together.

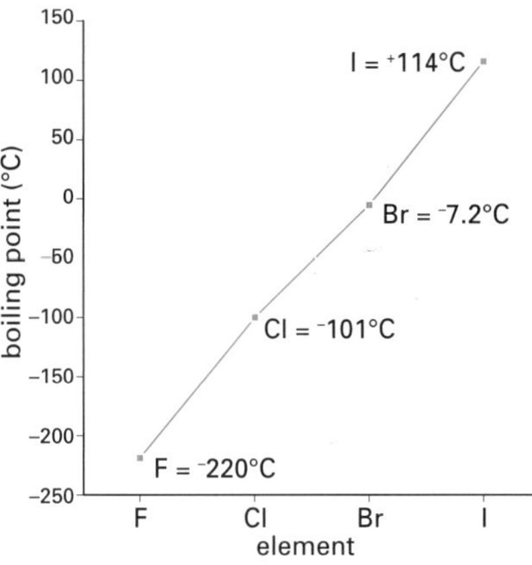

Down group 7, the halogen boiling point increases due to the stronger van der Waals' forces between the molecules.

How electronegative are the halogens?

All the halogen atoms are attractive to electrons, so all the halogens are strongly electronegative. There is a trend to the electronegativity. The most electronegative is fluorine and the least electronegative is iodine, so down group 7 the halogens become less electronegative.

Down group 7

- Successive atoms have more occupied energy levels.
- The distance between the nucleus and the pair of electrons in a covalent bond increases.
- The attraction of the nucleus for these electrons decreases.
- Electronegativity decreases.

An electronegativity reminder

Electronegativity is the power of an atom to withdraw electron density in a covalent bond

Electronegativity and the hydrogen halides

All the halogens react with hydrogen to make covalent hydrogen compounds called the hydrogen halides. It is likely you have met at least one, hydrogen chloride, before.

The fluorine atom is the most electronegative halogen atom, so in hydrogen fluoride the electron density in the covalent bond is strongly attracted towards the fluorine atom. Hydrogen fluoride is strongly polar.

An iodine atom has nearly the same electronegativity as a hydrogen atom. Hydrogen iodide is almost non-polar.

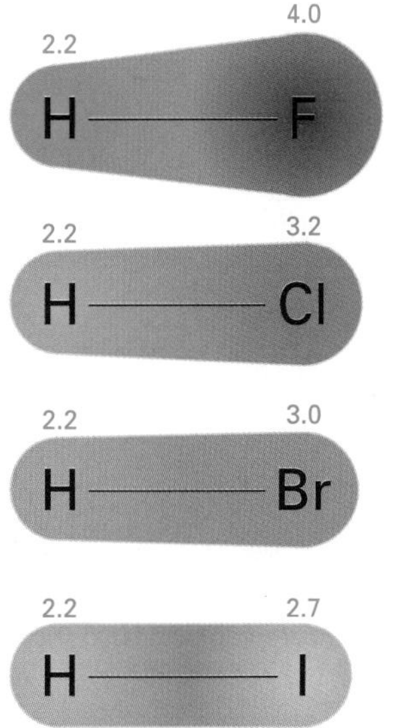

Down group 7 the hydrogen halides become less polar because the halogen atoms become less electronegative. The electronegativity of each atom is shown by the number (which is on the Pauling Scale of electronegativity).

Check your understanding

1. At the bottom of group 7 is the radioactive element astatine, At. Suggest
   a the state of astatine under normal laboratory conditions
   b the electronegativity of astatine compared with iodine
   c the colour of astatine
2. Why is HCl a more polar molecule than HBr?
3. State the structure and bonding of hydrogen chloride, HCl. Explain why HCl has a low boiling point at normal temperature and pressure.

9.02 Group 7: oxidizing power

OBJECTIVES

already from GCSE, you know

- what is meant by oxidation and reduction
- what is meant by an oxidizing agent and a reducing agent
- the colours of the halogens

and after this spread you should know

- that the halogens are oxidizing agents
- that some halogen elements can oxidize other halide ions
- that the oxidizing power of the halogens decreases down group 7
- what is meant by a displacement reaction

Chlorine atoms gain an electron to make chloride ions:

$Cl + e^- \rightarrow Cl^-$

The chlorine atom has gained an electron so is reduced, so the chlorine atom is an oxidizing agent.

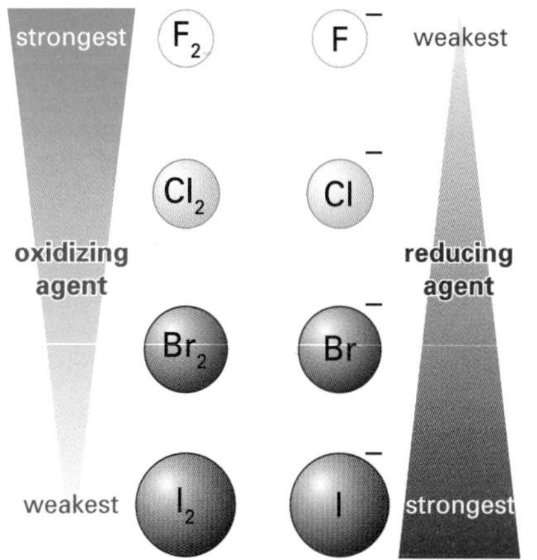

Down the group the halogens become weaker oxidizing agents. Also the halide ions become stronger reducing agents.

Powerful oxidizing agents

All halogens gain electrons to make halide ions, so all the halogens are oxidizing agents. You need to know the trend in the oxidizing power as you go down group 7.

A reminder

Oxidizing agents are electron acceptors.

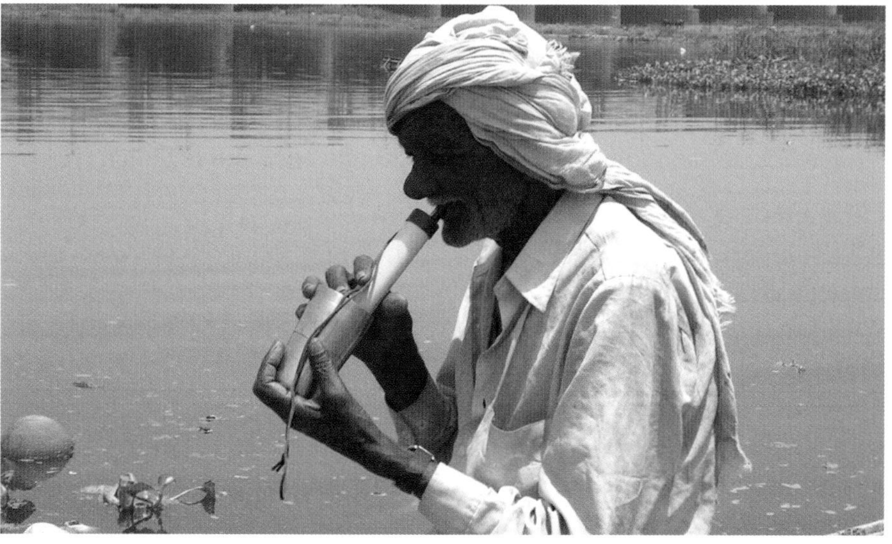

Travelling in the wilderness? This pen uses electricity and salt to kill harmful bacteria in water. It uses the oxidizing power of chlorine.

Halogen oxidizing power

Fluorine is the strongest oxidizing agent because fluorine atoms are the smallest halogen atoms, and so accept electrons most easily. Down group 7, as the halogen atoms get larger, they accept electrons less easily, and the oxidizing power becomes weaker.

Halide ions oxidized

When a halide ion loses an electron then a halogen atom is made. For example:

$$Cl^- \rightarrow Cl + e^-$$

This change is oxidation. The halide ion is an electron donor, so is a reducing agent.

Which halogen wins in a competition?

If a halogen element is mixed with a halide ion, will there be a reaction? It depends on which halogen element and which halide ion. Here are two examples.

Chlorine vs bromide ions

When green chlorine gas is mixed with aqueous potassium bromide then each chlorine atom gains an electron from a bromide ion, leaving red-brown bromine. The chlorine atoms accept the electrons:

$$Cl_2(g) + 2KBr(aq) \rightarrow Br_2(l) + 2KCl(aq)$$

$Cl_2(g)$	$2KBr(aq)$	$Br_2(l)$	$2KCl(aq)$
green chlorine	colourless potassium bromide	red-brown bromine	colourless potassium chloride

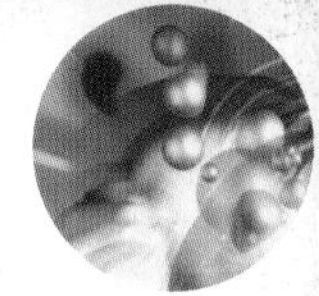

As chlorine is a stronger oxidizing agent than bromine, it removes the electron from the bromide ion.

A redox reaction in which one halogen replaces another is called a *displacement reaction*.

Bromine vs chloride ions?

What happens when bromine is mixed with chloride ions? Chlorine is the stronger oxidizing agent, so the chloride ions keep the electrons. A bromine atom is not sufficiently oxidizing to pull an electron away from a chloride ion.

$$Br_2(l) + 2KCl(aq) \rightarrow \text{no reaction}$$

Iodine

Iodine is obtained from seaweed. Chlorine oxidizes the iodide in the seaweed to give iodine.

Check your understanding

1. What would you see if the these chemicals were mixed (write 'no change' if you think no reaction happens):
   - **a** chlorine and sodium iodide
   - **b** potassium bromide and iodine
   - **c** sodium chloride and fluorine
   - **d** sodium iodide and bromine
2. Which halogens will oxidize sodium bromide?
3. Write balanced equations for the reactions between these chemicals (Write 'no reaction' if you think there is no reaction):
   - **a** sodium iodide and bromine
   - **b** potassium fluoride and iodine
   - **c** sodium chloride and fluorine
   - **d** chlorine and sodium iodide
4. Write half-equations for
   - **a** the oxidation of bromide ions to bromine
   - **b** the reduction of iodine to iodide ions

9.03 Group 7: reducing power of halide ions

OBJECTIVES

after this spread you should know

- that halide ions are reducing agents that can reduce other halogens
- that halide ions can reduce sulfuric acid to form different sulfur compounds
- that the reducing power of halide ions increases down group 7

A reminder

Reducing agents are electron donors.

Halide ions – powerful reducing agents

You already know how halogens can oxidize halide ions to give other halogens. For example, iodide ions lose electrons to give iodine atoms, which join to make iodine molecules:

$$2I^- \rightarrow I_2 + 2e^-$$

The iodide ions donate electrons to the oxidizing agent, so the iodide ions are themselves reducing agents.

All the halide ions are *reducing agents*.

The trend in reducing power

Down group 7

- The halide ions become larger because there are more occupied energy levels.
- The outer electrons are further from the nucleus.
- There is less attraction between the nucleus and the outer electrons.
- The electrons are donated more easily, so the halide ions become stronger reducing agents.

We have already discussed how halide ions react with halogens. Now we will look at whether halide ions reduce sulfuric acid.

Which halide ions reduce sulfuric acid?

Sulfuric acid reacts differently with different halide ions. We will look at their reactions starting from the top of the group.

Fluoride ions and sulfuric acid

A fluoride ion is a very weak reducing agent. This is not surprising as fluorine is very reactive, and so gains electrons very easily to make fluoride ions. To reverse this reaction would be difficult.

The reaction of fluoride ions, in solid sodium fluoride, with concentrated sulfuric acid is not a redox reaction. It is an acid-base reaction:

$$NaF + H_2SO_4 \rightarrow NaHSO_4 + HF$$

Chloride ions and sulfuric acid

A chloride ion is also a very weak reducing agent, so it reacts in a similar way to fluoride ions.

Solid sodium chloride reacts with concentrated sulfuric acid like this:

$$NaCl + H_2SO_4 \rightarrow NaHSO_4 + HCl$$

In this reaction, and the fluoride ion reaction above, a hydrogen ion (H^+) has been lost from the sulfuric acid and joined with the halide ion. The sodium ion is just a spectator to the reaction.

Bromide ions with sulfuric acid: A redox reaction

Bromide ions start off reacting in a similar way to chloride ions, so some misty HBr fumes may be seen:

$$NaBr + H_2SO_4 \rightarrow NaHSO_4 + HBr$$

Bromide ions in the hydrogen bromide are quite strong reducing agents, so they reduce the sulfuric acid and are oxidized to red-brown bromine:

$$2HBr + H_2SO_4 \rightarrow Br_2 + SO_2 + 2H_2O$$

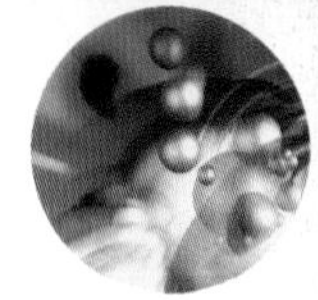

This is a redox reaction as there is a change in oxidation states:

+1 −1   +1 +6−2   0   +4 −2   +1 −2

$$2HBr + H_2SO_4 \rightarrow Br_2 + SO_2 + 2H_2O$$

Here is an equation for the reaction in one step;

$$2NaBr + 3H_2SO_4 \rightarrow Br_2 + SO_2 + 2H_2O + 2NaHSO_4$$

(a) On the left, the sodium bromide has reduced the sulfuric acid. The misty fumes coming off are HBr, and the red-brown colour is due to bromine. Also made are sulfur dioxide molecules, SO_2, which would be invisible, but they do change potassium dichromate paper from orange to green. (b) On the right, the iodide ions are oxidized to black iodine by the sulfuric acid. Also made is hydrogen sulfide, H_2S, which smells like bad eggs and is toxic.

The iodide ion: an even stronger reducing agent

An iodide ion is so large that it loses an electron easily, so it is an electron donor, a strong reducing agent. It is so strong that it reduces sulfuric acid to hydrogen sulfide:

1 HI is made:

$$NaI + H_2SO_4 \rightarrow NaHSO_4 + HI$$

2 Then the I in HI is oxidized, and so reduces the sulfuric acid:

$$8HI + H_2SO_4 \rightarrow 4I_2 + H_2S + 4H_2O$$

This time the change in oxidation states is this:

+1 −1   +1 +6−2   0   +1 −2   +1 −2

$$8HI + H_2SO_4 \rightarrow 4I_2 + H_2S + 4H_2O$$

How is bromine made?

Bromide ions are found in seawater. The Dead Sea is very concentrated so bromine is easily extracted from there using chlorine. The bromide ions act as a reducing agent and reduce the chlorine to chloride ions:

$$2Br^- + Cl_2 \rightarrow Br_2 + 2Cl^-$$

Check your understanding

1. What would you see if you added concentrated sulfuric acid to
   a potassium chloride
   b lithium bromide
   c potassium iodide?
2. Explain why iodide ions are stronger reducing agents than chloride ions.
3. State all the oxidation states for the elements in the reactants and products:

$$2NaBr + 3H_2SO_4 \rightarrow Br_2 + SO_2 + 2H_2O + 2NaHSO_4$$

9.04 Identifying halide ions

OBJECTIVES

already from AS level, you understand

- displacement reactions with halide ions in aqueous solution

and after this spread you should

- understand why acidified silver nitrate solution is used as a reagent to identify and distinguish between F^-, Cl^-, Br^-, and I^- ions
- know the trend in solubility of the silver halides in ammonia

Over 200 million containers are shipped around the world by sea every year. They are a very convenient way to transport goods of all kinds. But what do you do if your cargo arrives damaged by water? Is the damage a result of condensation inside the container, or has seawater been allowed to get in? Your insurance company will certainly need to know. Luckily there a simple chemical test they can do to find out.

About 90% of cargo travels in containers, stacked on the decks of container ships like this one.

The silver nitrate test

Halide ions in solution are detected using silver nitrate solution. The test is in two parts:

1 Add silver nitrate solution and look for a precipitate.

2 Add ammonia solution to see if the precipitate dissolves.

Adding silver nitrate solution

The first part of the test relies on differences between the different silver halides AgF, AgCl, AgBr, and AgI. Silver nitrate is soluble in water and reacts with halide ions in solution to form silver halides. Except for silver fluoride, the silver halides are insoluble. This is why they form precipitates. The precipitates are different colours, allowing you to tell them apart.

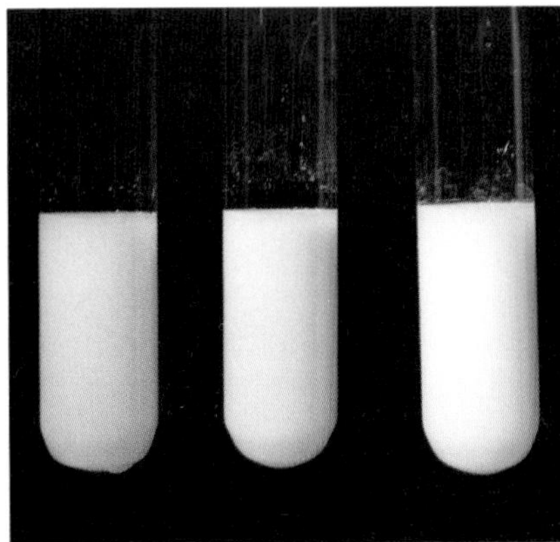

From left to right, precipitates of: silver chloride, silver bromide, and silver iodide.

halide ion	observation	ionic equation
F^-	no change	$Ag^+(aq) + F^-(aq) \rightarrow AgF(aq)$
Cl^-	white precipitate	$Ag^+(aq) + Cl^-(aq) \rightarrow AgCl(s)$
Br^-	cream precipitate	$Ag^+(aq) + Br^-(aq) \rightarrow AgBr(s)$
I^-	pale yellow precipitate	$Ag^+(aq) + I^-(aq) \rightarrow AgI(s)$

Chlorides, bromides, and iodides give characteristic precipitates.

Acidifying the sample

You should add a few drops of dilute nitric acid to your test solution before adding silver nitrate solution. This prevents the formation of other insoluble silver compounds, such as silver carbonate. The acid reacts with the carbonate ions forming carbon dioxide gas. You cannot use hydrochloric acid because it contains chloride ions, which would give a white precipitate with silver nitrate solution. Nitric acid contains nitrate ions, but all nitrates are soluble in water.

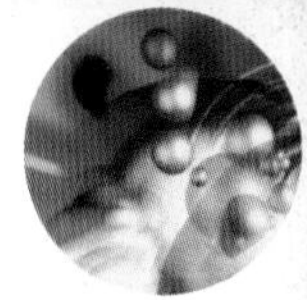

Adding ammonia solution

It can be difficult to tell the three different silver halide precipitates apart, especially if relatively little precipitate forms. Ammonia solution is used in a further test to confirm the identity of the halide present. This second part of the test relies on differences in solubility in ammonia between the three silver halide precipitates.

The solubility of the silver halides decreases in the order AgCl > AgBr > AgI. Silver chloride readily dissolves in dilute ammonia solution, silver bromide is soluble in concentrated ammonia solution, and silver iodide is insoluble in concentrated ammonia solution.

precipitate	addition of dilute ammonia solution	addition of concentrated ammonia solution
AgCl	precipitate redissolves	precipitate redissolves
AgBr	no change	precipitate redissolves
AgI	no change	no change

The results of confirmatory tests on silver halide precipitates. The silver iodide precipitate turns white in concentrated ammonia solution.

A complex ion

The silver chloride and silver bromide precipitates do not just dissolve in ammonia solution to give aqueous silver ions and halide ions. Aqueous silver ions form a stable complex ion with ammonia: $[Ag(NH_3)_2]^+(aq)$. A small concentration of aqueous silver ions is present in equilibrium with the silver halide precipitates. The addition of ammonia alters the position of equilibrium sufficiently to dissociate the silver chloride and silver bromide precipitates, but not the iodide precipitate.

Testing times

Seawater contains dissolved halide ions. Chloride ions are the most abundant of these, at around $0.55\,mol\,dm^{-3}$. Fluoride, bromide, and iodide ions are present at much smaller concentrations. The silver nitrate test gives a white precipitate with seawater. This is how the nature of any water damage to cargo can be determined. Cargo damaged by seawater will contain a high concentration of chloride ions, but cargo damaged by condensation will not.

Check your understanding

1. Describe the silver nitrate test for the presence of aqueous halide ions. Include practical details, observations, and relevant equations.
2. Explain why the absence of a precipitate in the silver nitrate test does not prove the presence of aqueous fluoride ions.
3. A sample of water-damaged cargo was tested for the presence of chloride ions. It was added to a measured amount of de-ionized water, and then silver nitrate solution was added.
   - **a** Explain why dilute nitric acid should also be added.
   - **b** Suggest why the test should be repeated on an undamaged sample of cargo.

9.05 Reactions of chlorine

OBJECTIVES

already from GCSE, you know

- that chlorine is produced industrially by the electrolysis of aqueous sodium chloride

already from AS level, you understand

- that the oxidizing ability of the halogens decreases down the group

and after this spread you should know

- the reaction of chlorine with water
- the reaction of chlorine with cold dilute aqueous sodium hydroxide, and the uses of the solutions formed

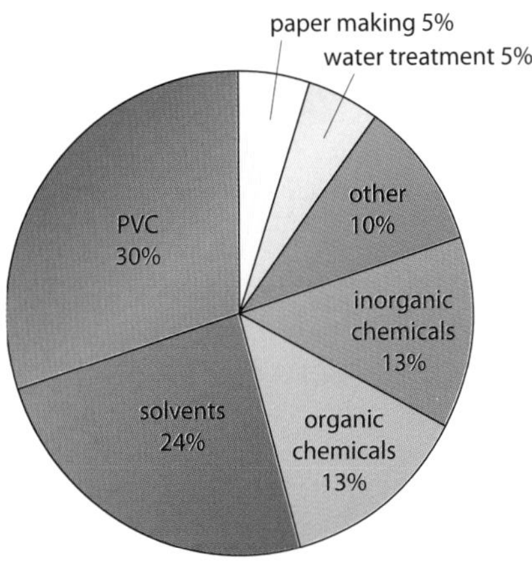

Chlorine has a wide range of industrial uses.

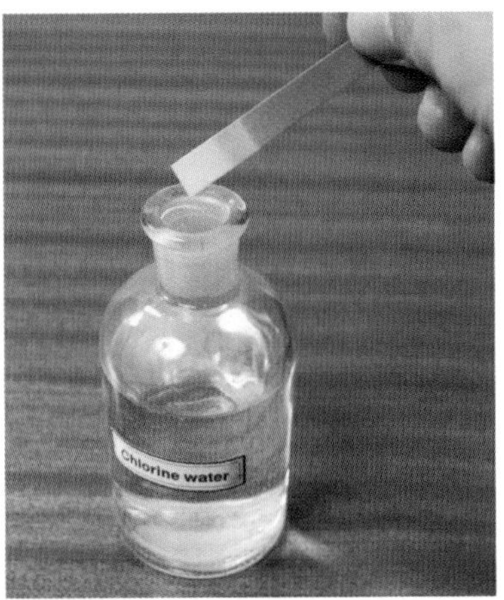

Chlorine released from chlorine water turns damp blue litmus paper red, then bleaches it white. The chlorine reacts with the water to form an acidic solution, and the chloric(I) acid bleaches the paper.

Chlorine is an important industrial chemical. About 44 million tonnes of it is produced around the world each year, with 1.6 million tonnes produced in the UK. Over half of this is used to make solvents and PVC. Most of the rest is used to make a wide range of chemicals such as disinfectants, bleaches, glues, medicines, and deodorants. Chlorine is also used directly to sterilize drinking water and water in swimming pools.

Reaction of chlorine with water

Chlorine dissolves in water and reacts with it to produce an acidic mixture. Two acids are formed in the reaction:

- hydrochloric acid, HCl
- **chloric(I) acid**, HClO, also called hypochlorous acid

The reaction is a **disproportionation** reaction because chlorine is both oxidized and reduced.

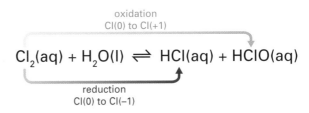

oxidation
Cl(0) to Cl(+1)

$$Cl_2(aq) + H_2O(l) \rightleftharpoons HCl(aq) + HClO(aq)$$

reduction
Cl(0) to Cl(−1)

Chlorine water in the laboratory

The mixture formed when chlorine dissolves in water is called chlorine water. There are two ways to prepare it safely in the laboratory. Chlorine gas is bubbled in to water until the solution turns light green. Or equal volumes of 2 M hydrochloric acid and 0.15 M **sodium chlorate(I)**, NaClO, are mixed together. Chlorine water must be prepared in the fume cupboard, and the container should be labelled as Toxic and Corrosive.

Chloric(I) acid

Chloric(I) acid is a weak acid. It ionizes in water to give the chlorate(I) ion ClO^-:

$$HClO(aq) \rightleftharpoons H^+(aq) + ClO^-(aq)$$

The chlorate(I) ion is also called the hypochlorite ion. It is an oxidizing agent. For example, it will oxidize the iodide ion to iodine:

$$2I^-(aq) + 2H^+(aq) + ClO^-(aq) \rightarrow I_2(aq) + H_2O(l) + Cl^-(aq)$$

The chlorate(I) ion is a powerful disinfectant. It kills bacteria by attacking their cell walls, and destroying enzymes and other substances inside the cell. It is also a bleach. It breaks down the structure of coloured organic chemicals so that they no longer appear coloured.

Reaction of chlorine with aqueous sodium hydroxide

The equilibrium mixture of chlorine and water is acidic:

$$Cl_2(aq) + H_2O(l) \rightleftharpoons 2H^+(aq) + Cl^-(aq) + ClO^-(aq)$$

The position of equilibrium moves to the right if an alkali is added. It moves so far to the right when cold dilute sodium hydroxide solution is added that the reaction goes to completion:

$$Cl_2(aq) + 2NaOH(aq) \rightarrow NaCl(aq) + NaClO(aq) + H_2O(l)$$

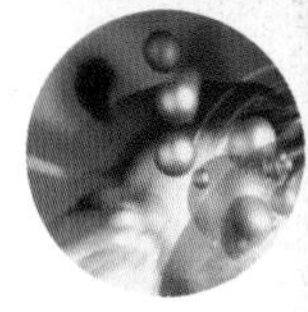

NaClO is sodium chlorate(I), also called sodium hypochlorite. A dilute solution of sodium chlorate(I) is a key ingredient of chlorine-based household bleaches.

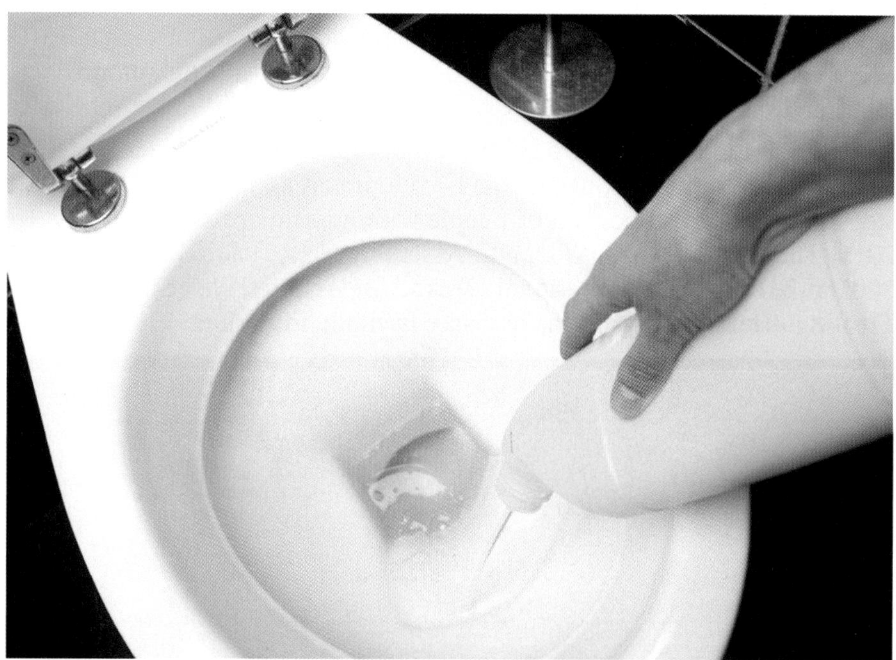

Household bleaches are used to clean and disinfect toilets, drains, and kitchen and bathroom surfaces.

Warmed sodium chlorate(I)

Above about 75°C, the chlorate(I) ions disproportionate to form chloride ions and chlorate(V) ions, ClO_3^-:

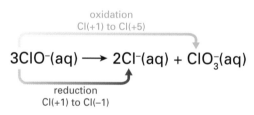

oxidation
Cl(+1) to Cl(+5)

$$3ClO^-(aq) \longrightarrow 2Cl^-(aq) + ClO_3^-(aq)$$

reduction
Cl(+1) to Cl(−1)

Sodium chlorate(V), $NaClO_3$, is a powerful weedkiller.

Check your understanding

1. The reaction between chlorine and water is a disproportionation reaction.
   a What is a disproportionation reaction?
   b Write an ionic half-equation for the reduction of chlorine in the reaction.
2. Describe a simple laboratory test for chlorine gas.
3. Suggest why chlorine water is able to kill bacteria.
4. a Write an equation for the reaction of chlorine with cold dilute sodium hydroxide solution.
   b Write ionic half-equations for the oxidation and reduction of chlorine in the reaction.

9.06 Chlorine and water treatment

OBJECTIVES

already from AS level, you know

- the reactions of chlorine with water

and after this spread you should

- know the use of chlorine in water treatment
- appreciate that the benefits to health of water treatment by chlorine outweigh its toxic effects

We need water from our food and drink to survive. But water can easily become contaminated by harmful micro-organisms. These can cause illnesses such as cholera and diarrhoea. Four billion cases of diarrhoea occur each year, resulting in over two million deaths, mostly among children under five years old.

At least a billion people in the world do not have access to clean water. The World Summit for Sustainable Development agreed in 2002 to take steps to halve the proportion of people without safe drinking water by 2015. The United Nations General Assembly later declared the years 2005 to 2015 as the International Decade for Action, 'Water for Life'. Treating drinking water to make it safe is vitally important.

It can be easy in the UK to take clean, safe water for granted. People in some other parts of the world are not so fortunate.

Killing micro-organisms

Water can be contaminated with faeces from animals and humans. Faeces may contain harmful micro-organisms such as bacteria and viruses. People drinking water contaminated with these are likely to fall ill.

micro-organism	disease caused
Salmonella enterica	salmonellosis
Shigella dysenteriae	dysentery
Vibrio cholerae	cholera
Yersinia enterocolitica	diarrhoea

Some bacteria that contaminate drinking water, and the diseases they cause. These bacteria are killed when chlorine is added to water.

Chlorine has been used to disinfect drinking water in the UK for over a hundred years, and about 98% of drinking water in Western Europe is chlorinated. Chlorine is added to the water at the water treatment works. Sodium chlorate(I) is also used. Chlorine reacts with the water to form a mixture of hydrochloric acid and chloric(I) acid:

$$Cl_2(aq) + H_2O(l) \rightleftharpoons HCl(aq) + HClO(aq)$$

Chloric(I) acid ionizes to form the chlorate(I) ion:

$$HClO(aq) \rightleftharpoons H^+(aq) + ClO^-(aq)$$

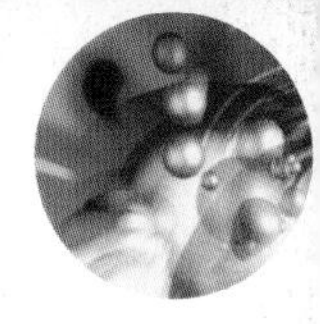

Chloric(I) acid is better at killing micro-organisms than chloric(I) ions. The position of equilibrium lies to the left in acidic conditions. At pH 7.5 the concentrations of HClO and ClO^- are about the same, but at pH 6.5 the concentration of HClO is about nine times more than the concentration of ClO^-. This means that chlorination is more effective when the pH of the water is kept low. The World Health Organization recommends that the pH of drinking water should be less than 8.

A matter of taste

Most people can smell or taste chlorine in water at concentrations below $5\,mg\,dm^{-3}$ and some people can still detect it at $0.3\,mg\,dm^{-3}$. The World Health Organization recommends that water is treated at a chlorine concentration of at least $0.5\,mg\,dm^{-3}$. Complaints from consumers about the taste of chlorine start to increase when it rises above $0.6\,mg\,dm^{-3}$. The concentration of chlorine in the water from your taps is usually less than this, around $0.2\,mg\,dm^{-3}$. This ensures that the water is still safe to drink after travelling through the pipes from the treatment works.

Swimming pool water typically contains $3\,mg\,dm^{-3}$ chlorine. The pH is adjusted to between 7.3 and 7.4 using sodium hydrogensulfate or sodium carbonate.

Safe to drink?

Chlorine is a toxic gas. It also reacts with organic compounds in water to form compounds such as trichloromethane and dichloroethanoic acid. Some of these compounds are potentially hazardous. For example, it is possible that chloromethane causes cancer. There is sufficient evidence in experimental animals that it does, but not enough evidence in humans. Scientists have estimated that the risk of death from harmful micro-organisms in untreated water is between a hundred and a thousand times more than the risk of cancer from by-products of chlorination.

The consequences of not chlorinating water supplies can be very serious. An outbreak of cholera began in Peru in 1991. It probably began when a ship dumped its sewage off the coast. The public water supplies were inadequate for the number of people who relied on them. The water was not properly chlorinated because of the cost and a lack of sufficient chlorine. Cholera spread across other countries in Latin America and several thousand people died.

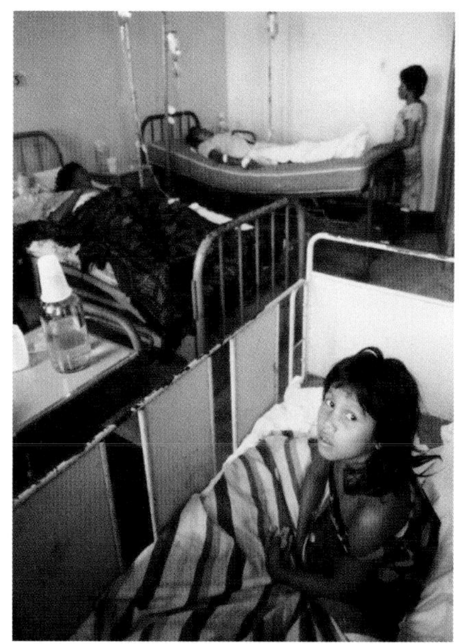

Patients in a cholera ward in Peru.

Check your understanding

1. a Explain why the amount of chlorine needed to disinfect drinking water increases as the pH increases.
   b Explain why it is undesirable to use high concentrations of chlorine in drinking water under normal circumstances.
   c Suggest why higher concentrations of chlorine in drinking water are acceptable where many people are living close together in an emergency camp after a disaster.
2. Outline some of the advantages and disadvantages of chlorinating water supplies, and explain why chlorine is used even though it may have toxic effects.

10.01 Group 2: physical properties

OBJECTIVES

already from AS level, you

- can explain the trend atomic radius of the period 3 elements
- understand how ionization energies give evidence for energy levels
- understand that metallic bonding involves a lattice of positive ions surrounded by delocalized electrons

and after this spread you should

- understand the trends in atomic radius, first ionization energy, and melting point of the elements in group 2

Period	Group 1	Group 2	
②	Li 3 Lithium	Be 4 Beryllium	
③	Na 11 Sodium	Mg 12 Magnesium	
④	K 19 Potassium	Ca 20 Calcium	Sc 21 Scandium
⑤	Rb 37 Rubidium	Sr 38 Strontium	Y 39 Yttrium
⑥	Cs 55 Caesium	Ba 56 Barium	La 57 Lanthanum
⑦	Fr 87 Francium	Ra 88 Radium	Ac 89 Actinium

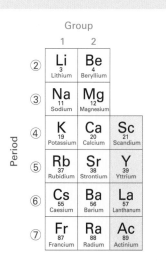

Group 2 is in the s block of the Periodic Table. It contains the elements Be to Ra.

Shielding

The number of protons increases down group 2. You might expect the attraction for the electrons in the highest occupied energy level to increase, leading to a decrease in the atomic radius. The electrons in the lower occupied energy levels become closer to the nucleus, but they shield the electrons in the highest occupied energy level from the attraction of the nucleus. So, the main factor in determining atomic redius becomes the number of occupied energy levels.

The elements of group 2 are also called the **alkaline earth metals**. This is because they form basic oxides. The group 2 elements have different uses, related to their physical and chemical properties.

- Beryllium alloys and magnesium alloys are light and strong. They are used in aircraft, spacecraft and missiles.
- Calcium carbonate is used in the manufacture of cement, concrete, and steel.
- Strontium compounds burn with a bright crimson flame. They are used in tracer bullets, fireworks, and signal flares.
- Many barium compounds are poisonous. Barium carbonate is used as rat poison. Barium sulfate is safe to use in medical X-ray photographs because it is insoluble and not absorbed.

The elements of group 2. From left to right: beryllium, magnesium, calcium, strontium, and barium. Radium is very radioactive and has little everyday practical use.

Trend in atomic radius

The atomic radius increases as you go down group 2. This is because there are more filled energy levels between the nucleus and the electrons in the highest occupied energy level.

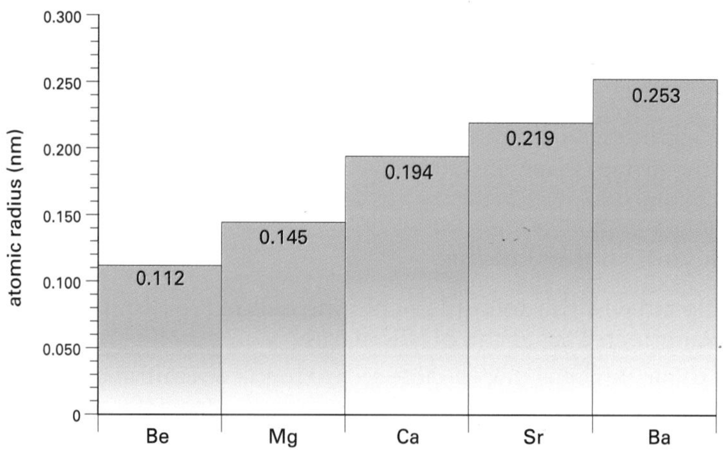

Atomic radius increases down group 2.

Trend in first ionization energy

The first ionization energy decreases as you go down group 2. In each case, an electron from the highest occupied a sub-level is being removed.

As you go down group 2, each successive element has more occupied energy levels. The distance between the outer electrons and the nucleus increases. Electrons in higher energy levels are less strongly attracted to the nucleus than electrons closer to it. So, even though the nuclear charge increases down the group, less energy is needed to remove an outer electron.

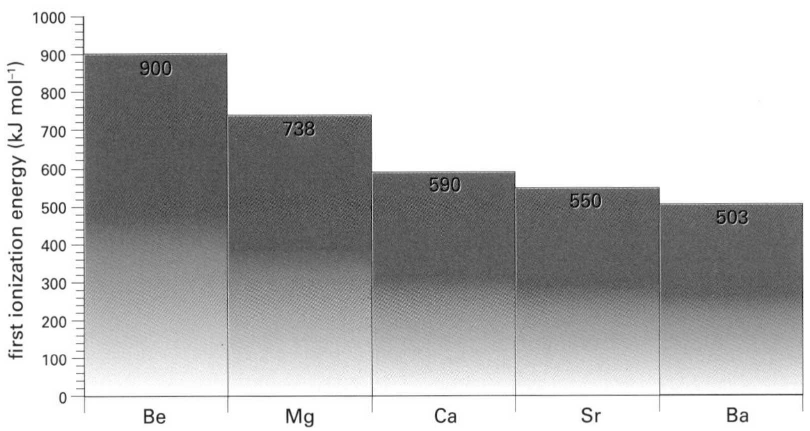

First ionization energy decreases down group 2.

First ionization energy

The first ionization energy is the energy needed to remove one electron from a gaseous atom. The general equation for this process is:

$M(g) \rightarrow M^{+}(g) + e^{-}$

where M stands for the chemical symbol of the element.

Trend in melting point

The melting point generally decreases as you go down group 2. The radius of the metal ions increases going down group 2 so their charge density decreases. The force of attraction between the metal ions and the delocalized electrons decreases, reducing the strength of the metallic bonding.

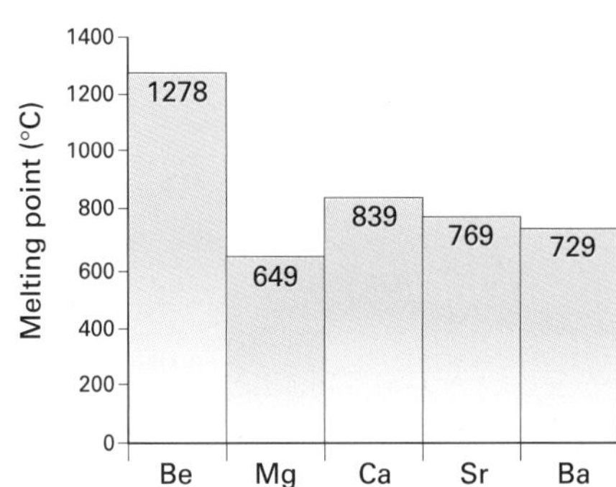

Melting point T_m generally decreases down group 2.

Magnesium

You may have noticed that magnesium has a lower melting point than might be expected. One possible explanation involves the structures of the solid metals. Beryllium and magnesium have a different structure from the other elements in the group. They have a hexagonal close-packed structure, but calcium and strontium have a face-centred cubic structure, and barium has a body-centred cubic structure. But magnesium also has a lower boiling point than expected. Boiling involves the state change from liquid to gas, so the different solid structures do not seem a likely explanation.

There should be two moles of delocalized electrons per mole of metal atoms in the group 2 elements. If there were fewer than two moles of delocalized electrons per mole of magnesium atoms, the metallic bonding would be weaker than expected, leading to a lower melting point. Is there any evidence for fewer delocalized electrons in magnesium? Metals conduct electricity because they have delocalized electrons, and magnesium has a lower electrical conductivity than beryllium or calcium.

Check your understanding

1. Describe and explain the trends down group 2 in
   - a atomic radius
   - b first ionization energy
   - c melting point

10.02 Group 2: reactions with water

OBJECTIVES

already from GCSE, you know that

- the group 1 elements react with water to form alkaline hydroxides and hydrogen gas
- the group 1 elements become more reactive as you go down the group

and after this spread you should know

- the reactions of the group 2 elements magnesium to barium with water, and recognize the trend

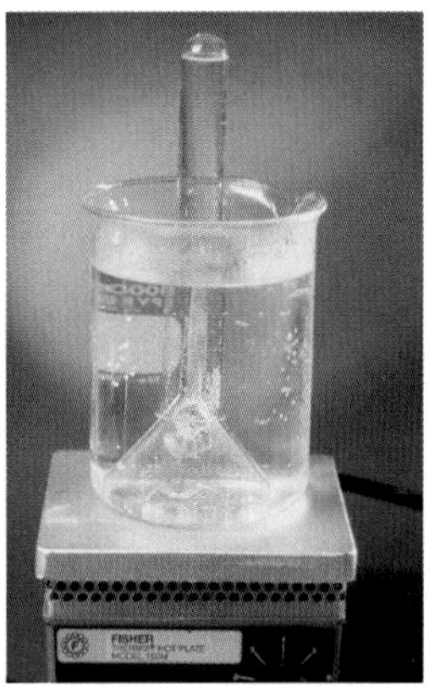

Magnesium reacts very slowly with water. Hydrogen is collected over several days by upward displacement through an inverted filter funnel.

Testing for hydrogen

There is a simple laboratory test for the presence of hydrogen. A lighted wooden splint will ignite a test tube of hydrogen gas with a popping sound.

Calcium reacts steadily with water.

Beryllium at the top of group 2 does not react with water or steam, even if it is heated until it glows. But the other group 2 elements react with water to produce the metal hydroxide and hydrogen:

$$M(s) + 2H_2O(l) \rightarrow M(OH)_2(aq) + H_2(g)$$

The reactions become more vigorous as you go down the group.

Period	Group 1	Group 2	
2	Li 3 Lithium	Be 4 Beryllium	
3	Na 11 Sodium	Mg 12 Magnesium	
4	K 19 Potassium	Ca 20 Calcium	Sc 21 Scandium
5	Rb 37 Rubidium	Sr 38 Strontium	Y 39 Yttrium
6	Cs 55 Caesium	Ba 56 Barium	La 57 Lanthanum
7	Fr 87 Francium	Ra 88 Radium	Ac 89 Actinium

Group 2 is in the s block of the Periodic Table. It contains the elements Be to Ra.

Magnesium

Magnesium burns vigorously with a bright white flame, so people often expect it to react vigorously with water, too. Instead, it only reacts very slowly with water. It may take several days to collect enough hydrogen to test. The magnesium hydroxide produced is sparingly soluble. Just enough dissolves to produce an alkaline solution.

Magnesium and steam

Magnesium burns with a bright white light when it is heated in steam. The hydrogen gas can be led out through a tube and, with care, it can be ignited. A white solid, magnesium oxide, is produced in this reaction instead of magnesium hydroxide.

$$Mg(s) + H_2O(g) \rightarrow MgO(s) + H_2(g)$$

Magnesium reacts vigorously with steam.

Calcium

Calcium reacts steadily with cold water. The reaction is exothermic, and the rate of reaction increases as the reaction mixture heats up. Calcium hydroxide is slightly soluble and produces an alkaline solution. This usually looks cloudy white because undissolved calcium hydroxide forms a suspension.

Strontium is stored in oil to keep air and water away.

Strontium and barium

Strontium and barium react with water more quickly than magnesium and calcium do. Barium, near the bottom of group 2, reacts immediately with water and produces hydrogen gas very quickly. Strontium and barium are stored in oil, just like sodium and other elements in group 1.

Barium reacts quickly with water.

Explaining the trend

The calculated standard enthalpy changes for the reactions between the group 2 elements and water are very similar to each other. This is also true of beryllium, even though this element does not react with water. So another factor must be responsible for the different reactivities.

group 2 element	$\Delta H^{\ominus}$ (kJ mol^{-1})
Be	–331
Mg	–353
Ca	–415
Sr	–387
Ba	–373

Standard enthalpy changes for the reactions of the Group 2 elements with water.

The group 2 hydroxides are ionic compounds. They contain the metal ion M^{2+} and the hydroxide ion OH^-. The metal atoms become metal ions in the reaction with water. You can calculate the energy needed to do this. It involves three processes:

- atomization energy (the energy needed to break up the metal lattice)
- first ionization energy (for the process $M(g) \rightarrow M^+(g) + e^-$)
- second ionization energy (for the process $M^+(g) \rightarrow M^{2+}(g) + e^-$)

The reactions between the group 2 metals and water are exothermic. These three processes are all endothermic. They represent an activation energy that must be overcome for the reactions to happen. The graph shows the standard enthalpy changes for the formation of the M^{2+} ions. Notice that the energy needed decreases down the group. The reactions happen faster when the activation energy is lower.

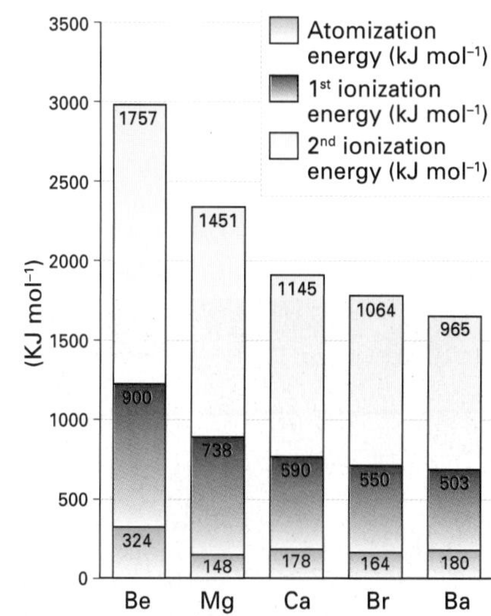

The standard enthalpy changes for forming M^{2+} ions from group 2 atoms.

Check your understanding

1. What is the trend in the reactivity of group 2 elements towards water?
2. Describe, with the help of relevant equations, the following reactions:
   a the reaction between calcium and water
   b the reaction between magnesium and steam
3. a Use the bar chart to estimate the standard enthalpy change for forming Ra^{2+} ions from radium atoms.
   b Predict what you would observe if a piece of radium were added to water, and explain how you answer to part **a** helps you to do this.

10.03 Group 2: hydroxides

OBJECTIVES

already from AS Level, you know

- the reactions of the group 2 elements magnesium to barium with water, and can recognize the trend

and after this spread you should know

- the relative solubilities of the hydroxides of the elements magnesium to barium
- the use of magnesium hydroxide in medicine
- the use of calcium hydroxide in agriculture

Solubility

The solubility of a substance is a measure of how much of it will dissolve in a certain solvent. It is the maximum mass of the substance that will dissolve under stated conditions. It is not how quickly a solute dissolves.

There are several factors that can affect the solubility of an individual substance in a solvent, including

- the volume of the solvent
- the temperature of the solvent

More solute will dissolve in a larger volume of solvent, and the solubility of most solids increases as the temperature increases. To take these factors into account, solubility is given for a specified temperature and volume of solvent.

IUPAC recommend that solubility *s* should be measured in $mol\,m^{-3}$. But you will usually see solubility measured in g solute per $100\,cm^3$ solvent. Note that if the solvent is water, the unit may instead be g solute per 100 g H_2O. This is because the density of water is approximately $1\,g\,cm^{-3}$, so $1\,cm^3$ has a mass of 1 g.

group 2 hydroxide	solubility (g per $100\,cm^3$ of water)
$Mg(OH)_2$	0.0012
$Ca(OH)_2$	0.113
$Sr(OH)_2$	0.410
$Ba(OH)_2$	2.57

The solubility of Group 2 hydroxides increases down the group.

Group 2 hydroxides

The solubility of the group 2 hydroxides increases down the group. Magnesium hydroxide is **sparingly soluble**: very little of it dissolves. Barium hydroxide is much more soluble, and it is possible to prepare a $0.1\,mol\,dm^{-3}$ solution of it.

Magnesium hydroxide

Magnesium hydroxide is used as an **antacid**. It forms a white suspension, usually called milk of magnesia. Stomach acid is hydrochloric acid. Indigestion is caused by excess stomach acid, and heartburn by stomach acid being pushed up into the oesophagus (gullet). Both conditions are painful. Magnesium hydroxide and other antacids relieve the symptoms by neutralizing the acid. Magnesium hydroxide reacts with hydrochloric acid to produce magnesium chloride and water:

$$Mg(OH)_2(s) + 2HCl(aq) \rightarrow MgCl_2(aq) + 2H_2O(l)$$

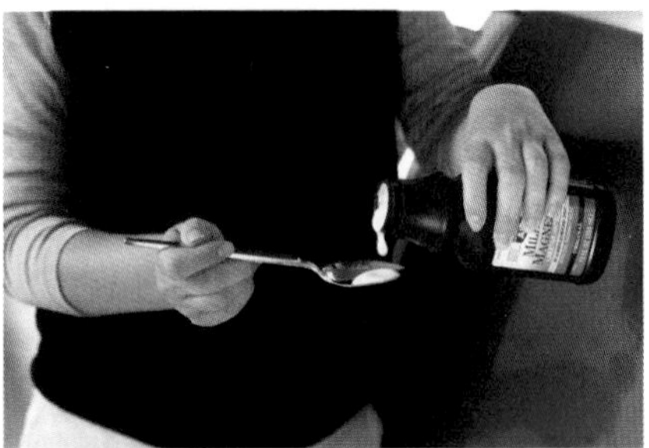

Magnesium hydroxide is used as a medicine to treat indigestion, heartburn, and constipation.

Calcium hydroxide

Limestone is a common rock made from calcium carbonate. This decomposes on heating to form calcium oxide:

$$CaCO_3(s) \rightarrow CaO(s) + CO_2(g)$$

Calcium oxide reacts with water to form calcium hydroxide:

$$CaO(s) + H_2O(l) \rightarrow Ca(OH)_2(s)$$

Farmers 'lime' their fields to control the acidity in the soil. They add powdered limestone, calcium oxide, or calcium hydroxide. This is important because different crops grow best in soils with different degrees of acidity. For example, potatoes grow best at pH 5.5 to 6.5, oilseed rape at pH 6.0 to 7.5, and barley at pH 6.5 to 7.5.

Farmers use a soil test kit to find the pH of the soil, then add sufficient lime to bring the pH of the soil into the required range. This needs to be repeated regularly because liming is not a permanent answer to the problem of soil acidity. Acids from acid rain, plants, and fertilizers react with the lime. This reduces its effectiveness over time.

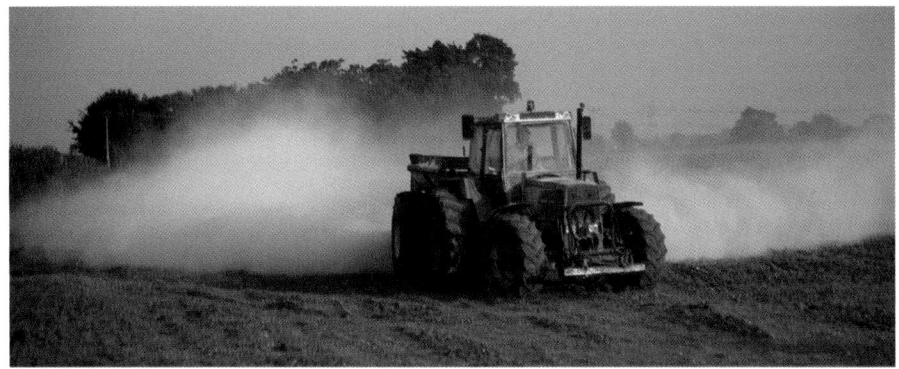

Soil acidity is controlled by spreading calcium hydroxide on fields.

Science @ Work

Laxative effect

Magnesium hydroxide also acts as a laxative. It relieves constipation and makes it easier to pass faeces. The magnesium ions are responsible for this action, rather than the hydroxide ions. Magnesium ions are not absorbed by the intestines very well, causing water to be drawn into the intestines. This makes the faeces softer and easier to pass. But it also means that indigestion sufferers need to make sure they do not swallow too much milk of magnesia.

Science @ Work

Limewater

Calcium hydroxide solution is also called limewater. This is used to detect the presence of carbon dioxide. When carbon dioxide is bubbled through limewater, a cloudy white suspension of calcium carbonate is formed:

$$Ca(OH)_2(aq) + CO_2(g) \rightarrow CaCO_3(s) + H_2O(l)$$

This gradually clears when excess carbon dioxide is bubbled through:

$$CaCO_3(s) + CO_2(g) + H_2O(l) \rightarrow Ca(HCO_3)_2(aq)$$

Limewater is commonly used to test for carbon dioxide in the breath.

Check your understanding

1. a What is the trend in the solubility of the group 2 hydroxides?
   b Predict the relative solubility of radium hydroxide, and give a reason for your answer.
2. Explain the use of
   a magnesium hydroxide in medicine
   b calcium hydroxide in agriculture

10.04 Group 2: sulfates

OBJECTIVES

already from AS level, you know

- the relative solubilities of the hydroxides of the elements magnesium to barium

and after this spread you should

- know the relative solubilities of the sulfates of the elements magnesium to barium
- know the use of barium sulfate in medicine
- understand why acidified barium chloride solution is used as a reagent to test for sulfate ions

The solubility of the group 2 sulfates decreases as you go down the group. Note that this is the opposite trend to the one shown by the group 2 hydroxides. Magnesium hydroxide is sparingly soluble but magnesium sulfate is soluble.

group 2 sulfate	solubility (g per 100 cm^3 of water)
$MgSO_4$	25.5
$CaSO_4$	0.24
$SrSO_4$	0.013
$BaSO_4$	0.00022

The solubility of group 2 sulfates decreases down the group.

Other sulfates

Magnesium sulfate-7-water, $MgSO_4.7H_2O$, is also called Epsom salts. It gets its name from mineral water at Epsom in Surrey, which naturally contains magnesium sulfate. It is used in bath salts, fireproofing, and artificial snow for film sets. It is a laxative, just like magnesium hydroxide.

Plaster of Paris is $2CaSO_4.H_2O$. When it is mixed with water, it hydrates to form calcium sulfate-2-water, $CaSO_4.2H_2O$. This expands slightly and sets hard. It is used to make plaster casts to keep broken bones still, and by forensic scientists to make plaster casts of shoeprints at crime scenes.

Plaster of Paris is used to make plasterboard, and for plastering walls and ceilings.

Barium sulfate

Barium compounds are toxic. Soluble barium compounds are particularly hazardous as they could be absorbed through the intestines if swallowed. Barium sulfate is very insoluble in water. It is also opaque to X-rays, which means that they cannot pass through it. These features make barium sulfate useful as a *contrast medium* in medical X-rays of the digestive system. The lower part of the digestive system shows up in X-rays if the patient is given an enema of barium sulfate suspension. The upper part shows up in X-rays if the patient swallows barium sulfate suspension. In both cases, the barium sulfate eventually passes harmlessly out of the body.

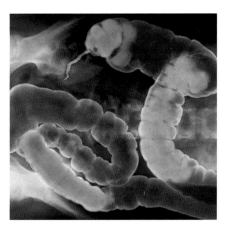

Barium sulfate shows up in medical X-rays, like this one of the large intestine.

Rat poison

Barium carbonate is used as rat poison. It is not soluble, but it reacts with stomach acid to produce barium chloride, which is soluble and also toxic:

$BaCO_3(s) + 2HCl(aq) \rightarrow BaCl_2(aq) + H_2O(l) + CO_2(g)$

Testing for sulfates

The insolubility of barium sulfate is the basis of a simple laboratory test for the presence of sulfate ions in solution. A white precipitate of barium sulfate forms when barium chloride solution is added to a solution containing sulfate ions. For example, when barium chloride solution is added to sodium sulfate solution the following reaction occurs.

$$BaCl_2(aq) + Na_2SO_4(aq) \rightarrow BaSO_4(s) + 2NaCl(aq)$$

This can be simplified to an ionic equation, since the Na^+ and Cl^- ions remain in solution as spectator ions:

$$Ba^{2+}(aq) + SO_4^{2-}(aq) \rightarrow BaSO_4(s)$$

The test works just as well if barium nitrate is used instead. But in both cases it is important to first acidify the test sample using hydrochloric acid or nitric acid. This is needed to prevent the false detection of sulfite ions SO_3^{2-} in the test. Barium sulfite $BaSO_3$ is insoluble. It would also form a white precipitate in the test. The acids react with the sulfite ion to form sulfur dioxide:

$$2H^+(aq) + SO_3^{2-}(aq) \rightarrow SO_2(g) + H_2O(l)$$

Acidified barium chloride solution produces a white precipitate if sulfate ions are present in a sample.

Check your understanding

1. a What is the trend in the solubility of the group 2 sulfates?
   b In what way is this trend different from the trend in the solubility of the group 2 hydroxides?
2. Explain the use of barium sulfate in medicine, even though barium compounds are toxic.
3. a Describe the laboratory test for the presence of aqueous sulfate ions.
   b Explain why the barium chloride used in the test can be acidified with hydrochloric acid, but not with sulfuric acid, H_2SO_4.

10.05 Testing for anions

OBJECTIVES

already from AS level, you understand

- why acidified silver nitrate solution is used as a reagent to identify and distinguish between F^-, Cl^-, Br^-, and I^- ions
- why acidified barium chloride solution is used as a reagent to test for sulfate ions

and after this spread you should know

- how to carry out anion tests competently

One of the recommended Practical Skills Assessment tasks is to carry out some inorganic tests. You may be asked to test some unknown solutions to find out which anions they contain. For example, they may contain chloride, bromide, iodide, or sulfate ions. You will be assessed on your ability to

- work safely and carefully
- use sensible volumes of the reagents
- obtain the correct observations for each test solution

Since you will be following a set of instructions given to you, it is possible that you might have to carry out tests for other anions. For example, you might also test for the presence of carbonate or nitrate ions.

Results

It is always wise to draw a blank results table before you start work. That way, you can be certain that you have completed all the tasks before you tidy away. If you are uncertain of an observation and repeat a certain test, do not erase your original record. Just put a line through your writing in such a way that you can still see your original observations. You may have been correct the first time around.

Watch the volumes

A common mistake in test tube reactions is to add too much reagent. When you do this, it is very difficult to mix the contents effectively. It is best to limit the volume of your test substance to about 1 cm^3, unless you are told otherwise.

If you are asked to add a reagent dropwise, this is exactly what it means.

- Add a drop using a teat pipette.
- Hold the tube at the top.
- Mix the contents by shaking the bottom of the tube from side to side.

Substance	Test	Observations	Inference
sodium chloride solution	A few drops of dil. HCl added, then a few drops of $AgNO_3(aq)$	White precipitate forms	Cl^- ions present
A			
B			
C			

You may wish to draw a table like this one. It has space to record what you did, your observations, and what you think your observations mean.

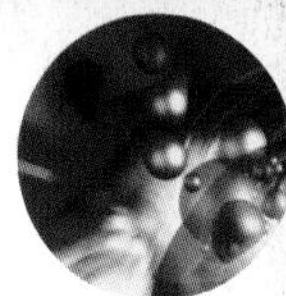

Anion tests – a reminder

Halide ions

You can test for the presence of Cl^-, Br^-, and I^- ions using silver nitrate solution.

- Add about 1 cm^3 of your test substance to a clean test tube.
- Add a few drops of dilute nitric acid and shake.
- Add a few drops of silver nitrate solution.

observations	inference
No precipitate	Does not contain Cl^-, Br^-, or I^- ions
White precipitate	Cl^- ions present
Cream precipitate	Br^- ions present
Yellow precipitate	I^- ions present

Typical observations and inferences for halide ion tests. Note that the absence of a precipitate may also mean that the concentration of halide ions is too low to detect.

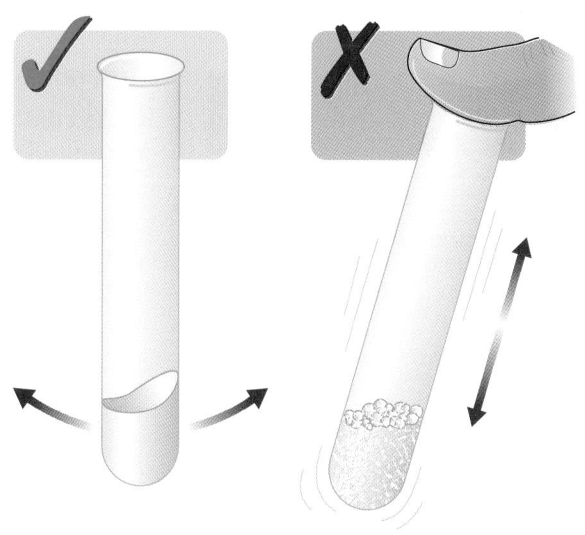

Do not put your thumb over the top of the test tube and shake it up and down.

You may need to carry out a confirmatory test on each precipitate using aqueous ammonia. Remember that ammonia is corrosive and has a sharp, irritating smell. Concentrated ammonia should be used in a fume cupboard. You may need to add an excess of ammonia.

- White precipitate of AgCl redissolves in dilute aqueous ammonia.
- Cream precipitate of AgBr redissolves in concentrated ammonia.
- Yellow precipitate of AgI does not redissolve in dilute or concentrated ammonia.

Sulfate ions

You can test for the presence of SO_4^{2-} ions using aqueous barium chloride or barium nitrate solution.

- Add about 1 cm^3 of your test substance to a clean test tube.
- Add a few drops of dilute hydrochloric acid and shake.
- Add a few drops of barium chloride or barium nitrate solution.

observations	inference
No precipitate	Does not contain SO_4^{2-} ions
White precipitate	SO_4^{2-} ions present

Typical observations and inferences for sulfate ion tests. Note that the absence of a precipitate may also mean that the concentration of sulfate ions is too low to detect.

Check your understanding

1. Suggest why the reagents should be made up in distilled or de-ionized water, rather than tap water.
2. Explain why you should be careful to
   - **a** use the correct acids in these tests
   - **b** use clean test tubes
   - **c** record your observations from one test before carrying out the next one

11.01 Metal ores

OBJECTIVES

already from AS level, you know

- that combustion of hydrocarbons containing sulfur
  leads to sulfur dioxide that causes air pollution

and after this spread you should

- know that metals are found in ores, usually as oxides or sulfides
- know that sulfide ores are usually converted into oxides by roasting in air
- understand the environmental problems associated with the conversion of sulfides into oxides
- know that the sulfur dioxide produced can be used to manufacture sulfuric acid

Around 80% of the world's copper is produced from chalcopyrite.

Sulfuric acid is the acid in car batteries. Old car batteries must be disposed of responsibly, as the acid can harm the environment.

Minerals and ores

Rocks contain minerals. Each mineral is an individual chemical compound with a particular crystal structure. Some rocks consist almost entirely of one mineral. For example, limestone and marble are both mainly calcite, which is a form of calcium carbonate. But most rocks contain a mixture of different minerals. An ore is a rock that contains enough of a certain mineral to make it worth mining. A rock is not an ore if it costs more to extract the desired metal from it than the metal is worth.

metal	ore mineral	formula of mineral
iron	haematite	Fe_2O_3
copper	chalcopyrite	$CuFeS_2$
manganese	pyrolusite	MnO_2
aluminium	bauxite	Al_2O_3
titanium	rutile	TiO_2
tungsten	scheelite	$CaWO_4$
zinc	zinc blende	ZnS

Some common metals and one of their main ore minerals.

Most ores contain metal oxides or metal sulfides. It is difficult to extract the metal from a metal sulfide. When an ore contains a metal sulfide, the sulfide is first converted to an oxide.

Roasting

Roasting in air is the method used to convert metal sulfides to metal oxides. The ore is usually crushed to increase its surface area, then it is heated strongly in a stream of air. The metal oxide and sulfur dioxide are produced in the reaction. For example, zinc blende contains zinc sulfide. This is roasted to produce zinc oxide:

$$ZnS(s) + 1\tfrac{1}{2}O_2(g) \rightarrow ZnO(s) + SO_2(g)$$

A similar thing happens with chalcopyrite:

$$2CuFeS_2(s) + 6\tfrac{1}{2}O_2(g) \rightarrow 2CuO(s) + Fe_2O_3(s) + 4SO_2(g)$$

Roasting reactions are strongly exothermic and help to reduce the energy costs of the process. They can also weaken hard ores, making it easier to crush them further for the next stage of metal extraction. But they produce sulfur dioxide as a by-product. As you discovered in Unit 1, this gas contributes to acid rain if it is allowed to escape into the atmosphere.

Fortunately, sulfur dioxide from roasting sulfide ores is used as a raw material for making sulfuric acid. This reduces the environmental impact of the process. It turns a harmful by-product into a useful substance. Sulfuric acid is used to make fertilizers, explosives, medicines, paints, dyes, detergents, and plastic.

Sulfuric acid

Sulfuric acid is manufactured using the *contact process*. Sulfur dioxide is oxidized in the presence of a vanadium(V) oxide catalyst:

$$2SO_2(g) + O_2(g) \xrightarrow[\text{catalyst}]{V_2O_5} 2SO_3(g)$$

The sulfur trioxide produced in the reaction is absorbed in concentrated sulfuric acid. This forms an oily liquid called oleum. The oleum is diluted with water to form a larger volume of concentrated sulfuric acid. Overall:

$$SO_3(g) + H_2O(l) \rightarrow H_2SO_4(aq)$$

Science @ Work

Acid mine drainage

Many underground mines are below the natural groundwater level. Pumps are needed to stop them flooding. When a mine is abandoned and its pumps turned off, the water level rises. The mine floods with water. Metal sulfides begin to oxidize once they are exposed to water. The sulfides are oxidized to sulfuric acid, which dissolves potentially toxic metals such as iron, cadmium, and copper.

The acidic pollution eventually runs out into nearby rivers. Dissolved Fe^{2+} ions oxidize to Fe^{3+} ions, and these produce an orange precipitate of iron(III) hydroxide. This damages the rivers. It coats the gills of fish and can kill them. Acid mine drainage can be combated using limestone and reed beds. The limestone neutralizes the acid, and reed beds contain bacteria that can reduce the sulfuric acid to hydrogen sulfide. This reacts with the metal ions to form insoluble sulfide minerals again.

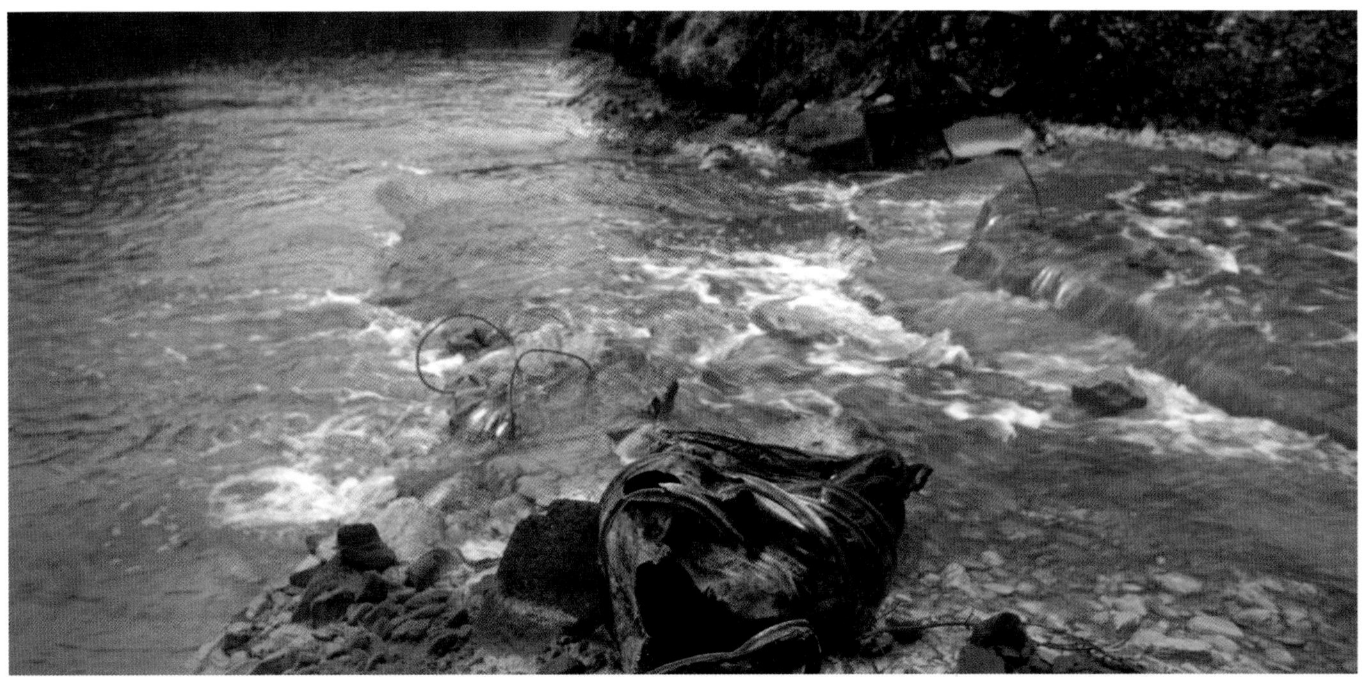

In 1992, acid mine drainage from the Wheal Jane tin mine heavily polluted rivers in Cornwall.

Check your understanding

1. a What is an ore?
   b Name three minerals found in ores. For each one, give its formula and name the metal extracted from it.
2. a Explain why sulfide ores are usually converted into oxides.
   b Describe, with the help of an equation, how zinc blende is converted into an oxide.
3. Discuss the environmental problems posed by sulfide ores, and how their impact can be decreased.

11.02 Metal extraction using carbon

OBJECTIVES

already from AS Level you know

- that metals are found in ores, usually as oxides or sulfides
- that sulfide ores are usually converted into oxides by roasting in air

and after this spread you should understand

- that extraction of metals involves reduction
- that carbon and carbon monoxide are cheap and effective reducing agents that are used in the extraction of iron, manganese, and copper

K	potassium	most reactive
Na	sodium	
Ca	calcium	
Mg	magnesium	
Al	aluminium	
Ti	titanium	
Mn	manganese	
Zn	zinc	
Fe	iron	
W	tungsten	
Cu	copper	
Ag	silver	
Au	gold	least reactive

The relative reactivity of some common metals.

Metal oxides must be reduced to extract the metal they contain. The way in which a metal oxide is reduced depends on how reactive the metal is. This in turn depends on factors such as the amount of energy involved in gaining electrons and breaking the ionic bonds in the metal oxide.

Reducing an oxide

Electrolysis

The most reactive metals are extracted using electrolysis. Aluminium, potassium, sodium, calcium, and magnesium are extracted this way. Electricity is expensive to produce. Metals produced using electrolysis are likely to be expensive, even if their ores are abundant.

Using a more reactive metal

Some metals are difficult to extract using electrolysis. Titanium is like this. It is extracted by reacting titanium(IV) chloride with a more reactive metal such as magnesium or sodium. Since the reactive metal itself must be extracted using electrolysis, titanium is an expensive metal.

Using hydrogen

Some of the least reactive metals, such as tungsten, are extracted using hydrogen. Hydrogen can be manufactured by the electrolysis of sodium chloride solution, or directly from fossil fuels:

- cracking of crude oil fractions
- reaction of methane (natural gas) with steam

$$CH_4(g) + H_2O(g) \rightarrow CO(g) + 3H_2(g)$$

- reaction of carbon (coke from coal) with steam

$$C(s) + H_2O(g) \rightarrow CO(g) + H_2(g)$$

These processes add to the expense of extracting the metal.

Using carbon

Carbon is also used to extract less reactive metals. Coal is a convenient source of carbon. It is usually treated first by heating it strongly in the absence of air. This drives off impurities and produces coke, which is almost pure carbon. Coke is a cheap and effective reducing agent. It is used in the extraction of metals such as iron, manganese, and copper. In the conditions used to reduce the metal oxide, the reducing agent may also be carbon monoxide.

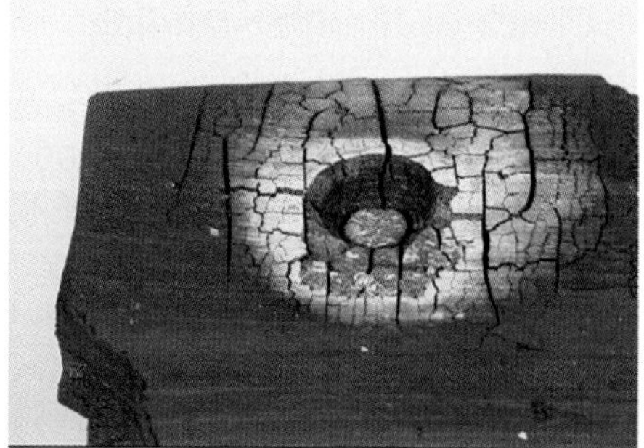

Lead oxide can be reduced to lead by heating it on a carbon block.

Extraction of iron

Iron is extracted from its ores in a blast furnace. The raw materials needed are

- iron ore such as haematite, Fe_2O_3
- coke, the source of the reducing agent
- limestone, a source of calcium carbonate
- air, a cheap source of oxygen

These are the main reactions that take place in the blast furnace:

1. Coke burns in the air, raising the temperature:

$$C(s) + O_2(g) \rightarrow CO_2(g)$$

2. Calcium carbonate decomposes in the high temperatures:

$$CaCO_3 \rightarrow CaO(s) + CO_2(g)$$

3. Coke reduces carbon dioxide to form carbon monoxide:

$$C(s) + CO_2(g) \rightarrow 2CO(g)$$

4. Carbon monoxide reduces the iron oxide at around 1500°C:

$$Fe_2O_3(s) + 3CO(g) \rightarrow 2Fe(l) + 3CO_2(g)$$

5. Impurities in the iron ore (mostly silica) react with calcium oxide from the decomposed limestone, forming a liquid slag:

$$CaO(s) + SiO_2(s) \rightarrow CaSiO_3(l)$$

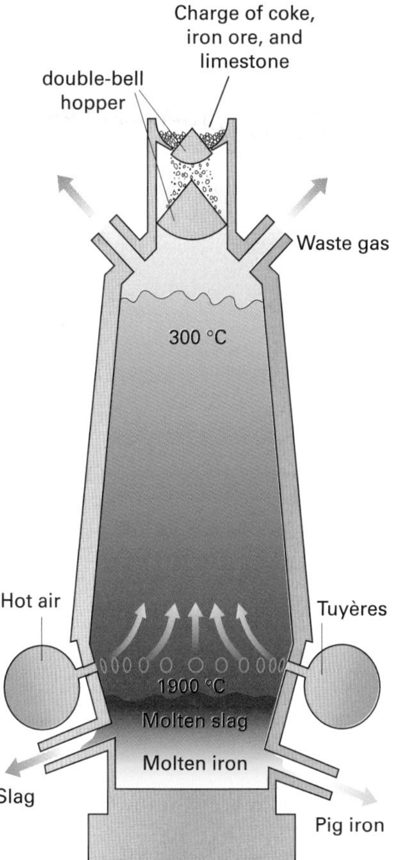

A modern blast furnace can be up to 30 m high and produces around 10 000 tonnes of iron per day. Iron straight from the blast furnace contains about 4% carbon. Steel is made from iron by lowering the carbon content by reaction with oxygen.

Manganese and copper

Manganese is similar to iron but is harder and more brittle. Its main ore is pyrolusite, which contains manganese(IV) oxide, MnO_2. This can be reduced to manganese(II) oxide using carbon monoxide:

$$MnO_2(s) + CO(s) \rightarrow MnO(s) + CO_2(g)$$

The manganese(II) oxide is reduced to manganese at high temperature by carbon:

$$MnO(s) + C(s) \rightarrow Mn(l) + CO(g)$$

Copper can be extracted from copper(II) oxide by heating it strongly with carbon:

$$CuO(s) + C(s) \rightarrow Cu(l) + CO(g)$$

Today, copper is more usually extracted straight from chalcopyrite in a single stage:

$$2CuFeS_2(s) + 3\tfrac{1}{2}O_2(g) \rightarrow 2Cu(l) + Fe_2O_3(s) + 2SO_2(g)$$

Science @ Work: Ferromanganese

Most manganese is used in steel making, but without bothering to extract it first. Iron ore and manganese ore are mixed in the required proportions and processed to make an alloy called ferromanganese. Most steel contains up to 1% manganese, which allows it to be rolled at high temperature without breaking.

Check your understanding

1. Explain the advantages of using carbon and carbon monoxide as reducing agents in the extraction of metals, compared with using electrolysis or hydrogen.
2. With the help of equations, explain how iron is extracted from iron ore.
3. It has been suggested that copper was discovered thousands of years ago when green malachite was used as a pottery glaze. Explain how the reaction between malachite $Cu_2CO_3(OH)_2$ and carbon from a potter's charcoal fire could produce copper.

11.03 Aluminium production

OBJECTIVES

already from GCSE you know

- that aluminium is extracted using electrolysis

after this spread you should

- know why carbon reduction is not used for the extraction of aluminium
- understand how aluminium is manufactured from purified bauxite

Aluminium has several properties that make it a very useful metal. Aluminium

- forms strong alloys
- is a good conductor of electricity
- has a low density
- resists corrosion

Aluminium is the most abundant metal in the Earth's crust, but it is too reactive to be found naturally as the free metal. It is also too reactive to be extracted using carbon as the reducing agent. Hans Christian Oersted produced impure aluminium in 1825 by heating aluminium chloride with potassium. Friedrich Wöhler improved the process two years later. The required potassium had to be extracted from potassium compounds by electrolysis. Electricity is expensive to produce. Aluminium was very expensive as a result.

The cost of the process fell in 1855 when Henri Étienne Sainte-Claire Deville improved it further. He used sodium instead of potassium, and sodium is cheaper to produce. But the problem still faced by scientists then was that an abundant metal with desirable properties was just too expensive to manufacture for everyday use.

The Hall–Héroult process

Charles Hall in the USA and Paul Héroult in France independently solved the problem in 1886. The Hall–Héroult process greatly reduced the cost of aluminium during the nineteenth century. It is the process used today. It still uses electrolysis because aluminium cannot be extracted using carbon.

Heat energy considerations

Aluminium oxide or *alumina* is purified from bauxite. Electrolysis only works when the substance is molten or dissolved, as its ions must be free to move. Aluminium oxide has a high melting point, 2050°C. Uneconomic amounts of energy would be needed to heat it to that temperature. So the aluminium oxide is dissolved in molten cryolite, Na_3AlF_6. This has a much lower melting point, and lets electrolysis happen at a much lower temperature, about 950°C.

This is a coloured anodized aluminium heat sink for a computer component.

A protective layer

Aluminium resists corrosion because it is protected by a transparent layer of aluminium oxide. This forms when the metal is exposed to air. Unlike a layer of rust on iron, this layer does not flake off. The natural layer is only about 10 nm thick but it can be thickened by a process called anodizing.

Anodizing involves electrolysis in sulfuric acid. The aluminium is the anode in the process, and it reacts with oxygen given off during electrolysis. The new thicker layer of aluminium oxide improves the corrosion resistance of the metal. It also absorbs dyes before a final treatment by heating. Tough, coloured aluminium items are produced this way.

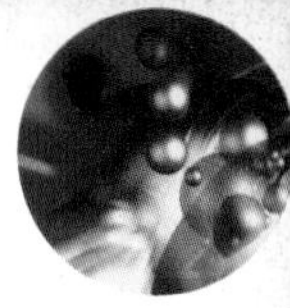

Electrical energy considerations

Electrolysis takes place in a large steel tank lined with graphite. The graphite lining acts as the cathode. A series of large graphite blocks act as the anode. A current of about 200 000 A is passed through the molten mixture of cryolite and aluminium oxide. This decomposes the aluminium oxide. It also maintains the temperature of the reaction mixture.

Aluminium ions are attracted to the cathode, where they are reduced to aluminium:

$$Al^{3+}(l) + 3e^- \rightarrow Al(l)$$

Oxide ions are attracted to the anode, where they are oxidized to oxygen:

$$2O^{2-}(l) \rightarrow O_2(g) + 4e^-$$

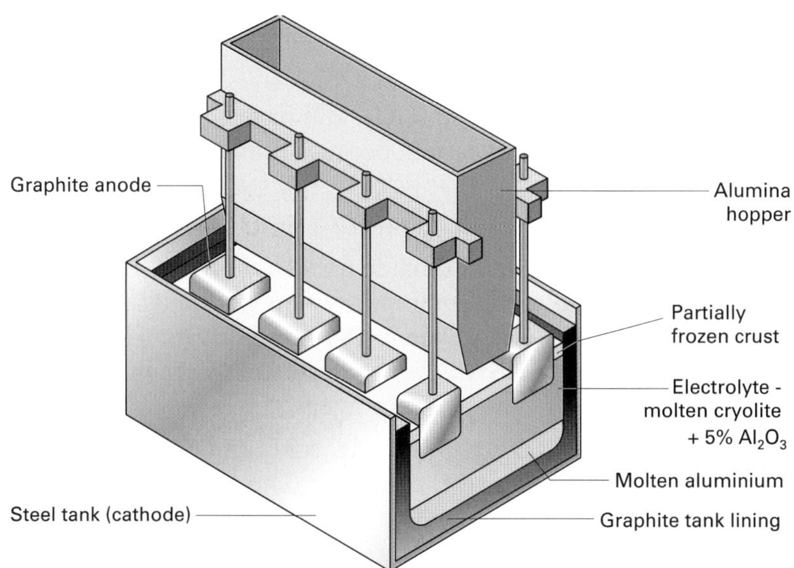

Aluminium is extracted from purified bauxite by electrolysis.

A typical electrolysis cell produces one tonne of aluminium each day. This needs 50 GJ (14 000 kWh) of electricity, enough to run a 100 W light bulb continuously for 16 years. Electricity is expensive to produce, so around 60% of the world's aluminium is produced using hydroelectric power. This is relatively cheap. It is a renewable energy resource and does not produce large amounts of carbon dioxide, unlike fossil fuels.

Science @ Work: Consumable anodes

The electrolysis process is a continuous process. The electrolysis cell works all the time, with regular additions of fresh aluminium oxide and removal of aluminium. But the graphite anodes must be replaced regularly. Oxygen is produced at their surface during electrolysis. This reacts with the anodes at the high temperatures in the electrolysis cell and erodes them.

The aluminium smelter at Fort William in the Scottish Highlands uses hydroelectric power.

Check your understanding

1. Why is carbon not used as the reducing agent in the extraction of aluminium?
2. What is the main aluminium ore, and what aluminium compound does it contain?
3. **a** Why is cryolite used in the production of aluminium?
   b Write ionic equations for the reactions occurring at the electrodes during aluminium production.
   c Use your answers to part **b** to write the full equation for the extraction of aluminium.

11.04 Titanium production

OBJECTIVES

already from AS level, you understand

- how aluminium is manufactured from purified bauxite

and after this spread you should

- know why carbon reduction is not used for the extraction of titanium
- understand how titanium is extracted from TiO_2 via $TiCl_4$

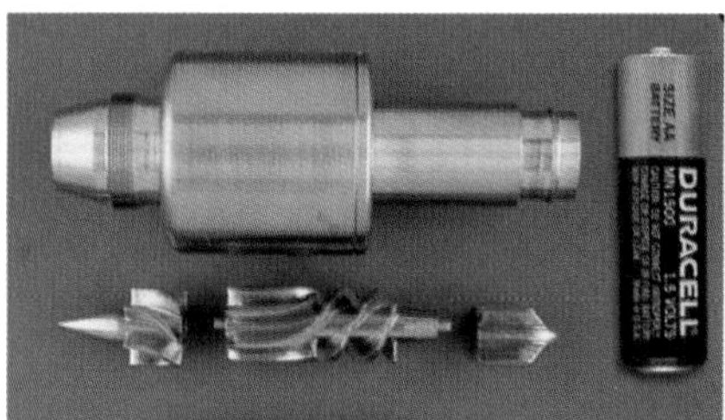

Titanium does not react with body tissues and liquids. It is used to make artificial hip joints and the tiny turbine in this heart bypass pump. (The battery is included for scale.)

Titanium is the seventh most abundant metal in the Earth's crust. Its density is half that of steel but it is just as strong as steel. Titanium alloys are even stronger. Titanium has a high melting point of 1668°C and keeps its strength even when hot. Like aluminium, titanium is a reactive metal that is protected from corrosion by a layer of its oxide. These properties make titanium a very useful metal.

Titanium is expensive to extract. It is used to make upmarket everyday goods such as bicycle frames and pans, but it is most widely used to make aircraft parts. Today's scientists face the same problem with titanium as scientists in the nineteenth century faced with aluminium: an abundant metal with desirable properties that is just too expensive to manufacture for everyday use.

The Hunter process

The main titanium ores are rutile, TiO_2, and ilmenite, $FeO.TiO_2$. These ores cannot be reduced using carbon because titanium carbide is formed instead of titanium:

$$TiO_2(s) + 3C(s) \rightarrow TiC(s) + 2CO(g)$$

It is possible to reduce titanium dioxide using a more reactive metal. But this remains in the titanium as an impurity that is difficult and expensive to remove. This problem was solved in 1910 by Matthew Hunter, a New Zealander working in the USA. His *Hunter process* works in two main stages and uses sodium as the reducing agent.

Stage 1 A mixture of titanium(IV) oxide, coke, and chlorine is heated to about 900°C in a steel container. The reaction produces titanium(IV) chloride:

$$TiO_2(s) + 2C(s) + 2Cl_2(g) \rightarrow TiCl_4(g) + 2CO(g)$$

Stage 2 The titanium(IV) chloride is heated to around 1000°C with sodium, a more reactive metal:

$$TiCl_4(g) + 4Na(l) \rightarrow Ti(s) + 4NaCl(l)$$

The sodium chloride is removed using dilute hydrochloric acid.

The Kroll process

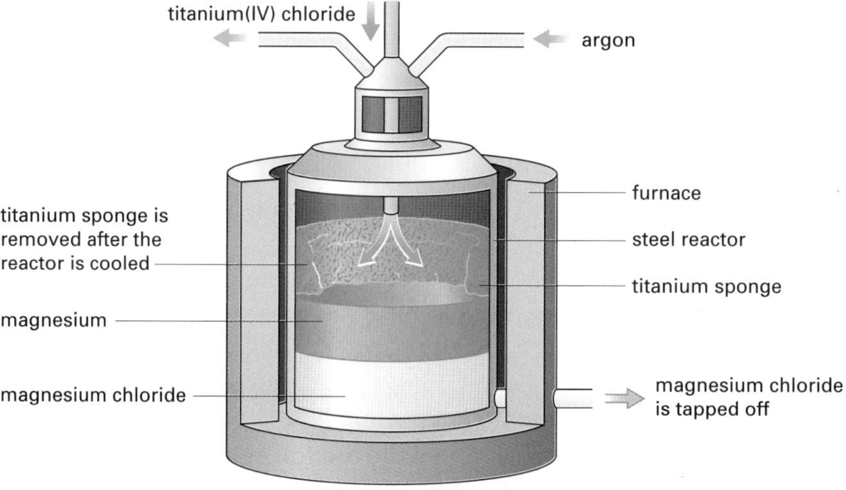

The Kroll reactor produces up to three tonnes of titanium at a time. The titanium sponge is removed from the reactor when it has cooled down.

The Hunter process was improved by William Kroll from Luxembourg in 1932. Commercial production of titanium using his *Kroll process* began in 1946. It is similar to the Hunter process but the second stage uses magnesium as the reducing agent instead of sodium:

$$TiCl_4(g) + 2Mg(l) \rightarrow Ti(s) + 2MgCl_2(l)$$

Argon, the cheapest noble gas, is used to provide an inert atmosphere. This stops the hot titanium reacting with oxygen from the air to form titanium(IV) oxide. Magnesium is cheaper than sodium, so the Kroll process is more economic than the Hunter process. Even so, titanium is expensive to extract from its ores.

- The process is a **batch process**. The Kroll reactor must be cooled down and the titanium removed before it can be set off again.
- High temperatures of around 1000°C are needed.
- A lot of energy is needed to produce the magnesium by electrolysis.
- Titanium(IV) chloride is a hazardous substance. It reacts with moisture in the air to form acidic fumes of hydrogen chloride.
- The titanium is produced as a "sponge" containing impurities that must be removed by heating.

There are some economies in the extraction of titanium. The reaction between the titanium(IV) chloride and magnesium is exothermic. This helps to raise the temperature in the reactor as the reaction proceeds. The waste magnesium chloride is electrolysed to produce magnesium for the second stage, and chlorine for the first stage.

FFC Cambridge process

In 1998 scientists working at the University of Cambridge patented a new, simpler way to produce titanium. The process is called the *FFC-Cambridge process*, after the surnames of the scientists (Fray, Farthing, and Chen). It promises to substantially reduce the cost of extracting titanium in the future. The scientists discovered how to convert titanium(IV) oxide directly to titanium.

Electrolysis is carried out in a graphite or titanium container. A pellet of titanium(IV) oxide is the cathode, and a graphite rod is the anode. Molten calcium chloride at 950°C is added. Oxygen forms at the anode. Titanium is deposited at the cathode as a result of two reactions:

$$Ca^{2+}(l) + 2e^- \rightarrow Ca(l)$$

$$TiO_2(s) + 2Ca(l) \rightarrow Ti(s) + 2CaO(s)$$

The process was unexpected because titanium(IV) oxide is an electrical insulator rather than a conductor. It appears that just enough oxide ions leave the titanium(IV) oxide for it to become a conductor.

Titanium could be used more often for everyday items in the future.

Check your understanding

1. Why is carbon not used as the reducing agent in the extraction of titanium?
2. Describe the Kroll process for extracting titanium from titanium(IV) oxide using magnesium. Include the conditions and relevant equations in your answer.

11.05 Tungsten production

OBJECTIVES

already from AS level, you understand

- how aluminium and titanium are extracted from their ores

and after this spread you should

- know why carbon reduction is not used for the extraction of tungsten
- understand how tungsten is extracted from WO_3 by reduction with hydrogen

Tungsten is the eighteenth most abundant metal in the Earth's crust. Its density is 70% greater than the density of lead. The melting point of tungsten is 3422°C, the highest melting point of any metal. Like aluminium and titanium, tungsten is protected from corrosion by a layer of its oxide. However, this only protects it up to about 400°C, after which it begins to oxidize rapidly in air.

The main uses of tungsten are a result of its high density. These include

- gyroscope rotors
- radiation shielding
- counterweights in aircraft

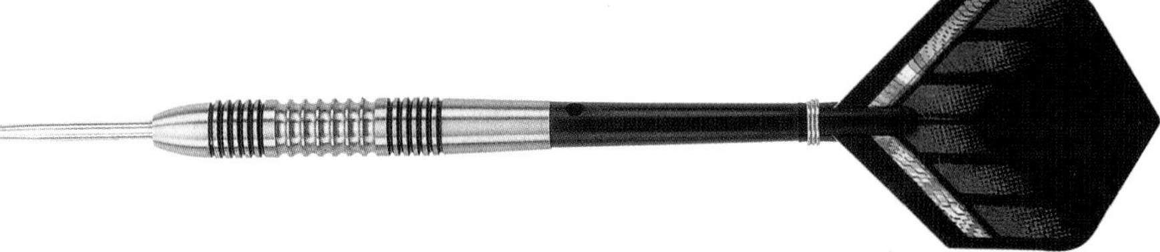

Darts with dense tungsten tips are slimmer than ordinary darts.

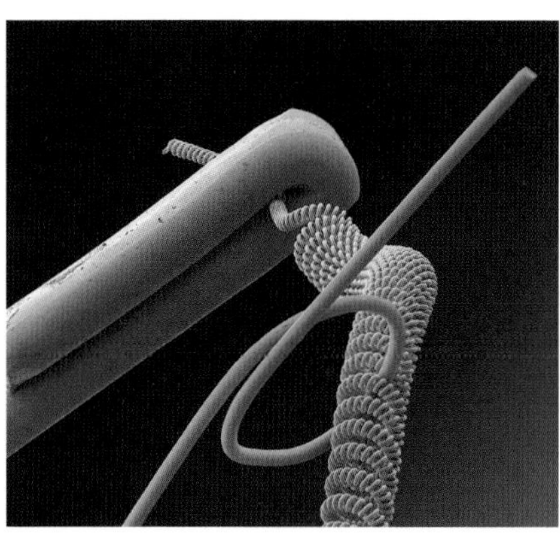

The most familiar use of tungsten is in the coiled filament of ordinary incandescent electric light bulbs. The demand for this use of tungsten will fall as energy-saving compact fluorescent lamps become more widespread.

Tungsten's high melting point makes it ideal for

- heat shields
- electrodes for arc welding
- the filament wires for electric light bulbs

Reduction with hydrogen

Tungsten ores are processed to produce tungsten(VI) oxide, WO_3, which is then reduced to tungsten. Carbon could be used as the reducing agent. But this also produces tungsten carbide, WC. This impurity would adversely affect the properties of the tungsten, making it much less useful. So hydrogen is used as the reducing agent instead.

Tungsten(VI) oxide powder is loaded into narrow metal tubes. It is heated in a furnace to about 850°C. Hydrogen gas is passed through the tubes, where it reacts with the powder:

$$WO_3(s) + 3H_2(g) \rightarrow W(s) + 3H_2O(g)$$

After passing through the furnace, any unreacted hydrogen gas is recycled. It is dried to remove the water vapour and mixed with fresh hydrogen.

Risks of using hydrogen

The use of hydrogen as the reducing agent has some risks. Hydrogen is a very flammable gas. It forms an explosive mixture with air when its concentration is just 4%. Hydrogen is odourless and colourless. It burns with an invisible flame, so a leak of hydrogen is difficult to detect.

Tungsten ores

There are four main ores of tungsten:

- ferberite, $FeWO_4$
- huebnerite, $MnWO_4$
- wolframite, $(Fe,Mn)WO_4$ (the symbol for tungsten comes from the name of this ore)
- scheelite, $CaWO_4$ (named after Carl Wilhelm Scheele, a scientist who played a major part in the discovery of tungsten)

In each case, the ore is chemically treated to produce ammonium paratungstate $(NH_4)_{10}(W_{12}O_{41}).5H_2O$. This is decomposed to WO_3 by heating it to 600°C.

Tungsten carbide

Carbides are often undesirable impurities formed when metals are reduced using carbon. But tungsten carbide is deliberately manufactured because it is extremely hard. It is made by reacting tungsten and carbon together at 1500°C. Titanium carbide may be added to it to make it even harder.

Tungsten carbide is used for a wide range of tough, wear-resistant tools. These include drills, grinders, and mining tools. Around 65% of tungsten is used as tungsten carbide, rather than as the metal itself.

Tungsten carbide wheel cutter tools are used to cut glass and ceramic tiles.

Check your understanding

1. Why is carbon not used as the reducing agent in the extraction of tungsten?
2. Describe how tungsten is manufactured from tungsten(VI) oxide. Include the conditions, risks, and relevant equations in your answer.

11.06 Recycling metals

OBJECTIVES

already from AS level, you understand

- oxidation and reduction reactions
- that extraction of metals involves reduction
- how several metals are manufactured

and after this spread you should

- understand the environmental and economic advantages and disadvantages of recycling scrap metals compared with the extraction of metals
- understand the environmental advantages of using scrap iron to extract copper from aqueous solutions compared with the high temperature carbon reduction of copper oxide
- know that the usual source of such aqueous solutions is low-grade ore

In 2005 over 400 million tonnes of metal was recycled in the world, 13 million tonnes in the UK alone. Metal ores are finite resources. There is only a limited amount of them in the Earth's crust. At the present rate of consumption, aluminium would last for about a thousand years. But the situation is much worse for many other metals. Copper would last for 60 years, lead and platinum for 40 years. Metals must be recycled, otherwise many of them will run out during your lifetime.

Mining and processing ores

Recycling reduces the need to mine metal ores. Modern opencast mines are huge, noisy, and dusty places. There may be a danger of acid mine drainage, particularly after a mine has been abandoned. This happens when metal sulfides are exposed to water and begin to oxidize. The sulfuric acid produced dissolves toxic metals, and water carries these out of the mine.

Recycling also reduces the need to process metal ores. These need to be crushed and transported, and energy and chemicals are used to process them. Sulfide ores are roasted to convert them into oxides. There may be a lot of waste material left over. For example, bauxite is processed using sodium hydroxide solution. An alkaline red mud is left, which is difficult to dispose of and is usually stored on-site.

Up to two tonnes of red mud may be produced for every tonne of aluminium oxide extracted from bauxite.

Reducing the need to mine and process ores has economic and environmental advantages. But mining and ore processing may be a significant source of employment, often in very poor parts of the world. There may be few other employment opportunities should a mine or processing plant close down.

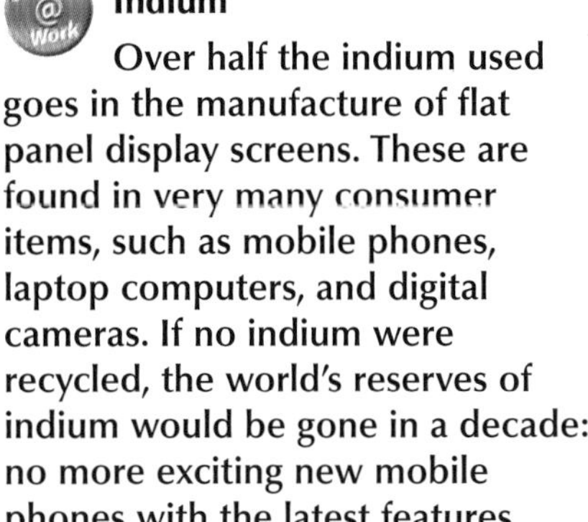

Indium

Over half the indium used goes in the manufacture of flat panel display screens. These are found in very many consumer items, such as mobile phones, laptop computers, and digital cameras. If no indium were recycled, the world's reserves of indium would be gone in a decade: no more exciting new mobile phones with the latest features after that.

Indium is used in flat panel displays for mobile phones and other electronic devices.

Extracting metals by reduction

Large amounts of energy are needed to extract metals from their oxides. Saving energy means that less fossil fuel is used. This may be a direct saving, as in iron and steel making, which uses coke from coal. It may be an indirect saving, as in aluminium manufacture, which uses large amounts of electricity. These savings result in less carbon dioxide being released into the atmosphere. This helps to reduce global warming.

metal	% energy saved
aluminium	95
copper	85
steel	74

Significant amounts of energy are saved by recycling metals.

There are disadvantages to recycling, including:

- resistance by households and businesses to change from simply throwing used items into the refuse
- the costs involved in collecting, transporting, and sorting scrap
- the need to remove other material from the scrap metal, such as the tin coating inside steel cans
- the market for scrap may vary with changes in demand

The UK recycles about 39% of its aluminium, 32% of its copper, and 42% of its iron and steel. But the proportions are increasing.

Using scrap iron to extract copper

Most high-grade copper ores have been used up. The remaining ores are low grade. When they are processed, the liquid wastes may still contain some copper. Scrap iron can be used to recover copper from these dilute copper solutions. Iron is a stronger reducing agent than copper. It can reduce copper(II) ions to copper:

$$Cu^{2+}(aq) + Fe(s) \rightarrow Cu(s) + Fe^{2+}(aq)$$

Notice that aqueous iron(II) ions are produced, so the iron appears to dissolve in the process. But copper is more valuable than iron. So a relatively cheap and abundant metal is used to recover a relatively expensive and increasingly scarce metal. Less energy and fewer resources are needed to extract copper this way than by extracting it from copper oxide using carbon at high temperatures.

Check your understanding

1. Outline some of the economic and environmental advantages and disadvantages to recycling metals.
2. Tin is removed from steel cans by treating them with an oxidizing agent in hot sodium hydroxide solution. The tin is then recovered by electrolysis. Suggest some advantages and disadvantages of this process.
3. Aluminium is a stronger reducing agent than iron. Suggest why it is not used instead of scrap iron to recover copper from copper solutions.

12.01 Making haloalkanes

OBJECTIVES

already from AS level, you

- know that alkanes are saturated hydrocarbons
- can apply IUPAC rules for naming haloalkanes with up to six carbon atoms

and after this spread you should

- understand the mechanism for the reaction between methane and chlorine

Silicones are used in bathroom sealants and artificial body parts.

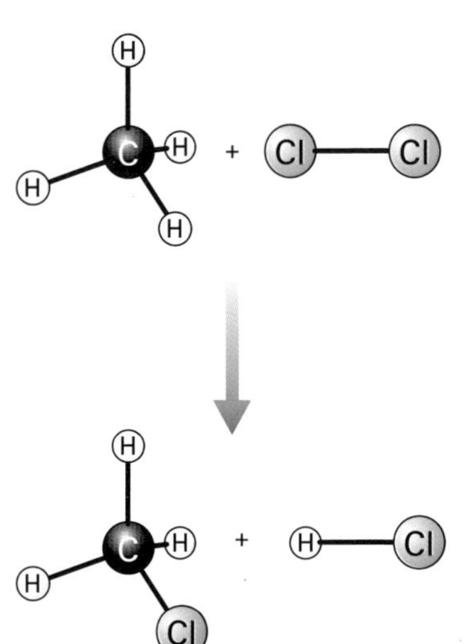

Methane and chlorine produce chloromethane in a free radical substitution reaction.

Chloromethane, CH_3Cl, is a haloalkane. It is a colourless gas that is used almost entirely to make other industrial chemicals. These include quaternary ammonium salts, which in turn are used to make fabric softeners and hair conditioners. Around 2% of chloromethane is used as a solvent in the manufacture of artificial rubber, but about 75% of it is used to make silicones. Chloromethane and other haloalkanes can be manufactured by reacting alkanes with halogens.

The only bonds in alkanes are C–C bonds and C–H bonds. These have relatively high mean bond enthalpies. Alkanes do not react easily with many substances because these strong bonds must be broken. Reactions involving alkanes have high activation energies.

Bond breaking

A covalent bond is a shared pair of electrons. When the covalent bond between two atoms breaks, one electron can be transferred to each atom. This is called homolytic fission. In general:

$$X{:}Y \rightarrow X\bullet + Y\bullet$$

The two products X• and Y• are called **free radicals**. The dots show that each one has an unpaired electron. The unpaired electron is available for bonding, so free radicals are very reactive. Alkanes can react with halogens using free radicals.

A substitution reaction

Methane and chlorine do not react together in the dark. But in ultraviolet light they react to produce chloromethane and hydrogen chloride:

$$CH_4(g) + Cl_2(g) \xrightarrow{\text{UV light}} CH_3Cl(g) + HCl(g)$$

During the reaction between methane and chlorine, a hydrogen atom in the methane molecule is replaced by a chlorine atom from the chlorine molecule. This sort of change is called a **substitution** reaction. The chlorination of methane involves free radicals, so it is called a *free radical substitution reaction*. It is possible to find out how it works, step by step. This is called a **reaction mechanism**.

The reaction mechanism

There are three main steps in the free radical substitution reaction between methane and chlorine. These are called **initiation**, **propagation**, and **termination**.

Initiation

Existing bonds must be broken for the reaction to begin. The Cl–Cl bonds in chlorine are weaker than the C–H bonds in methane. So the energy in ultraviolet light is more likely to break them:

$Cl_2 \rightarrow Cl\bullet + Cl\bullet$

This is an example of homolytic fission. The chlorine radicals are very reactive and can react with methane molecules.

bond	mean bond enthalpy ($kJ\,mol^{-1}$)
Cl–Cl	243
C–H	435

Mean bond enthalpies in chlorine and methane.

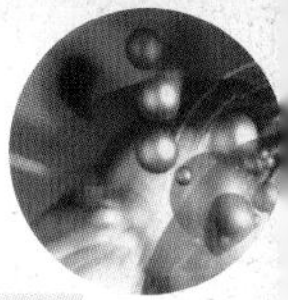

Propagation

When a free radical reacts with a molecule, a new free radical and a new molecule are formed. This is what can happen when a chlorine radical from the initiation stage reacts with a molecule of methane:

$$CH_4 + Cl\bullet \rightarrow \bullet CH_3 + HCl$$

The methyl radical $\bullet CH_3$ can then react with another chlorine molecule:

$$\bullet CH_3 + Cl_2 \rightarrow CH_3Cl + Cl\bullet$$

Notice that the products are chloromethane and another chlorine radical. This is why it is called a propagation step: it keeps using and making chlorine radicals until all the chlorine is used up. Also notice that the dot in the methyl radical $\bullet CH_3$ is shown next to the atom that carries the unpaired electron.

Termination

A termination reaction happens when two free radicals react together to form a molecule. For example, two chlorine radicals react to form a chlorine molecule:

$$Cl\bullet + Cl\bullet \rightarrow Cl_2$$

But this would probably not gain you a mark in an examination because it is just the reverse of the initiation reaction. You need to choose at least one other free radical, as in these two examples:

$$\bullet CH_3 + Cl\bullet \rightarrow CH_3Cl$$

$$\bullet CH_3 + \bullet CH_3 \rightarrow C_2H_6$$

One of these termination reactions produces chloromethane, the target product. The other produces ethane, C_2H_6, which is not an intended product.

Further substitution

Further substitution reactions can happen to chloromethane, producing dichloromethane, trichloromethane, and finally tetrachloromethane. The chance of these forming can be decreased if the methane is in excess (a lot of it compared to chlorine). Removing by-products like these increases costs and reduces the efficiency of the process. There are other ways to make chloromethane that are better for sustainable development.

```
     H              H              H              Cl
     |              |              |              |
 H — C — H     H — C — Cl    Cl — C — Cl    Cl — C — Cl
     |              |              |              |
     Cl             Cl             Cl             Cl
```

chloromethane	dichloromethane	trichloromethane	tetrachloromethane
CH_3Cl	CH_2Cl_2	$CHCl_3$	CCl_4

Methane and chlorine react to form chloromethane. Further substitution can take place, forming other products.

Check your understanding

1. Explain why the following reaction involves homolytic fission:

$$Br_2 \rightarrow Br\bullet + Br\bullet$$

2. Bromine, Br_2, reacts with ethane to form bromoethane, CH_3CH_2Br. The mechanism is similar to the one for the chlorination of methane. Write equations for the initiation, propagation, and termination steps of the reaction mechanism.

12.02 Damaging the ozone layer

OBJECTIVES

already from AS Level you understand

- the mechanism for the reaction between methane and chlorine

and after this spread you should

- know that chloroalkanes and chlorofluoroalkanes can be used as solvents
- understand that ozone, formed naturally in the upper atmosphere, is beneficial
- know that chlorine atoms are formed in the upper atmosphere when energy from ultraviolet radiation causes C–Cl bonds in chlorofluorocarbons to break
- be able to use equations to explain why chlorine atoms catalyse the decomposition of ozone and contribute to the formation of a hole in the ozone layer

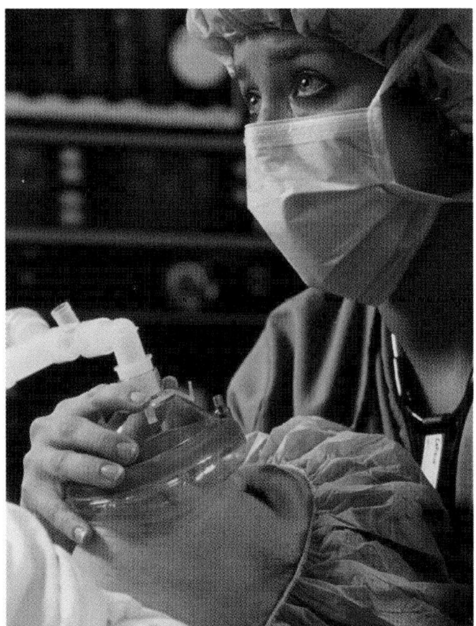

Sevoflurane, $C_4H_3F_7O$, and desflurane, $C_3H_2F_6O$, are two haloalkanes used as general anaesthetics.

Haloalkanes

Haloalkanes are alkanes in which one or more hydrogen atoms are substituted by a halogen atom. Chloromethane, CH_3Cl, is a chloroalkane, but it is not the only one. More than one hydrogen atom in a carbon chain may be substituted by chlorine. This means that a large range of chloroalkanes is possible. Simple chloroalkanes are good solvents. For example, trichloromethane, $CHCl_3$, and tetrachloromethane, CCl_4, were used in the past as solvents for dry cleaning clothes. They are toxic and tetrachloromethane is no longer used at all. Modern dry cleaning solvents include tetrachloroethene, C_2Cl_4, which is less toxic.

Haloalkanes containing more than one halogen are possible, too. BCF or bromochlorodifluoromethane, $CBrClF_2$, is good at extinguishing fires. It releases bromine radicals in the heat of a fire, and these inhibit combustion.

Haloalkanes that contain chlorine and fluorine are called **chlorofluorocarbons**. These are called **CFCs** for short. CFCs and chloroalkanes damage the Earth's ozone layer when they are released into the environment.

The ozone layer

Ozone, O_3, is an allotrope of oxygen. It is formed in the upper atmosphere by free radical reactions. Ultraviolet light from the Sun provides the energy needed to break the O=O bond in oxygen molecules:

$$O_2 \xrightarrow{\text{UV light}} \bullet O \bullet + \bullet O \bullet$$

The oxygen atoms formed are free radicals with two unpaired electrons. They can react with other oxygen molecules to form ozone:

$$\bullet O \bullet + O_2 \rightarrow O_3$$

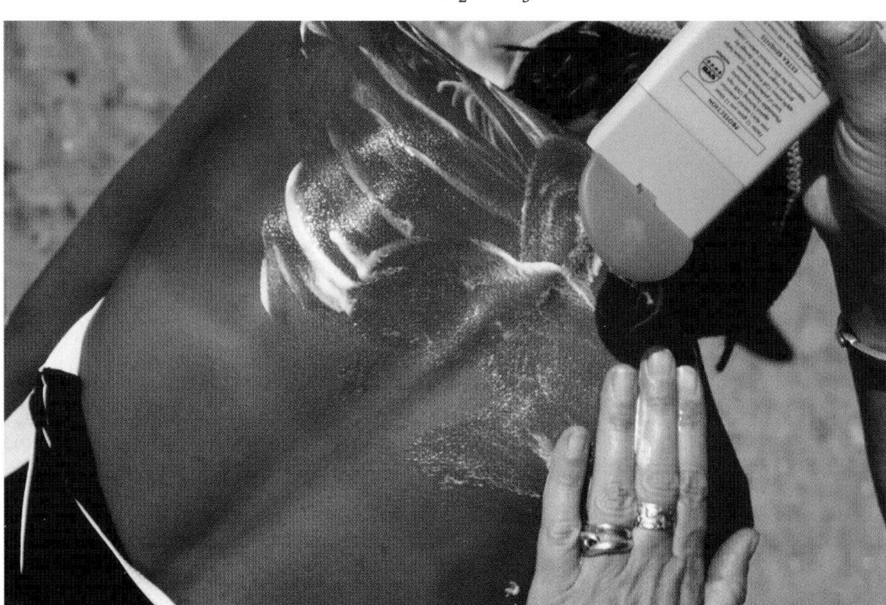

Overexposure to ultraviolet light from the sun can cause skin cancer. People with dark skin are less susceptible than those with fair skin, but they still need to apply sunscreen for protection.

The **ozone layer** is in the **stratosphere**, between 15 km and 40 km above the Earth's surface. But it is not really a layer at all. It is just the part of the atmosphere where ozone is found in the highest concentrations, between 2 parts per million and 8 parts per million. Ozone absorbs ultraviolet light and so prevents most of it reaching the ground. This is important because ultraviolet light harms living things. It damages DNA and can lead to skin cancer.

When an ozone molecule absorbs ultraviolet light, it dissociates to form an oxygen molecule and an oxygen atom:

$$O_3 \rightarrow \bullet O \bullet + O_2$$

Both reactions, the one that creates ozone and the one that decomposes it, naturally continue all the time. In the natural situation, the rate of ozone production is the same as the rate of ozone decomposition. So overall, the concentration of ozone in the stratosphere stays the same. Unfortunately, the production and release of CFCs has increased the rate of ozone decomposition. The concentration of ozone in the ozone layer has decreased as result.

Termination

Oxygen atoms do not have to produce ozone. They may react with each other or with ozone itself to produce O_2:

$$\bullet O \bullet + \bullet O \bullet \rightarrow O_2$$

$$\bullet O \bullet + O_3 \rightarrow 2O_2$$

Notice that the second reaction decomposes ozone.

CFCs and the ozone layer

Ultraviolet light provides the energy to break C–Cl bonds in CFC molecules when they reach the stratosphere. For example, in trichlorofluoromethane:

$$CCl_3F \xrightarrow{\text{UV light}} \bullet CCl_2F + \bullet Cl$$

The chlorine radicals react with ozone and decompose it:

$$\bullet Cl + O_3 \rightarrow ClO\bullet + O_2$$

The chlorine oxide radical itself can react with ozone and decompose it:

$$ClO\bullet + O_3 \rightarrow 2O_2 + \bullet Cl$$

Notice that a chlorine radical is released in this reaction. This can decompose more ozone. Reactions like these have decomposed sufficient ozone to cause a 'hole' in the ozone layer. This is a region in the stratosphere where the concentration of ozone is much less than it should be.

Check your understanding

1. Name two haloalkanes that have been used as solvents.
2. With the help of equations, explain how ozone is naturally created and decomposed.
3. a Chloromethane decomposes to form a methyl radical and a chlorine radical. Write an equation for this reaction, and explain why it is an initiation reaction for a free radical reaction.
   b With the help of two equations, show how chlorine and chlorine oxide radicals can decompose ozone.
   c Explain why the equations in part **b** show a propagation step in a free radical reaction.

12.03 Repairing the ozone layer

OBJECTIVES

already from AS Level, you

- know that chloroalkanes and chlorofluoroalkanes can be used as solvents
- understand that ozone, formed naturally in the upper atmosphere, is beneficial
- know that chlorine atoms are formed in the upper atmosphere when energy from ultraviolet radiation causes C–Cl bonds in chlorofluorocarbons to break
- can use equations to explain why chlorine atoms catalyse the decomposition of ozone and contribute to the formation of a hole in the ozone layer

after this spread you should

- appreciate that legislation to ban the use of CFCs was supported by chemists and that they have now developed alternative chlorine-free compounds

Lovelock's discovery

James Lovelock invented a very sensitive device to measure organic compounds in the atmosphere. He discovered in 1973 that trichlorofluoromethane, CCl_3F, an artificial haloalkane, was present wherever he took measurements. Not only that, the total amount in the atmosphere was similar to the total amount ever made. CFCs do not disappear quickly. They last for decades or more.

The scientific community was divided at the time. There was criticism of the two scientists' conclusions but also great concern. As a result, there was some restriction of CFCs during the 1970s.

Paul Crutzen, Mario Molina, and Sherwood Rowland won the 1995 Nobel Prize in Chemistry for 'their work in atmospheric chemistry, particularly concerning the formation and decomposition of ozone'. The work of these scientists led to the consensus of scientific and political opinion about the ozone layer and the need to protect it.

Crutzen showed in 1970 that nitrogen monoxide, NO, could catalyse the decomposition of ozone:

$$NO + O_3 \rightarrow NO_2 + O_2$$

$$NO_2 + O \rightarrow NO + O_2$$

The following year Crutzen and other scientists pointed out that supersonic airliners flew so high that they would release nitrogen monoxide and nitrogen dioxide straight into the ozone layer. This was a possible threat to the ozone layer.

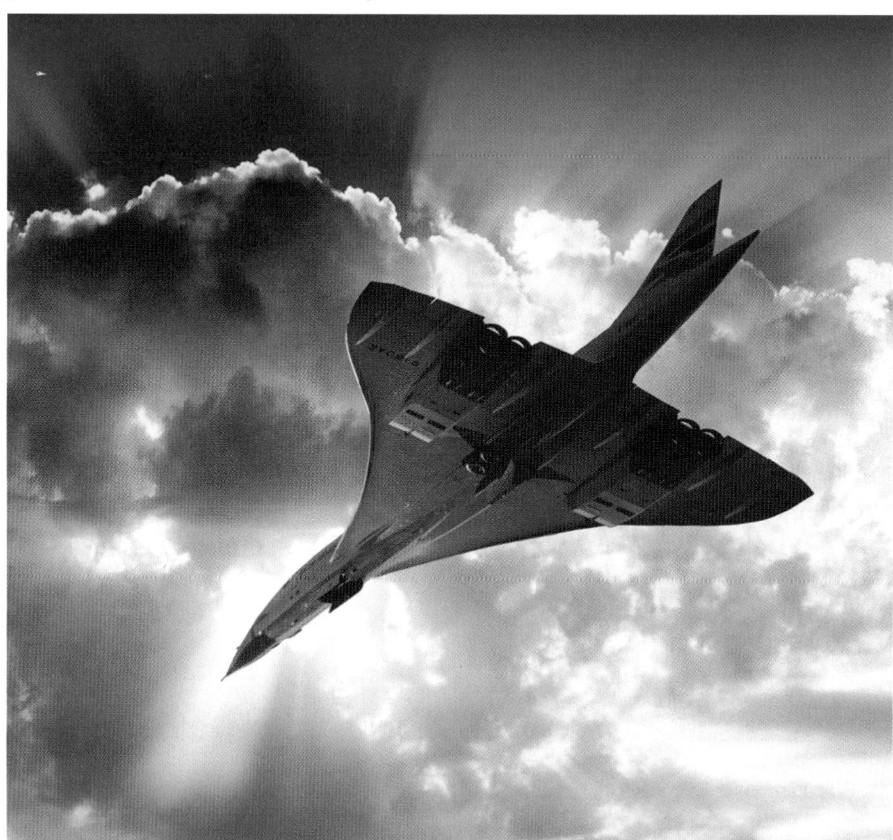

Concorde was a supersonic airliner jointly designed and built by Britain and France. It first flew in 1969 and made scheduled flights between 1976 and 2003.

In 1974 Molina and Rowland published a scientific paper on the threat to the ozone layer from CFCs. At the time CFCs were widely used in aerosol spray cans, refrigerators, and plastic foam packaging. The two chemists realized that CFCs would eventually reach the ozone layer after release because they are unreactive. Once there, ultraviolet light would decompose them to form chlorine radicals. They calculated that each chlorine radical could catalyse the decomposition of 100 000 ozone molecules. With about one million tonnes of CFCs being produced each year, they predicted a large reduction in the ozone layer within decades.

The hole over the Antarctic

The conditions over the Antarctic naturally lead to seasonal changes in the ozone layer there. But in 1985 it decreased far more than expected. Research eventually ruled out natural causes and blamed CFCs. Other ozone holes have appeared since, not just over the Antarctic.

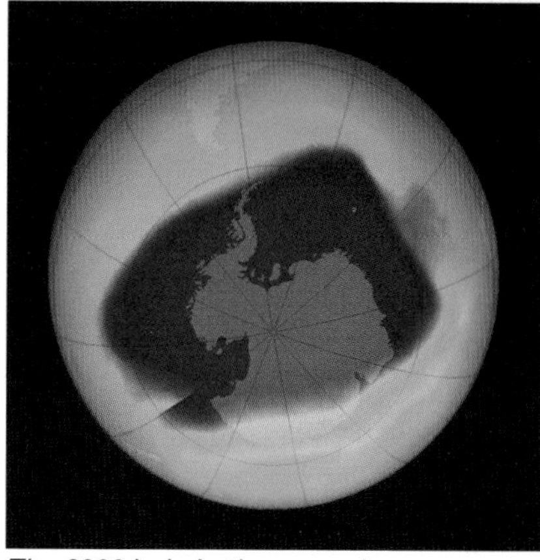

The 2006 hole in the ozone layer over the Antarctic, detected using a satellite. Purple and blue represent the lowest concentrations of ozone, green and yellow represent the highest.

Treaties and replacements

The work of scientists has led to a good understanding of how CFCs decompose ozone. This has greatly informed the decisions of governments around the world. Early action was taken in 1978 when four countries banned the non-essential use of CFCs in aerosol cans. At a meeting in Montreal in 1987, twenty-four countries signed the *Montreal Protocol on Substances that Deplete the Ozone Layer*. This introduced strict limits on the production and use of CFCs. It has been revised and extended several times since then, with almost all countries signing the treaty.

Chemists have now developed alternative compounds to replace CFCs. Hydrochlorofluorocarbons, HCFCs, contain less chlorine than CFCs. They have less potential to decompose ozone. But chemists have also developed chlorine-free compounds. These are called hydrofluorocarbons, or HFCs. For example, tetrafluoroethane, CF_3CH_2F, is used as a solvent and in refrigerators, and pentafluoroethane, CF_3CHF_2, is used in fire suppression systems. HFCs have no known effects on the ozone layer.

Old refrigerators may contain CFCs in their insulating foam and coolant. The CFCs must be removed and destroyed. If the refrigerator was just crushed, the CFCs would escape into the atmosphere.

Check your understanding

1. a Outline the work of Crutzen, Molina, and Rowland.
   b Suggest how Lovelock's discovery in 1973 might have influenced the work of Molina and Rowland.
2. Concorde did not achieve its predicted sales. Tickets to fly on it were very expensive and it was noisy. To what extent do you think that the work of Crutzen and others made Concorde less successful than hoped?
3. CFCs, HCFCs, and HFCs are greenhouse gases. The *Kyoto Protocol to the UN Framework Convention on Climate Change* seeks to reduce the overall emissions of six greenhouse gases, including HFCs, between 2008 and 2012. Suggest some implications this decision has for chemists.

12.04 Alcohols from haloalkanes

OBJECTIVES

already from AS level, you

- know the definition of electronegativity
- understand that the electron distribution in a covalent bond may not be symmetrical

and after this spread you should understand

- that haloalkanes contain polar bonds
- that haloalkanes are susceptible to nucleophilic attack by OH^- ions
- the mechanism of nucleophilic substitution in primary haloalkanes

Primary haloalkanes

Haloalkanes such as bromoethane and 1-chlorobutane are **primary haloalkanes.** The carbon atom that their halogen atom is attached to is directly bonded to just one other carbon atom.

Haloalkanes contain a carbon–halogen bond. Apart from astatine at the bottom of group 7, all the halogens are more electronegative than carbon. This means that the carbon–halogen bond is a polar bond. The carbon atom has a partial positive charge δ^+ and the halogen atom has a partial negative charge δ^-. The carbon atom is **electron-deficient** because the halogen atom withdraws negative charge from it. The presence of the electron-deficient carbon atom makes haloalkanes liable to attack by nucleophiles.

Nucleophiles

In chemistry, **species** is a general term for an atom, ion, or molecule. A **nucleophile** is a species with a lone pair of electrons that is available to form a co-ordinate bond. Nucleophiles are attracted to regions of positive charge. Their name means **nucleus-loving**, since a nucleus is positively charged.

Nucleophiles can be

- negatively charged ions, such as OH^- and CN^-
- molecules with a lone pair of electrons, such as H_2O and NH_3

Nucleophiles like these can attack the electron-deficient carbon atom in a haloalkane molecule. They can bring about a reaction in which they replace the halogen atom. Reactions like these are called **nucleophilic substitution** reactions.

Nucleophilic substitution reactions

Bromoethane reacts with warm dilute sodium hydroxide solution or dilute potassium hydroxide solution to form ethanol:

$$CH_3CH_2Br(aq) + OH^-(aq) \rightarrow CH_3CH_2OH(aq) + Br^-(aq)$$

This is a substitution reaction. The bromine atom has been replaced by OH. Similar reactions happen with other haloalkanes. For example, 1-chlorobutane reacts with warm dilute sodium hydroxide solution to form butan-1-ol:

$$CH_3(CH_2)_3Cl(aq) + OH^-(aq) \rightarrow CH_3(CH_2)_3OH(aq) + Cl^-(aq)$$

These reactions are nucleophilic substitution reactions because they involve a nucleophile, in this case the hydroxide ion OH^-.

Reaction mechanism

The reaction mechanism is shown with the help of curly arrows. A curly arrow indicates the movement of a pair of electrons from its tail to its head. A curly arrow can show the movement of

- a lone pair of electrons
- the pair of electrons in a covalent bond

The reaction mechanism for nucleophilic substitution in 1-bromopropane by hydroxide ions is explained on the opposite page. It is shown broken down into steps for simplicity, but the events actually happen all at once.

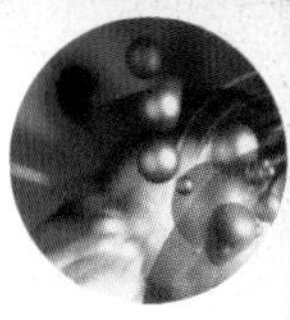

```
     H   H   H
     |   |   |δ+      δ−
 H — C — C — C — Br
     |   |   |
     H   H   H

              :OH⁻
```

1. In the first step, the nucleophile is attracted to the electron-deficient carbon atom. Notice that you should draw a lone pair of electrons in the nucleophile :OH⁻. The curly arrow is drawn from one side of the lone pair of electrons and points towards the electron-deficient carbon atom.

```
     H   H   H
     |   |   |δ+      δ−
 H — C — C — C — Br
     |   |   |
     H   H   H

              :OH⁻
```

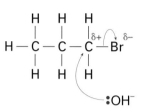

2. In the next step, the pair of electrons in the C–Br bond moves towards the bromine atom with its partial negative charge. Notice that the tail of the curly arrow starts from the bond and is not in mid-air above it.

```
     H   H   H
     |   |   |
 H — C — C — C — OH   +   :Br⁻
     |   |   |
     H   H   H
```

3. In the final step, the hydroxide ion has donated one its lone pairs of electrons to form a new bond with the carbon atom. The C–Br bond has broken to release a bromide ion.

The complete mechanism is drawn like this:

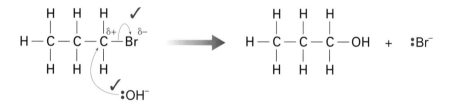

```
     H   H   H   ✓                            H   H   H
     |   |   |δ+      δ−                      |   |   |
 H — C — C — C — Br         ———>          H — C — C — C — OH   +   :Br⁻
     |   |   |                                |   |   |
     H   H   H                                H   H   H
           ✓
              :OH⁻
```

4. The complete reaction mechanism, with ticks to show the features an examiner is likely to look for in an examination.

The reaction mechanism for other haloalkanes will be just the same: you only need to change the halogen or carbon chain in your diagram.

Check your understanding

1. Define the term 'nucleophile' and give two examples of nucleophiles.
2. What does the curly arrow in a reaction mechanism show?
3. A substitution reaction happens when chloroethane reacts with warm sodium hydroxide solution.
   - **a** Name the organic compound formed in the reaction.
   - **b** Outline the mechanism for the reaction.

12.05 Nitriles and amines from haloalkanes

OBJECTIVES

already from AS level, you understand

- that haloalkanes are susceptible to nucleophilic attack by OH^- ions
- the mechanism of nucleophilic substitution in primary haloalkanes

and after this spread you should

- understand that the carbon–halogen bond enthalpy influences the rate of hydrolysis
- understand that haloalkanes are susceptible to nucleophilic attack by CN^- ions and ammonia
- appreciate the usefulness of these reactions in organic synthesis

Haloalkanes and reactivity

The rate of nucleophilic substitution depends upon the halogen in the haloalkane. It would be tempting to think that fluoroalkanes would be the most reactive, because fluorine is the most reactive halogen. But this is not the case. The reactivity of the haloalkanes depends upon the strength of the carbon–halogen bond. The lower its bond enthalpy, the weaker it is and the more easily it is broken.

bond	mean bond enthalpy (kJ mol^{-1})
C–F	484
C–Cl	338
C–Br	276
C–I	238

The carbon–fluorine bond is the strongest carbon–halogen bond.

The carbon–halogen bond is weakest in iodoalkanes, so these are the most reactive. The reactivity decreases going to bromoalkanes and then to chloroalkanes. The carbon–fluorine bond is so strong that the fluoroalkanes are not readily attacked by nucleophiles. This trend can be seen in the reaction of haloalkanes with water.

Hydrolysis of haloalkanes

The oxygen atom in the water molecule has two lone pairs of electrons. So water can act as a nucleophile. It is a weaker nucleophile than the hydroxide ion but it can still produce the same products in reactions with haloalkanes. The reaction is called a **hydrolysis** reaction because water is used to break down the haloalkane. The rate of hydrolysis can be followed using silver nitrate solution.

As the water and haloalkane react, halide ions are formed. These react with the silver ions to form precipitates. When the hydrolysis of chloroethane, bromoethane, and iodoethane is followed using silver nitrate solution, silver iodide forms faster than silver bromide. Silver chloride is the slowest to form.

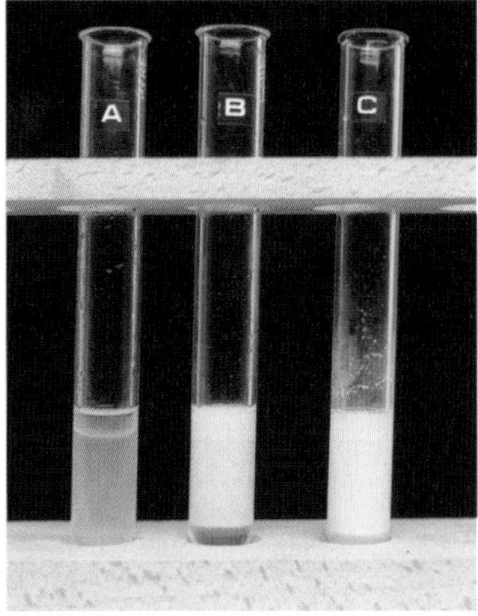

A contains a faint white precipitate of AgCl.
B contains a significant cream precipitate of AgBr.
C contains a distinct yellow precipitate of AgI.

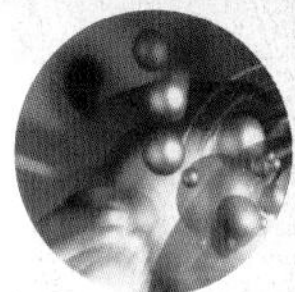

Reaction with cyanide ions

The cyanide ion, CN^-, acts as a nucleophile. It reacts with haloalkanes to produce nitriles. The cyanide ion is provided by sodium cyanide, NaCN, or potassium cyanide, KCN. These need to be dissolved in ethanol and heated with the haloalkane for the reaction to work.
1-Bromopropane reacts with hot ethanolic sodium hydroxide solution to form butanenitrile:

$$CH_3CH_2CH_2Br + CN^- \rightarrow CH_3CH_2CH_2CN + Br^-$$

This is a nucleophilic substitution reaction. The reaction mechanism is the same as the one where OH^- is the nucleophile.

The nitrile formed in the reaction has one more carbon atom than the original haloalkane. This is very useful when making new substances.

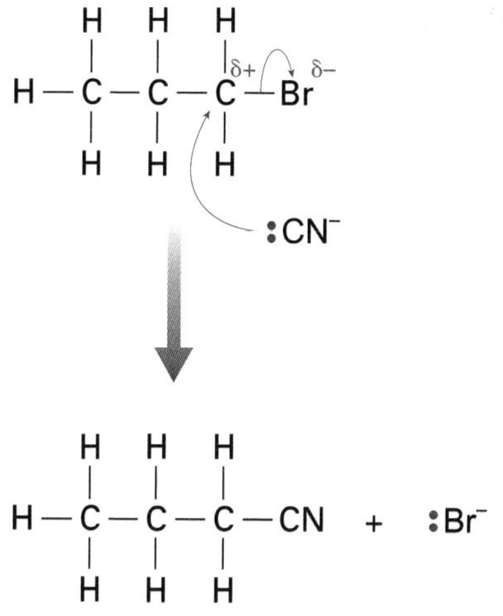

The reaction mechanism for nucleophilic substitution by cyanide ions.

Naming nitriles

The nitrile group can be shown as $-C{\equiv}N$ or just $-CN$. Nitriles are named using the suffix nitrile. The carbon atom in the nitrile group is included when counting the carbon atoms in the main chain. So, CH_3CN is ethanenitrile, not methanenitrile.

Reaction with excess ammonia

Ammonia, NH_3, acts as a nucleophile as its nitrogen atom has a lone pair of electrons. Ammonia reacts with haloalkanes to produce amines. These have an amine group NH_2. For example 1-bromopropane reacts with ammonia to form 1-aminopropane:

$$CH_3CH_2CH_2Br + 2NH_3 \rightarrow CH_3CH_2CH_2NH_2 + NH_4Br$$

Two ammonia molecules are needed to react with one haloalkane molecule. One acts as a nucleophile and the other acts as a base.

Step 1 Ammonia acting as a nucleophile:

$$CH_3CH_2CH_2Br + NH_3 \rightarrow CH_3CH_2CH_2NH_3^+ + Br–$$

Step 2 Ammonia acting as a base:

$$CH_3CH_2CH_2NH_3^+ + NH_3 \rightarrow CH_3CH_2CH_2NH_2 + NH_4^+$$

Here the ammonia molecule accepts a hydrogen ion from the positively charged intermediate.

The ammonium ion from step 2 and the bromide ion from step 1 form ammonium bromide.

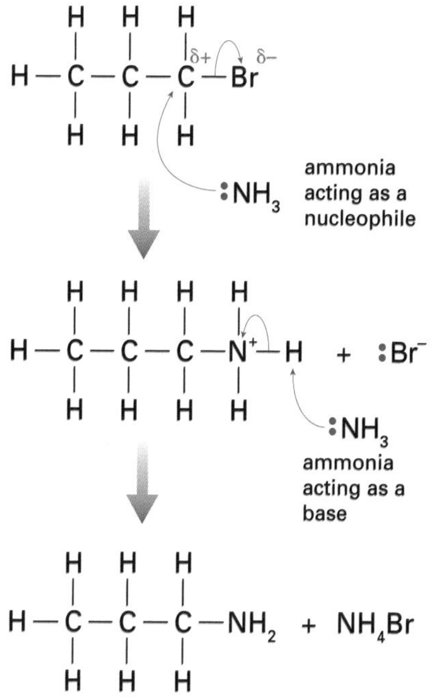

ammonia acting as a nucleophile

ammonia acting as a base

The reaction mechanism for nucleophilic substitution by excess ammonia.

Why excess ammonia?

One example of a primary amine is 1-aminopropane. Its nitrogen atom has a lone pair of electrons. Once it has formed in the reaction mixture, the primary amine can act as a nucleophile, just like ammonia. It can react with a haloalkane to form a secondary amine. This can react further to form a tertiary amine, and eventually a quaternary ammonium salt can be formed. The use of excess ammonia reduces these reactions and ensures that the main product is a primary amine.

Check your understanding

1. Explain why iodoalkanes are more reactive than chloroalkanes.
2. A substitution reaction happens when chloroethane reacts with hot ethanolic potassium cyanide solution.
   a Name the organic compound formed in the reaction.
   b Outline the mechanism for the reaction.
3. Outline the mechanism for the substitution reaction between chloroethane and excess ammonia.

12.06 Alkenes from haloalkanes

OBJECTIVES

already from AS level, you understand

- the mechanism for nucleophilic substitution in haloalkanes

and after this spread you should

- understand the role of the hydroxide ion as a nucleophile and as a base
- understand concurrent substitution and elimination in haloalkane reactions
- appreciate the usefulness of this reaction in organic synthesis

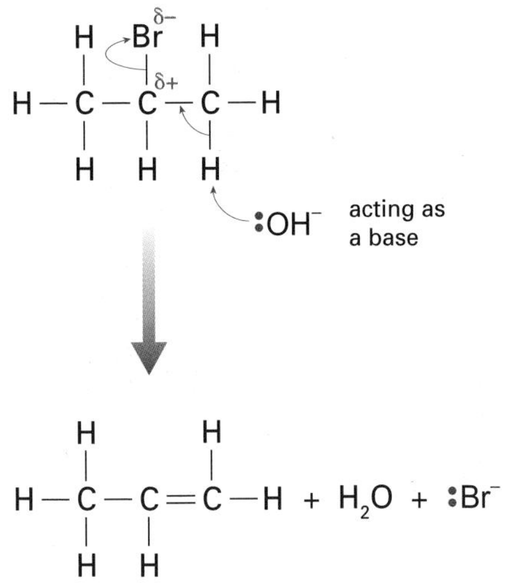

H δ− Br H
H—C—C δ+—C—H
H H H
:OH⁻ acting as a base

H H
H—C—C=C—H + H_2O + :Br⁻
H H

The reaction mechanism for elimination in 2-bromopropane to form propene.

When a haloalkane reacts with excess ammonia, the ammonia can act as a nucleophile and as a base. The hydroxide ion can also act as a nucleophile and as a base. A substitution reaction happens when the hydroxide ion acts as a nucleophile, producing an alcohol. But when the hydroxide ion acts as a base, an elimination reaction happens. It produces an alkene instead of an alcohol.

Elimination from haloalkanes

If 2-bromopropane is heated strongly with concentrated sodium or potassium hydroxide in ethanol, propene is formed:

$$CH_3CH_2CH_2Br + OH^- \rightarrow CH_3CH{=}CH_2 + H_2O + Br^-$$

The mechanism needs three curly arrows to describe it. They show that

- the hydroxide ion forms a co-ordinate bond with a hydrogen atom
- the bonding pair of electrons from the carbon–hydrogen bond forms a second covalent bond between two carbon atoms
- the carbon–bromine bond breaks, releasing a bromide ion

Note that all three pairs of electrons move simultaneously and there is no **intermediate** species. The hydrogen atom that is attacked is joined to a carbon atom next to the carbon atom with the carbon–halogen bond. The hydrogen atom is removed as a hydrogen ion, H^+, which is accepted by the hydroxide ion to form water. This is how the hydroxide ion acts as a base here.

Concurrent substitution and elimination

When haloalkanes react with hydroxide ions, both types of reaction can happen together. This is called *concurrent substitution and elimination*. The reaction conditions can be adjusted to favour one type of reaction over the other.

- Substitution is favoured by using warm, dilute aqueous sodium hydroxide or potassium hydroxide.
- Elimination is favoured by using hot, concentrated ethanolic sodium hydroxide or potassium hydroxide.

To some extent the two reactions are always in competition with each other because hydroxide ions can act both as a nucleophile and as a base.

type of reaction	reaction conditions			hydroxide ion acts as a:
	temperature	hydroxide	solvent	
substitution	low	dilute	water	nucleophile
elimination	high	concentrated	ethanol	base

A summary of the conditions needed to favour substitution or elimination reacts in haloalkanes.

The favoured type of reaction is also influenced by whether the haloalkane is a primary, secondary, or tertiary haloalkane. This is often shortened to 1°, 2°, and 3°. Look at the carbon atom to which the halogen atom is attached. This is directly attached to

- one other carbon atom in a primary haloalkane
- two other carbon atoms in a secondary haloalkane
- three other carbon atoms in a tertiary haloalkane

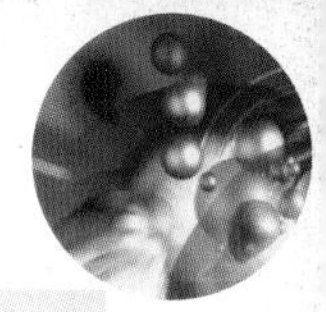

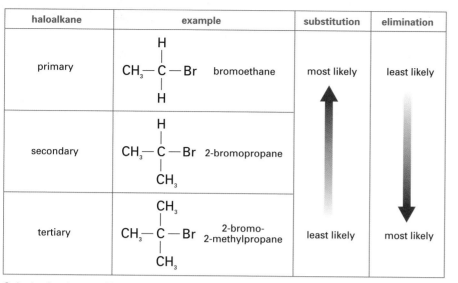

haloalkane	example		substitution	elimination
primary	CH_3-CH_2-Br	bromoethane	most likely	least likely
secondary	$CH_3-CH(CH_3)-Br$	2-bromopropane	↑	↓
tertiary	$CH_3-C(CH_3)_2-Br$	2-bromo-2-methylpropane	least likely	most likely

Substitution is most likely in primary haloalkanes and elimination is most likely in tertiary haloalkanes.

Two elimination products

Only one product of elimination is possible from 2-bromopropane. But two products of elimination are possible from longer, unsymmetrical secondary haloalkanes. For example, the bromine atom in 2-bromobutane is joined to the second carbon atom of four. As a result, elimination produces but-1-ene and but-2-ene. The diagram shows how this works.

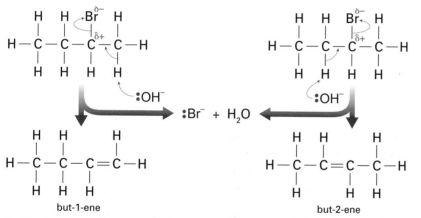

$CH_3CH_2CH^{\delta+}Br^{\delta-}CH_3$ + $:OH^-$ → $:Br^-$ + H_2O + but-1-ene ($CH_3CH_2CH=CH_2$) or but-2-ene ($CH_3CH=CHCH_3$)

but-1-ene

but-2-ene

Elimination in 2-bromobutane produces but-1-ene and but-2-ene.

Check your understanding

1. 2-Bromopentane reacts with hot ethanolic potassium hydroxide to produce pent-2-ene:

$$CH_3(CH_2)_2CHBrCH_3 + KOH \rightarrow CH_3CH_2CH{=}CHCH_3 + H_2O + KBr$$

   a What type of reaction is this?

   b Explain why the potassium ion, K^+, is a spectator ion in this reaction.

   c Outline a mechanism for this reaction.

   d Give the structural formula and name of another alkene that could be formed by the reaction mixture.

   e Suggest the conditions needed to favour the production of pentan-2-ol in the reaction instead of pentene.

Refluxing

Organic reactions are often slow, so the reaction mixture may need to be heated for several hours. During this time, the volatile substances may well boil away. A method called refluxing is used to stop this happening. A condenser is fitted into the neck of the reaction vessel. As vapours rise from the boiling reaction mixture, they are cooled and condensed, so they fall back into the reaction vessel.

Refluxing is often used for haloalkane substitution reactions with cyanide ions, and haloalkane elimination reactions with hydroxide ions.

13.01 Alkenes and *E-Z* isomerism

OBJECTIVES

already from AS Level, you can

- apply IUPAC rules for naming alkenes with up to six carbon atoms

and after this spread you should

- know that alkenes are unsaturated hydrocarbons with a double covalent bond
- know that the arrangement >C=C< is planar
- understand that *E-Z* isomers exist owing to restricted rotation about the C=C bond
- know that the alkenes can exhibit *E-Z* stereoisomerism
- be able to draw the structures of *E* and *Z* isomers

The alkenes form a homologous series of hydrocarbons with the general formula C_nH_{2n}. They are **unsaturated** because they have carbon–carbon double bonds. Like alkanes, alkenes burn in air to form carbon dioxide and water vapour. But the presence of the double bond means that alkenes are more reactive than the alkanes. They are generally too useful just to be burnt as fuels. Instead, they are the feedstock for making haloalkanes, alcohols, and polymers.

The double bond

Ethene is a planar molecule. Each carbon atom has four bonding pairs of electrons. You might expect that this would lead to a tetrahedral arrangement of bonds around each carbon atom. Instead the arrangement is trigonal planar because of the double bond. When determining the shape of ethene, the four bonding pairs of electrons behave like three pairs of electrons.

The >C=C< arrangement is planar in all alkenes. For example, the methyl group in propene has a tetrahedral shape but the region around the double bond is still planar. The movement around the double bond is restricted: it resists being rotated. This leads to the existence of **isomers**.

The displayed formula of ethene C_2H_4, the simplest alkene.

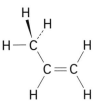

The shape of the propene molecule. Remember that solid lines represent bonds in the plane of the paper, wedges represent bonds that come out of the paper, and dotted lines represent bonds that go into the paper

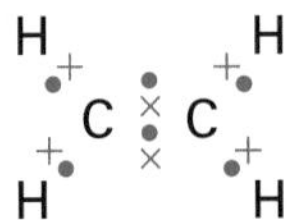

A dot and cross diagram for the bonding in ethene.

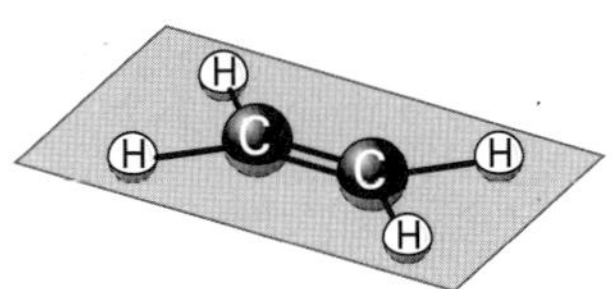

Ethene is a planar molecule.

Diastereoisomers

Stereoisomers have the same molecular and structural formulae, but a different arrangement of their atoms in space. Their three-dimensional shape is different. Some stereoisomers are mirror images of each other. The ones that are not mirror images are called **diastereoisomers**. An alkene will have diastereoisomers if both carbon atoms involved in the double bond have different groups attached to them.

Butene exists as but-1-ene and but-2-ene. These are position isomers because the double bond is in different positions. Does butene have diastereoisomers? But-1-ene does not. Only one of the two carbon atoms involved in its double bond has two different groups attached to it. But-2-ene does have diastereoisomers. Both carbon atoms involved in its double bond have two different groups attached to them.

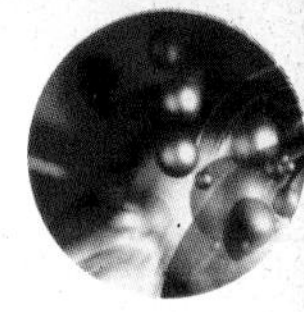

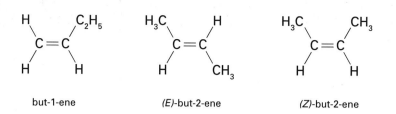

but-1-ene (E)-but-2-ene (Z)-but-2-ene

But-1-ene does not have stereoisomers, but but-2-ene does have diastereoisomers. The E and Z in brackets show which diastereoisomer is which.

The *E-Z* naming system

Diastereoisomers are identified using the *E-Z* naming system. *E* comes from the German word *entgegen*, which means *opposite*. *Z* comes from the German word *zusammen*, which means *together*. To decide whether an alkene is the *E* isomer or the *Z* isomer, you need to work out the priority of the groups attached to each carbon atom involved in the double bond. This is simple for the alkenes you should meet at A Level: priority increases in the order $H < CH_3 < C_2H_5$. Then work out whether the highest priority group on each carbon atom is above or below the double bond.

- If one of the highest priority groups is above the double bond and the other is below, it is an *E* isomer.
- If both of the highest priority groups are above the double bond, or both are below, it is the *Z* isomer.

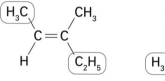

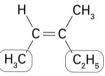

(E)-3-methylpent-2-ene (Z)-3-methylpent-2-ene

*The two E-Z isomers of 3-methylpent-2-ene. The highest priority groups on each side have been circled for clarity. The ones in the E isomer are on '**e**ither half' and the ones on the Z isomer are on the '**z**ame half' of the double bond.*

Cis-trans isomerism

You will come across the terms *cis*, *trans*, and *geometrical isomerism* when you read about chemistry. The *cis-trans* naming system is similar to the *E-Z* naming system. It works well when only two or three different groups involved, but it struggles when there are four different groups. Isomers in which the same group is above and below the double bond are *trans*. A *trans* isomer would also be an *E* isomer. Isomers in which the same group is above, or below, on both sides of the double bond are *cis*. A *cis* isomer would also be a *Z* isomer. *Geometrical isomerism* is another name for *cis-trans* isomerism. IUPAC discourages the use of this term.

Check your understanding

1. Explain why hex-3-ene has *E-Z* isomers but hex-1-ene does not.
2. Show the displayed formulae for the *E-Z* isomers of but-2-ene.
3. Is this an *E* isomer or a *Z* isomer? Explain how you know.

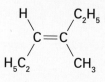

13.02 Haloalkanes from alkenes

OBJECTIVES

already from AS Level, you understand

- the mechanism for elimination in haloalkanes to form alkenes

and after this spread you should

- understand that the double bond in an alkene is a centre of high electron density
- understand the mechanism of electrophilic addition of alkenes with hydrogen bromide and with bromine
- know that bromine can be used to test for unsaturation

The double bond in alkenes consists of two shared pairs of electrons. It is a centre of high electron density. The presence of the double bond makes alkenes much more reactive than alkanes. It can be attacked by electrophiles.

Electrophiles

An **electrophile** is a species that can accept a pair of electrons. The name means *electron-loving*, as electrophiles are attracted to regions of negative charge. Common electrophiles include

- positively charged ions such as H^+ and NO_2^+
- atoms that have a partial positive charge δ^+ because they are covalently bonded to a more electronegative atom, for example hydrogen in HBr

An electrophile can attack the centre of high electron density in the double bond in an alkene molecule. It accepts a pair of electrons to form a new covalent bond.

Electrophilic addition reactions

Hydrogen bromide

Ethene and hydrogen bromide readily react together to form bromoethane:

$$CH_2{=}CH_2 + HBr \rightarrow CH_3CH_2Br$$

The reaction is an **addition** reaction because the H and Br atoms from hydrogen bromide are added to the ethene, one to each of the two carbon atoms involved in the double bond. It is an **electrophilic addition** reaction because the attacking species is an electrophile, in this case H^+ from the hydrogen bromide.

As with other organic reactions, it is possible to study the reaction mechanism. Remember that a curly arrow indicates the movement of a pair of electrons from its tail to its head. Here is the reaction mechanism for electrophilic addition of hydrogen bromide to ethene:

$$H_2C{=}CH_2 + H^{\delta+}{-}Br^{\delta-} \longrightarrow H_3C{-}CH_2^{+} + {:}Br^-$$

In the first step, a pair of electrons from the double bond forms a co-ordinate covalent bond with the hydrogen atom in hydrogen bromide. The H—Br bond breaks to release a bromide ion. Notice that a positively charged intermediate is formed in this step.

$$H_3C{-}CH_2^{+} + {:}Br^- \longrightarrow H_3C{-}CH_2Br$$

In the final step, a lone pair of electrons in the bromide ion forms a co-ordinate covalent bond with the positively charged intermediate.

The positively charged intermediate is called a **carbocation** (pronounced 'carbo-cat-ion'). It is a very reactive species that is readily attacked by nucleophiles such as bromide ions. The complete mechanism is drawn like this:

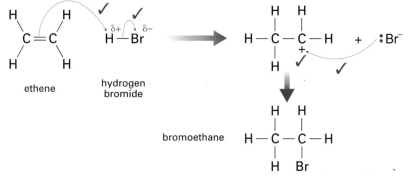

ethene hydrogen bromide bromoethane

The complete reaction mechanism, with ticks to show the features an examiner is likely to look for in an examination.

The reaction mechanism for other alkenes and electrophiles is very similar. When you draw them, take care to draw the correct carbocation intermediate, and include its positive charge.

Bromine

Bromine reacts slowly with ethene to form 1,2-dibromoethene:

$$CH_2{=}CH_2 + Br_2 \rightarrow BrCH_2CH_2Br$$

The Br–Br bond is not polar. But when a bromine molecule comes close to the double bond in an alkene molecule, the bonding pair of electrons is repelled. This gives the nearest bromine atom a partial positive charge, so it can act as an electrophile.

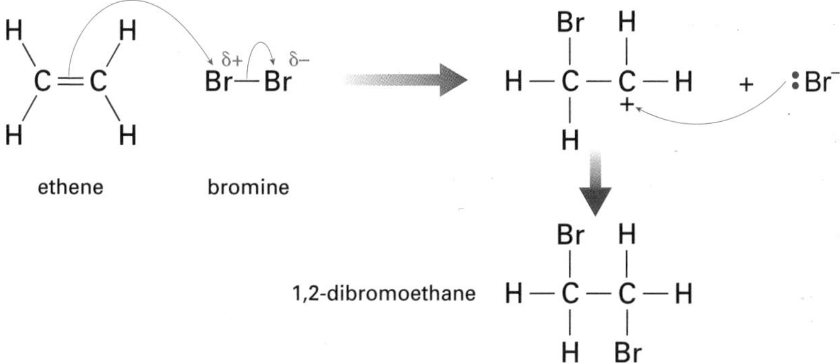

The complete reaction mechanism for the electrophilic addition of bromine to ethene. Notice how similar it is to the mechanism for the electrophilic addition of hydrogen bromide.

Check your understanding

1. a Outline the reaction mechanism for the reaction between but-2-ene and hydrogen bromide.
   b Name the organic product formed in the reaction.
   c What type of reaction is it?
2. Explain how bromine, a molecule without a permanent dipole, can take part in electrophilic substitution reactions.
3. When a solution of iodine in ethanol is added to sunflower oil, the brown colour of the iodine solution disappears. Explain this observation.

A test for unsaturation

Bromine is used in a simple laboratory test to see if a substance is unsaturated. It is dissolved in water to make a red-brown solution called *bromine water*. A stoppered test tube containing the test substance and a few drops of bromine water is shaken. The mixture becomes colourless if the test substance contains carbon–carbon double bonds. Alkenes decolourize bromine water.

Hex-1-ene on the right decolourizes bromine water but hexane on the left does not.

13.03 Sulfuric acid and alkenes

OBJECTIVES

already from AS Level, you understand

- the mechanism of electrophilic addition of alkenes with hydrogen bromide and with bromine

and after this spread you should

- understand the mechanism of electrophilic addition of alkenes with sulfuric acid
- understand that alcohols are produced industrially by hydration of alkenes in the presence of an acid catalyst
- know the typical conditions for the industrial production of ethanol from ethene

Ethanol is the liquid in spirit levels. Ethanol has a low freezing point of −114°C so the spirit level continues to work in cold weather.

Alkenes can be used to manufacture alcohols such as ethanol. This is important because alcohols are useful substances. For example, ethanol is a solvent for perfumes and paints, and the starting material for making other industrial substances such as ethyl esters. Alcohols can be made from alkenes in two stages using sulfuric acid, and directly using steam.

Ethene and sulfuric acid

Ethanol is produced from sulfuric acid on a laboratory scale in two stages:

1. Reaction of cold concentrated sulfuric acid with ethanol. This is an electrophilic addition. It forms ethyl hydrogensulfate:

$$CH_2{=}CH_2 + H_2SO_4 \rightarrow CH_3CH_2OSO_3H$$

2. Hydrolysis of ethyl hydrogensulfate by warming it with water:

$$CH_3CH_2OSO_3H + H_2O \rightarrow CH_3CH_2OH + H_2SO_4$$

Notice that sulfuric acid is used in the first stage and then produced in the second stage. It acts as a catalyst, so the overall equation is:

$$CH_2{=}CH_2 + H_2O \rightarrow CH_3CH_2OH$$

In the reaction mechanism for the first stage, electrophilic addition to form ethyl hydrogensulfate, sulfuric acid is shown as H–OSO_3H. This is analogous to H–Br for hydrogen bromide: the reaction mechanism is the same but you show :$^-OSO_3H$ instead of Br^-.

$H_2C{=}CH_2$ + $\overset{\delta+}{H}$—$\overset{\delta-}{OSO_3H}$ → $H_3C{-}\overset{+}{C}H_2$ + $:\bar{O}SO_3H$

↓

ethyl hydrogensulfate $CH_3CH_2OSO_3H$

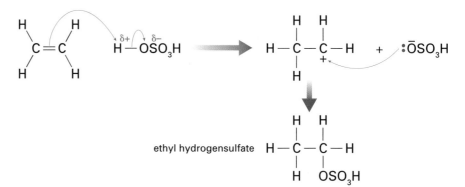

The full reaction mechanism for the electrophilic addition of sulfuric acid to ethene.

Ethene and steam

Ethene is produced on an industrial scale by the direct hydration of ethene. Ethene and steam react together at about 300°C and a high pressure of around 7 MPa. Phosphoric acid H_3PO_4 is used as a catalyst, held on an inert silica support. Ethanol is the only product:

$$CH_2{=}CH_2(g) + H_2O(g) \rightleftharpoons CH_3CH_2OH(g)$$

The reaction is reversible. The use of a high pressure moves the position of equilibrium to the right and increases the yield of ethanol. The forward reaction is exothermic. This means that it would be favoured by a low temperature. But a low temperature means a slow reaction, so a compromise temperature is chosen. It is hot enough to achieve a good rate of reaction without significantly reducing the yield of ethanol.

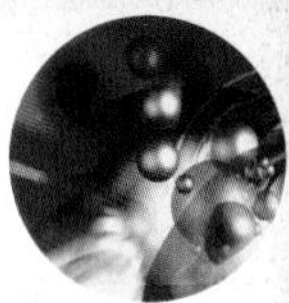

Phosphoric acid as a catalyst

Phosphoric acid supplies hydrogen ions in the direct hydration of ethene. The hydrogen ion is the electrophile that attacks the double bond in the ethene molecule. The steam provides the nucleophile that attacks the carbocation intermediate, forming a second intermediate. This intermediate forms ethanol when it releases a hydrogen ion. If you study the reaction mechanism, you will see that the hydrogen ion that joins the ethene molecule is not the one that leaves at the end.

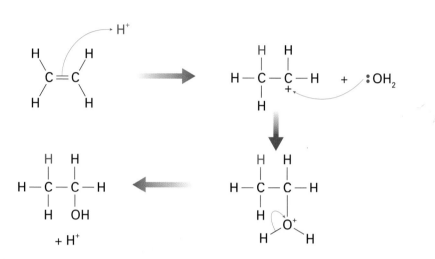

The mechanism for the direct hydration of ethene. Follow the hydrogen ion (in red) from the phosphoric acid catalyst.

Check your understanding

1. **a** Outline the reaction mechanism for the reaction between cold, concentrated sulfuric acid and ethene.
   b Name the organic product formed in the reaction.
   c Write an equation to show how ethanol is formed from the organic product in part **b**.
2. Ethanol can be produced on an industrial scale by direct hydration of ethene.
   a Outline the typical conditions needed for this process.
   b Identify the electrophile in the process and explain where it comes from.
3. In the 1930s, ethanol was produced on an industrial scale by the reaction of ethene with sulfuric acid, as in the laboratory method. Suggest why the direct hydration of ethene gradually replaced this process during the twentieth century.

13.04 Unsymmetrical alkenes

OBJECTIVES

already from AS Level, you understand

- the mechanism of electrophilic addition of alkenes

and after this spread you should be able to

- predict the products of addition to unsymmetrical alkenes by reference to the relative stabilities of primary, secondary, and tertiary carbocation intermediates

Ethene and (*Z*)-but-2-ene are symmetrical alkenes. Only one organic product forms when symmetrical alkenes undergo electrophilic addition with electrophiles such as hydrogen bromide.

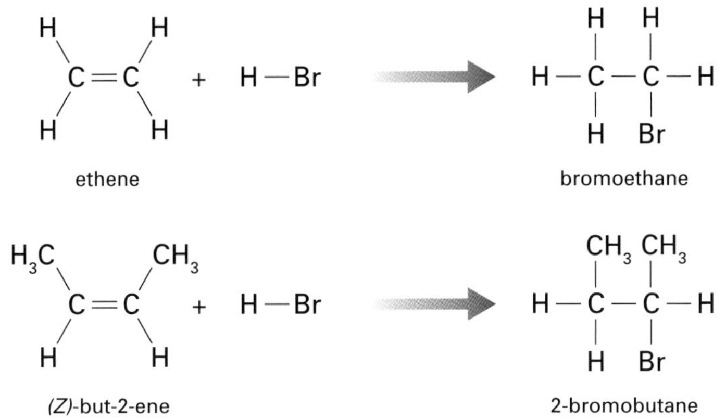

ethene + H—Br → bromoethane

(*Z*)-but-2-ene + H—Br → 2-bromobutane

Electrophilic addition reactions in symmetrical alkenes. Note that 2-bromobutane would still be obtained if (E)-but-2-ene were used instead of (Z)-but-2-ene.

The situation is different with unsymmetrical alkenes such as propene. Two organic products form when they undergo electrophilic addition with electrophiles such as hydrogen bromide. These products are not formed in equal amounts. Instead, one product is the minor product and the other is the major product. In the reaction between propene and hydrogen bromide

- 1-bromopropane is the minor product
- 2-bromopropane is the major product

The reason for this lies with the intermediate carbocation formed in the reaction.

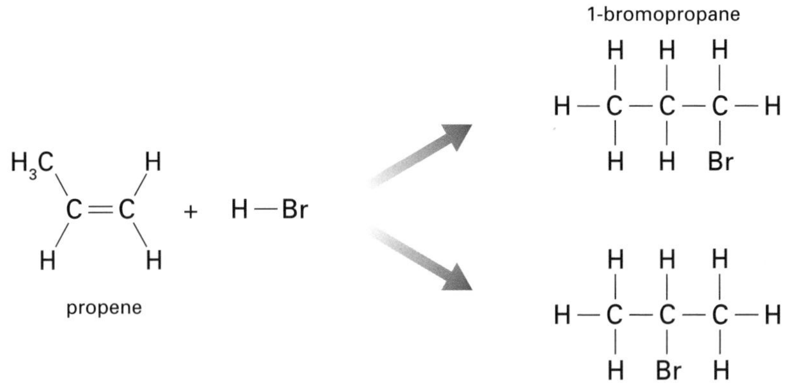

propene + H—Br → 1-bromopropane / 2-bromopropane

There are two possible products in the reaction between propene and hydrogen bromide. The major product is 2-bromopropane.

Different carbocations

The reaction mechanism for electrophilic addition has two steps. In the first one, the electrophile forms a co-ordinate bond with one of the carbon atoms involved in the double bond. Either carbon atom can be involved. This means that two different carbocations are possible in the reactions with unsymmetrical alkenes. Each will lead to a different

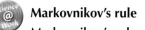

product in the second step, when a negatively charged ion forms a co-ordinate bond with the carbocation. This is why two products form in the reaction between propene and hydrogen bromide.

```
   H   H   H                        H   H   H
   |   |   |                        |   |   |
H— C — C — C⁺— H       ——→      H — C — C — C — H
   |   |                            |   |   |
   H   H         :Br⁻               H   H   Br

   H   H   H                        H   H   H
   |   |   |                        |   |   |
H— C — C⁺— C — H       ——→      H — C — C — C — H
   |       |                        |   |   |
   H       H     :Br⁻               H   Br  H
 2° carbocation                  2-bromopropane
```

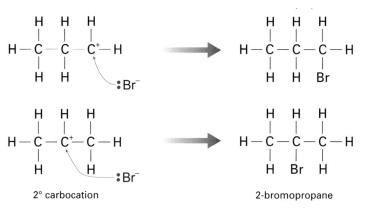

The primary carbocation leads to the formation of 1-bromopropane, the minor product. The secondary carbocation leads to the formation of 2-bromopropane, the major product.

The different carbocations are described as being primary, secondary, or tertiary. This is often shortened to 1°, 2°, and 3°.

type of carbocation	structure	the positively charged carbon atom is directly attached to
primary, 1°	$R_1-\underset{+}{\overset{H}{C}}-H$	one other carbon atom
secondary, 2°	$R_1-\underset{+}{\overset{H}{C}}-R_2$	two other carbon atoms
tertiary, 3°	$R_1-\underset{+}{\overset{R_3}{C}}-R_2$	three other carbon atoms

The different types of carbocation and how to recognize them. R_1, R_2, and R_3 stand for alkyl groups such as methyl $-CH_3$ and ethyl $-C_2H_5$. They can be all the same or different.

Stability of carbocations

Alkyl groups such as the methyl group are electron-releasing groups. The pairs of electrons in their C–H bonds are attracted towards the positive charge. This helps to stabilize the charge on the carbocation. The more alkyl groups there are attached to the positively charged carbon atom, the more stable the carbocation. So tertiary carbocations are the most stable, primary carbocations are the least stable, and the stability of secondary carbocations is in between.

The most stable carbocations will exist for the longest time, and the least stable carbocations will exist for the shortest time. This means that the major product in an electrophilic addition reaction comes from the most stable carbocation. This is why 2-bromopropane is the major product in the reaction between propene and hydrogen bromide: it is formed from a secondary carbocation, and this is more stable than a primary carbocation. The less stable primary carbocation leads to the formation of the minor product, in this case 1-bromopropane.

Markovnikov's rule

Markovnikov's rule was devised by the Russian chemist Vladimir Markovnikov in 1870, following his observations of electrophilic addition reactions. It lets you predict the major product in the electrophilic addition of an unsymmetrical alkene. It states that the hydrogen ion from reactants such as hydrogen bromide and sulfuric acid attaches to the carbon atom with the most hydrogen atoms. The negatively charged ion from the reactant attaches to the other carbon atom. Markovnikov's rule predicts the outcome of a reaction, but importantly it does not explain the reasons. You must always predict and explain the outcome of electrophilic addition reactions in terms of the relative stabilities of the possible carbocations.

Check your understanding

1. Explain why a tertiary carbocation is more stable than a secondary carbocation.
2. Show the displayed formulae for the possible carbocations formed in electrophilic addition reactions by
   - a but-1-ene
   - b but-2-ene
   - c 2-methylpropene
3. Name and then draw the displayed formulae for the major organic products formed in the reactions between the alkenes in question 2 and hydrogen bromide.

13.05 Polymers from alkenes

OBJECTIVES

already from GCSE, you know that

- many small molecules (monomers) join together to form very large molecules (polymers)
- alkenes can be used to make polymers such as poly(ethene) and poly(propene)

and after this spread you should

- know how addition polymers are formed from alkenes
- be able to recognize the repeating unit in a poly(alkene)
- recognize that poly(alkene)s are unreactive, like alkanes
- know some typical uses of poly(ethene) and poly(propene)

Alkene molecules can join together to form longer molecules called **polymers**. When alkene molecules do this, they are called **monomers**. Many thousand monomer molecules may join end to end to form a single polymer molecule. The reaction is an addition reaction. There is only one product, just like the addition reactions between alkenes and bromine. So the polymers formed from alkenes are called **addition polymers**.

Different monomers make different addition polymers. For example, ethene makes poly(ethene), and propene makes poly(propene). You may know these polymers as polythene and polypropylene. They have many applications because of their useful properties.

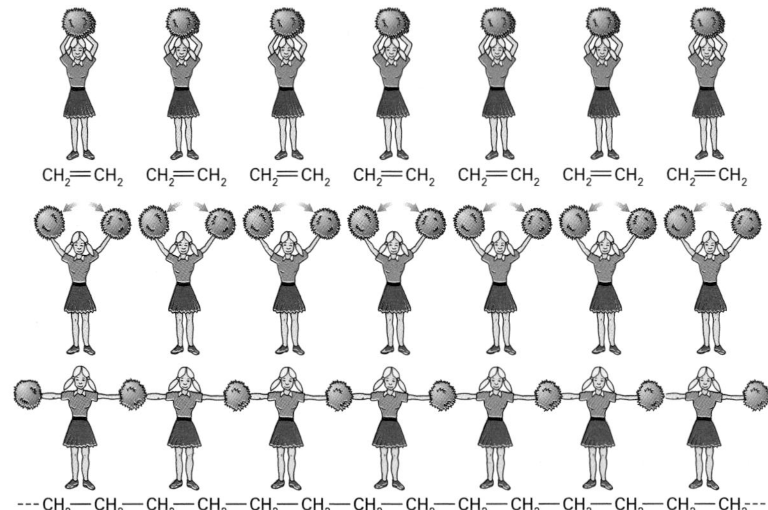

Ethene monomers take part in an addition reaction across the carbon–carbon double bond to form poly(ethene).

Repeating units

The diagram shows the reaction between three propene monomers to form a poly(propene) molecule.

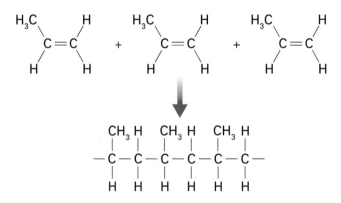

$$\mathrm{(H_3C)HC{=}CH_2 + (H_3C)HC{=}CH_2 + (H_3C)HC{=}CH_2}$$

$$\downarrow$$

$$\mathrm{-C(CH_3)H-CH_2-C(CH_3)H-CH_2-C(CH_3)H-CH_2-}$$

The formation of poly(propene).

Notice that the same structure is repeated three times in the polymer. This structure is called the **repeating unit** of poly(propene). The idea of a repeating unit makes it easier to represent reactions to form polymers and the structure of the polymer itself.

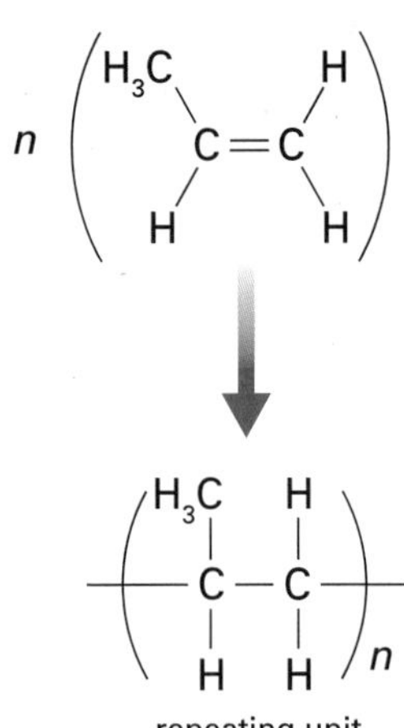

The formation of poly(propene) shown using its repeating unit.

To draw a repeating unit, starting with the displayed formula of the monomer

- change the double bond to a single bond
- draw the bond angles at 90°
- draw a longer bond to the left and right, and draw brackets through these
- add any remaining atom symbols

If you know the repeating unit you can easily work out the monomer. The repeating unit is the monomer molecule drawn with an extra bond to the left and right, and a single bond instead of a double bond. The name of the polymer comes from its monomer. An alkene makes a poly(alkene), so ethene makes poly(ethene) and propene makes poly(propene).

Uses of poly(ethene) and poly(propene)

Poly(alkenes) no longer contain a carbon–carbon double bond, so they are saturated and unreactive, like alkanes. They resist attack by acids and bases. The physical properties of polymers depend on the structure and bonding of the polymer chains.

- Longer chains have stronger van der Waals' forces between them than shorter chains, so they form stronger polymers.
- Chains with few branches pack together more closely than highly branched chains, so they have stronger van der Waals' forces between them and form stronger polymers.
- Chains that are linked together by covalent bonds form polymers that are harder and more difficult to melt than those without cross-links.

The reaction conditions during manufacture are adjusted to obtain a polymer with the desired properties. For example, low-density poly(ethene), LDPE, is made at a lower pressure than high-density poly(ethene), HDPE.

Tunnels made from poly(ethene) film protect crops such as lettuce from cold, wind, and heavy rain.

Poly(ethene)

LDPE has relatively long, branched chains. It is soft and flexible, so it is used to make polythene bags, bottles, cling film, and insulation for cables. HDPE has relatively few branches. It is stronger and more rigid than LDPE. It is used to make bottle tops and lids, bowls, and refuse bins.

Poly(propene)

Poly(propene) is less flexible than LDPE but it is tough and strong. It is used to make chairs, buckets, crates, and carpets. Poly(ethene) melts at around 100°C to 125°C, so it can become soft in hot water. Poly(propene) melts at around 160°C. This makes it robust for plastic kitchenware.

Fishing nets and ropes are made from poly(propene).

Check your understanding

1. Poly(ethene) and poly(propene) are addition polymers.
   a Explain why they are called addition polymers.
   b State two typical uses for each of them.
   c Explain why they are unreactive, like alkanes.
2. Draw the repeating unit of poly(ethene).
3. Polyvinyl chloride or PVC is made from chloroethene monomers. What is the systematic name for PVC?

13.06 Recycling polymers

OBJECTIVES

already from GCSE, you know that

- many polymers are not biodegradable – they are not broken down by micro-organisms and this can lead to problems with waste disposal

already from AS Level, you

- know how addition polymers are formed from alkenes
- can recognize the repeating unit in a poly(alkene)
- recognize that poly(alkene)s are unreactive, like alkanes
- know some typical uses of poly(ethene) and poly(propene)

and after this spread you should

- know that poly(propene) is recycled

What do Australian bank notes and flower pots have in common? They are both made from poly(propene). In 1988, Australia became the first country to introduce plastic bank notes. The plastic notes have several advantages over traditional paper bank notes. They are much more difficult to forge and they last longer. Old paper bank notes are taken out of circulation and then burned or buried. But old poly(propene) notes can be ground up and recycled to make pipes, compost bins, and flower pots. What makes poly(propene) suitable for recycling in this way?

Australian bank notes are made from two thin layers of poly(propene) joined under heat and pressure. They are more durable than paper notes and survive being washed in a pocket.

The fate of polymers

The manufacture of polymers such as poly(propene) needs large amounts of crude oil. Around 4% of the world's oil is used to provide the raw materials needed, and about the same amount to provide energy for the manufacturing process. Just throwing away used items made from polymers does not support sustainable development. Oil is a non-renewable resource, so this would be a waste of resources. In addition, polyalkenes such as poly(propene) are unreactive. This makes them useful to us, but it also means that they do not rot away easily in landfill sites. There are other solutions to the fate of used polymer items.

Other solutions include

- reusing polymers, for example by using more durable plastic bags that can be used for many shopping trips, not just one.
- reprocessing scrap from manufacturing processes. For example, pieces of polymer are left over after it is moulded or cut to shape. These can be melted down for re-use in the manufacturing process. Nearly all such poly(ethene) and poly(propene) scrap is reprocessed now.
- incineration. Burning used polymer items can release useful amounts of thermal energy for heating and generating electricity. Some polymers need additional fuel to burn them. But poly(propene) does not, so the incinerator can run more efficiently. Incineration releases carbon dioxide into the atmosphere and so contributes to global warming.
- recycling the used polymer

Poly(ethene) and poly(propene) form around two-thirds of household waste. There are different ways to recycle used polymers such as these.

Poly(ethene) and poly(propene) items can be recycled.

Recycling poly(propene)

Items made from polymers have recycling symbols moulded into them. High-density poly(ethene) HDPE and poly(propene) PP are particularly suitable for recycling.

Poly(propene) can be recycled by *mechanical recycling*. Waste plastic must be sorted to separate out the different types and colour of polymer. This can be done by hand. Increasingly the task is being done automatically using infrared spectroscopy or flotation. This involves separating the different polymers on the basis of their density.

After sorting, the poly(propene) is shredded into flakes, then processed to form granules. These can be melted down and moulded into new items. This is what happens with the Australian bank notes. Poly(propene) can be recycled in this way over fifty times without losing its strength. By contrast, paper can only be recycled about seven times before its fibres become too small to be useful.

Poly(propene) can also be recycled by *feedstock recycling*. This involves decomposing the polymer in a similar way to the cracking of heavy oil fractions. The polymer breaks down in the absence of air at high temperatures of around 500°C. This produces a fraction similar to naphtha, which is used as a feedstock for the chemical industry.

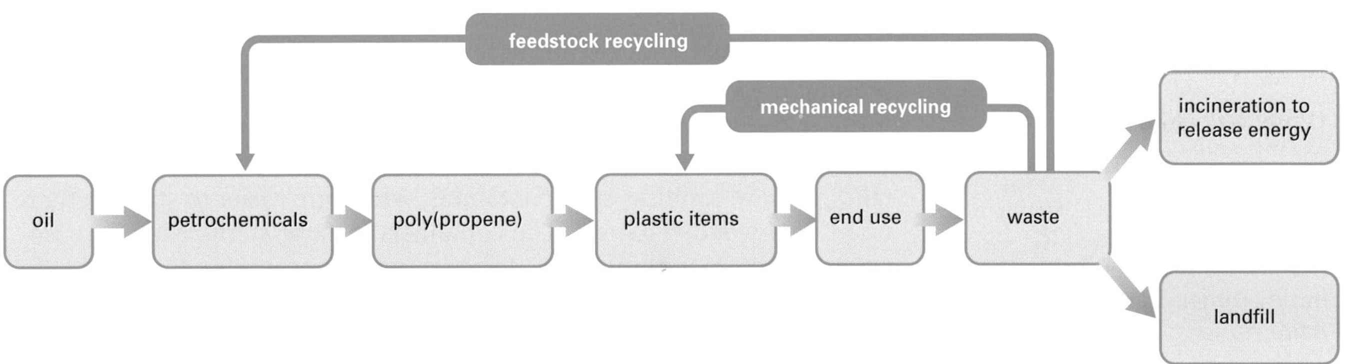

A summary of how poly(propene) can be recycled.

Bottle tops

Always take the tops off waste plastic bottles before putting the bottles in the refuse bin or recycling box. The bottles can be crushed more easily without their tops so they take up less volume.

Check your understanding

1. Outline some reasons why it is increasingly unacceptable to dispose of used polymers such as poly(propene) in landfill sites.
2. **a** PVC is poly(chloroethene). It releases harmful chlorine compounds when it is incinerated. Explain why this is not a problem for poly(propene).
   b State one advantage and one disadvantage of incinerating poly(propene).
3. Describe two ways in which poly(propene) can be recycled.

14.01 Ethanol – a biofuel

OBJECTIVES

already from AS Level, you know

- that alcohols can be produced industrially by hydration of alkenes in the presence of an acid catalyst

and after this spread you should

- know how ethanol is produced industrially by fermentation
- know the conditions for this reaction
- understand the meaning of the term biofuel

Fermentation

Fermentation has been used for thousands of years to produce ethanol CH_3CH_2OH. The process uses sugars from plants and microscopic single-celled fungi called yeast. The yeast contains **enzymes**, proteins that are natural catalysts. They catalyse the decomposition of sugar to ethanol and carbon dioxide:

$$C_6H_{12}O_6 \xrightarrow[\text{in yeast}]{\text{enzymes}} 2CH_3CH_2OH + 2CO_2$$

Fermentation works at normal atmospheric pressure and temperatures of around 10–40°C. The rate of fermentation increases as the temperature increases. It reaches its optimum rate at about 36°C but stops above 43°C. High temperatures denature the enzymes and kill the yeast.

Fermentation is an anaerobic process – it takes place in the absence of oxygen. It continues for several days until the concentration of ethanol reaches about 15%. This is toxic to the yeast, so it dies and fermentation stops.

Alcoholic drinks

Ethanol is produced in beer and wine on an industrial scale. The sugars for beer come from barley. This is germinated to convert its stored starch to sugars. The sugars for wine are contained in grape juice. Fermentation takes place in stainless steel containers, which are easier to sterilize than traditional wooden barrels. The containers must be sterilized before use so that other micro-organisms do not grow. For example, *Acetobacter* produces ethanoic acid from sugar in the early stages of the fermentation process. After fermentation, the reaction mixture is filtered to remove the dead yeast.

Modern wine-making takes place on an industrial scale.

Using the leftovers

The by-products of brewing beer are put to good use. The yeast is processed to make Marmite™, and the barley remains are fed to farm animals.

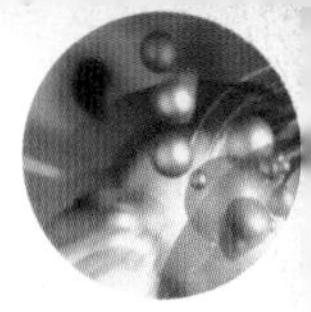

Industrial alcohol

The process for producing ethanol for fuel and industrial purposes is very similar to beer and wine making. The sugars can come from a wide variety of sources, including:

- sugar beet and sugar cane
- molasses, a by-product of processing sugar beet and sugar cane
- starch from wheat, potatoes, or maize

Fermentation is carried out at about 36°C. The pH is maintained at about 4.5 to prevent growth by contaminating bacteria.

A bioethanol factory and a field of maize. Maize is one of the sources of sugar for producing industrial ethanol by fermentation.

Processing ethanol for biofuel

A **biofuel** is a fuel made from the products of living things. Wood is a biofuel and so is methane produced by the bacterial digestion of sewage. Ethanol produced by fermentation of plant sugars is a biofuel. It can be used to fuel vehicles on its own or when mixed with petrol, but it must be processed first.

Fermentation produces a mixture containing about 15% ethanol. This is too dilute to be used as a fuel, so it must be distilled to concentrate the ethanol. Water boils at 100°C and ethanol boils at 78.4°C. The difference in boiling points allows ethanol to be separated from water. But ethanol and water form a mixture called an *azeotrope*. This boils at 78.1°C, which is below the boiling points of both ethanol and water. It contains 95.6% ethanol and 4.4% water. Ethanol cannot be made any more concentrated than this by distillation alone.

Ethanol can be mixed with petrol or used as a fuel itself. The ethanol azeotrope is usually dehydrated to remove the water content before it is mixed with petrol. E numbers tell you how much of the mixture is ethanol and how much is petrol. There are several mixtures, such as E5 and E20. E5 contains 5% ethanol and may be sold as high octane petrol, and E20 contains 20% ethanol. E100 is the azeotropic mixture of ethanol and water. Vehicles usually need different engines from the usual petrol engines to run on this fuel. They are popular in Brazil.

This car is designed to run on E85, a mixture of 85% ethanol and 15% petrol.

Check your understanding

1. a Write an equation for fermentation to produce ethanol.
   b State the typical conditions needed for fermentation.
2. a Explain why the concentration of ethanol from fermentation is about 15%.
   b Explain why there is a limit to the concentration of ethanol produced by distillation.
3. Explain why ethanol is described as a biofuel.

14.02 Renewable and non-renewable ethanol

OBJECTIVES

already from AS Level, you

- know how ethanol is produced industrially by fermentation, and by hydration of ethene in the presence of an acid catalyst
- understand the meaning of the term biofuel
- know the meaning of the term carbon neutral

and after this spread you should

- understand the economic and environmental advantages, and disadvantages, of fermentation compared with the industrial production of ethanol from ethene
- appreciate the extent to which ethanol, produced by fermentation, can be considered to be a carbon-neutral biofuel

The bubbles in sparkling mineral water and other fizzy drinks come from dissolved carbon dioxide.

Each of the two methods for producing ethanol has economic, and environmental, advantages and disadvantages.

Disadvantages of fermentation

	fermentation	hydration of ethene
speed of reaction	slow	fast
yield	15%	95%
atom economy	51.1%	100%
type of process	batch	continuous

Some disadvantages of fermentation compared to direct hydration of ethene.

Both processes use distillation to produce ethanol at a high concentration, and this needs a lot of heat energy. Ethanol produced by fermentation is less pure than ethanol produced by the hydration of ethene. Extra steps are needed to separate the ethanol solution from dead yeast, which add to the overall cost of the process. Fermentation has a low yield and atom economy:

- fermentation, $C_6H_{12}O_6 \rightarrow 2C_2H_5OH + 2CO_2$ is a decomposition reaction
- hydration of ethene, $C_2H_4 + H_2O \rightarrow C_2H_5OH$ is an addition reaction

Addition reactions always have an atom economy of 100%. A low atom economy like that of fermentation is usually seen as less helpful to sustainable development. But the carbon dioxide by-product of fermentation has a ready market putting the fizz into fizzy drinks.

Fermentation is a batch process and the direct hydration of ethene is a continuous process. Batch processes are difficult to automate. They need more labour than continuous processes, and the equipment must be cleaned and recharged with fresh reactants each time. But the cost of the equipment needed for batch processes is usually much less than the cost for continuous processes.

Advantages of fermentation

	fermentation	hydration of ethene
pressure (MPa)	0.1	7
temperature (°C)	36	300
catalyst	enzymes in yeast	phosphoric acid
energy use	low	high
raw materials	renewable sugars	non-renewable ethene

Some advantages of fermentation compared to direct hydration of ethene.

Fermentation is carried out at much lower temperatures and pressures than is the case with the hydration of ethene. As a result, its energy use is much lower. This reduces the use of fossil fuels. The low temperatures and pressures also mean that the cost of equipment needed for fermentation can be much lower. Indeed, you can carry out fermentation to make beer and wine in a warm place at home, if you wish.

The raw materials for fermentation are renewable resources, sugars from plant material. This makes ethanol important as a potential carbon-neutral fuel.

Home wine-making does not need complex and expensive equipment.

Carbon-neutral ethanol?

A carbon-neutral process has no net annual carbon (greenhouse gas) emissions to the atmosphere. Is the use of bioethanol as a fuel a carbon-neutral process? Here are the main chemical processes involved:

1. Photosynthesis in plants produces sugars from carbon dioxide and water:

$$6CO_2 + 6H_2O \xrightarrow[\text{from the Sun}]{\text{light energy}} C_6H_{12}O_6 + 6O_2$$

2. Fermentation produces ethanol and carbon dioxide from sugars:

$$C_6H_{12}O_6 \xrightarrow[\text{in yeast}]{\text{enzymes}} 2CH_3CH_2OH + 2CO_2$$

3. Combustion of ethanol produces carbon dioxide and water vapour:

$$2CH_3CH_2OH + 6O_2 \rightarrow 4CO_2 + 6H_2O$$

If equations 2 and 3 are combined, you get:

$$C_6H_{12}O_6 + 6O_2 \rightarrow 6CO_2 + 6H_2O$$

This is the reverse of the photosynthesis equation. The energy released by the fuel was originally light energy from the Sun. The carbon in the carbon dioxide released was originally absorbed from the atmosphere by plants. On this basis, ethanol is a carbon-neutral fuel. But there are other factors to consider, too. These include

- manufacturing, and running, farm equipment and bioethanol factories
- manufacturing and transporting fertilizers
- distilling and dehydrating the ethanol
- distributing the fuel

If these requirements are met using fossil fuels, bioethanol cannot be a carbon-neutral fuel. The manufacture of steel and concrete releases carbon dioxide into the atmosphere. Fertilizers are manufactured from ammonia. The hydrogen needed for making ammonia comes from the reaction of steam with natural gas or coal. This can also release carbon dioxide. Additional carbon dioxide will be released if fossil fuels are used to provide the heat for distillation and the fuel for tankers.

Wheat is used to make bioethanol in the UK. This competes with the use of wheat as a food crop.

The renewable transport fuel obligation

The UK government announced in 2005 that it wanted 5% of all car fuel in the UK to come from renewable sources by 2010. This was aimed at reducing emissions of carbon dioxide by one million tonnes a year.

Check your understanding

1. Outline three advantages and three disadvantages of manufacturing ethanol by fermentation.
2. A scientific study concluded that making bioethanol from maize needs 29% more energy than it releases. Another study concluded that bioethanol from maize releases 34% more energy than it takes to make it. Suggest why the two studies might have given such different conclusions.

14.03 Investigating the combustion of alcohols

OBJECTIVES

already from AS Level, you can

- recall the definition of standard enthalpy of combustion $\Delta H_c^{\ominus}$
- calculate the enthalpy change from the heat change in a reaction using the equation $q = mc\Delta T$

and after this spread you should know

- how to competently use a calorimetric method to measure the enthalpies of combustion in an homologous series of alcohols

One of the recommended Practical Skills Assessment tasks is to measure the enthalpies of combustion of some alcohols. You may be asked to calculate the enthalpies of combustion and to compare them with accepted values. You will be assessed on your ability to

- work safely and carefully
- measure masses and volumes precisely and within the range asked for
- measure the starting and finishing temperatures precisely
- obtain results that show the expected trend in enthalpy, and are within the range of expected values

Amounts

You will be assessed on your ability to make measurements correctly. These include measuring the

- volume of water used
- mass of a spirit burner before and after igniting the alcohol
- temperature of the water before and after heating

You need to be particularly careful when you measure the masses and temperatures. The values you will use in your calculations depend on the *differences* between the starting and finishing values.

Precision

Make sure you record your results to the precision of the measuring instrument. For example, if you use a 2 decimal place balance, record all the readings to 2 decimal places. So 125.00 would be correct, but 125.0 and 125 would not.

Results

It is always wise to draw a blank results table before you start work. That way, you can be certain that you have completed all the tasks before you tidy away. If you are uncertain of an observation and repeat a certain test, do not erase your original record. Just put a line through your writing in such a way that you can still see your original observations. You may have been correct the first time around.

Alcohol	Methanol	Ethanol	Propan-1-ol	Butan-1-ol
Initial mass of burner and fuel (g)		128.84		
Final mass of burner and fuel (g)		128.31		
Mass of alcohol burned (g)		0.53		
Initial temperature of water (°C)		21		
Final temperature of water (°C)		41		
Change in temperature (°C)		20		

Volume of water in calorimeter = 100 cm^3

You may wish to draw a table like this one. It has space to record what you did, your observations, and what you think your observations mean.

Measuring an enthalpy of combustion

This is a typical method for simple calorimetry to measure an enthalpy of combustion.

- Weigh a spirit burner with its alcohol and record the mass.
- Measure 100 cm^3 of water using a measuring cylinder and add it to the calorimeter.
- Record the temperature of the water.
- Place the burner under the calorimeter and light the alcohol.
- Stir the water with the thermometer until the temperature has increased by 20K.
- Replace the lid on the burner.
- Reweigh the spirit burner with its alcohol and record the mass.

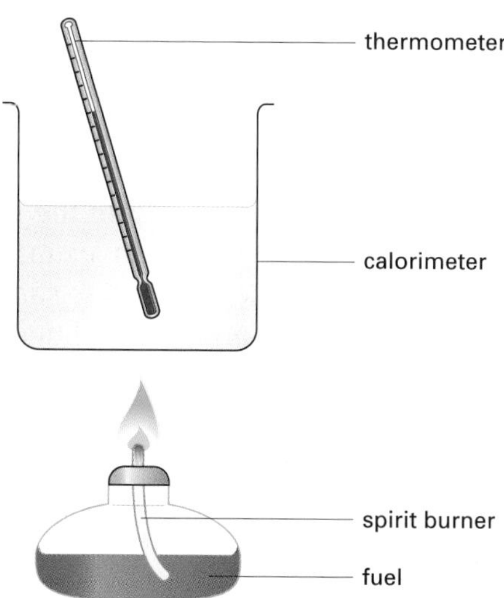

A simple laboratory experiment to measure an enthalpy of combustion.

Analysing the results

You will recall that you calculate the enthalpy change from the heat change in a reaction using the equation $q = mc\Delta T$, where

- m is the mass of water (100 cm^3 has a mass of 100 g)
- c is the specific heat capacity of water (4.18 K^{-1} g^{-1})
- ΔT is the temperature change

Once you have calculated the heat energy absorbed by the water, calculate the number of moles of alcohol burned. Divide the mass of alcohol burned by its M_r. Finally, calculate the enthalpy change by dividing the heat energy in kJ by the number of moles.

Heat energy is released from the alcohol to warm the water. This will be the enthalpy of combustion of the alcohol, assuming that all the heat energy warms the water. It will have a negative value because heat energy has left the alcohol.

The expected trend

You should expect the enthalpy of combustion to increase as the chain length of the alcohol increases. But it is very unlikely that your experimental results will be the same as the accepted values. The biggest source of error in the experiment is heat loss to the surroundings. This can be a large proportion of the heat energy released.

alcohol	number of C atoms	M_r	enthalpy of combustion $\Delta H_c^{\ominus}$ (kJ mol^{-1})
methanol	1	32.0	–726
ethanol	2	46.0	–1367
propan-1-ol	3	60.0	–2021
butan-1-ol	4	74.0	–2676

Accepted enthalpies of combustion for four alcohols.

Check your understanding

1. Apart from aiming for a small temperature rise, suggest two ways in which heat losses to the surroundings could be reduced in the calorimetry experiment.
2. Explain why it is important to keep the distance from the spirit burner to the bottom of the calorimeter constant.
3. A small temperature rise reduces the rate of heat loss from the calorimeter, but makes it difficult to carry out the experiment because the mass of fuel burned is very small. Discuss the problems associated with aiming for a large temperature rise.

14.04 Classifying alcohols

OBJECTIVES

already from AS Level, you know

- how alcohols can be obtained from haloalkanes and alkenes

and after this spread you should

- be able to apply IUPAC rules for naming alcohols with up to six carbon atoms
- understand that alcohols can be classified as primary, secondary, or tertiary

A functional group is a particular atom, or group of atoms, in a molecule that is responsible for the how the molecule reacts. The members of a homologous series will contain the same functional group. The alcohols contain the **hydroxyl group** –OH. Take care not to confuse this with the hydroxide ion OH^-. The general formula of the alcohols is $C_nH_{2n+1}OH$.

Naming alcohols

Alcohols are named after the number of carbon atoms they contain, and their names end in *ol*. The structural formula of methanol is CH_3OH and the structural formula of ethanol is CH_3CH_2OH.

```
     H               H   H
     |               |   |
  H—C—O—H        H—C—C—O—H
     |               |   |
     H               H   H
```

methanol ethanol

The displayed formulae of methanol and ethanol.

Alcohols with more than two carbon atoms can have position isomers. Position isomerism is a particular type of structural isomerism. It occurs when the functional group can be in different positions on the same carbon chain. Position isomers of alcohols are named in a similar way to position isomers of alkenes. The position of the hydroxyl is shown using a number. For example, butanol has two position isomers, butan-1-ol and butan-2-ol. You keep the total number as small as possible, so butan-3-ol would be incorrect.

```
    H  H  H  H            H  H  H  H
    |  |  |  |            |  |  |  |
 H—C—C—C—C—H          H—C—C—C—C—H
    |  |  |  |            |  |  |  |
    H  H  H  OH           H  H  OH H
```

butan-1-ol butan-2-ol

Position isomers of butanol.

When you write out the shortened structural formula for an alcohol where the hydroxyl group is not on an end carbon atom, show the OH group in brackets. For example, the shortened structural formula for butan-1-ol is

$$CH_3CH_2CH_2CH_2OH$$

and for butan-2-ol it is

$$CH_3CH_2CH(OH)CH_3.$$

Multiple hydroxyl groups

An alcohol molecule can have more than one hydroxyl group. If it does, you use the suffix –diol if there are two hydroxyl groups and –triol if there are three. You number the positions of the hydroxyl groups so that the total is as small as possible. So butane-1,2-diol would be correct but butane-3,4-diol would be incorrect. Notice that an e is added to the first part of the name when there are two or more hydroxyl groups.

```
    H  H  H
    |  |  |
 H—C—C—C—H
    |  |  |
    OH OH OH
```

Propane-1,2,3-triol is a sweet-tasting viscous liquid used in cough syrup and toothpaste. Its common name is glycerol.

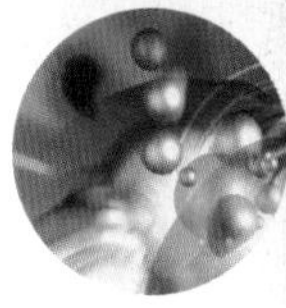

Primary, secondary, and tertiary alcohols

Alcohols are classified as primary, secondary, or tertiary alcohols. This is often shortened to 1°, 2°, and 3°. The classification is very similar to the one used for carbocations and haloalkanes.

type of alcohol	structure	the carbon atom with the hydroxyl group is directly attached to
primary, 1°	H | R_1—C—OH | H	one other carbon atom
secondary, 2°	R_2 | R_1—C—OH | H	two other carbon atoms
tertiary, 3°	R_2 | R_1—C—OH | R_3	three other carbon atoms

The different types of alcohol and how to recognize them. R_1, R_2, and R_3 stand for alkyl groups such as methyl $-CH_3$ and ethyl $-C_2H_5$. They can be all the same or different. Methanol is a primary alcohol.

```
                              H                         H
                              |                         |
                           H—C—H                     H—C—H
  H  H  H                  H  |  H                   H  |  H
  |  |  |                  |  |  |                   |  |  |
H—C—C—C—H               H—C—C—C—H               H—C—C—C—OH
  |  |  |                  |  |  |                   |  |  |
  H  OH H                  H  OH H                   H  H  H

propan-2-ol             methylpropan-2-ol        methylpropan-1-ol
```

Propan-2-ol is a 2° alcohol and methylpropan-2-ol is a 3° alcohol. Methylpropan-1-ol is a 1° alcohol, even though it is branched.

Check your understanding

1. There are four isomers of pentanol $C_5H_{11}OH$ that have one branch.
   - **a** Draw the displayed formulae for these isomers.
   - **b** Name the isomers you have drawn. When you do this, name the longest chain with the hydroxyl group attached first, then number the position of the methyl group.
   - **c** Classify each isomer as a primary, secondary, or tertiary alcohol.
   - **d** There is an isomer of pentanol with two branches.

     Draw the displayed formula for this isomer. Name it and explain why it is a primary alcohol.
2. Explain the advantages of using a systematic naming system for alcohols.

Science @ Work — Non-systematic names

The IUPAC naming system makes it easy to work out the structure of an alcohol, or name a given alcohol. But you may come across non-systematic names for alcohols when you read about chemistry. They are based on the name of the alkyl group attached to the hydroxyl group. For example, methanol is methyl alcohol, and ethanol is ethyl alcohol. Position isomers and chain isomers are shown by letters. For example, *n*-butyl alcohol is butan-1-ol, *s*-butyl alcohol is butan-2-ol, and *t*-butyl alcohol is methylpropan-2-ol.

14.05 Aldehydes, ketones, and carboxylic acids

OBJECTIVES

already from AS Level, you know

- how to apply IUPAC rules for naming alkanes, alkenes, haloalkanes, and alcohols with up to six carbon atoms

and after this spread you should know

- how to apply IUPAC rules for naming simple aldehydes, ketones, and carboxylic acids
- a simple chemical test to distinguish between aldehydes and ketones

The **carbonyl** functional group is >C=O. It is planar like the >C=C< group in alkenes. **Aldehydes** and **ketones** both contain the carbonyl group but in different places on the carbon chain:

- at the end of the chain in aldehydes
- not at the end of the chain in ketones

These differences in position are enough for aldehydes and ketones to have different properties from each other.

Aldehydes

Aldehydes are named after the number of carbon atoms they contain, including the carbon atom in the carbonyl group, and their names end in *al*. The carbonyl group in aldehydes is shown as –CHO in shortened structural formulae. The structural formula of methanal is HCHO and the structural formula of ethanal is CH_3CHO. Take care when you write the names of aldehydes. It is can be easy for 'ethanal' to look like 'ethanol'.

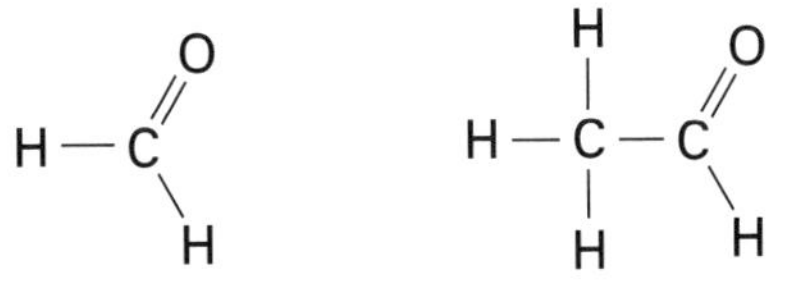

methanal

ethanal

The displayed formulae of methanal and ethanal.

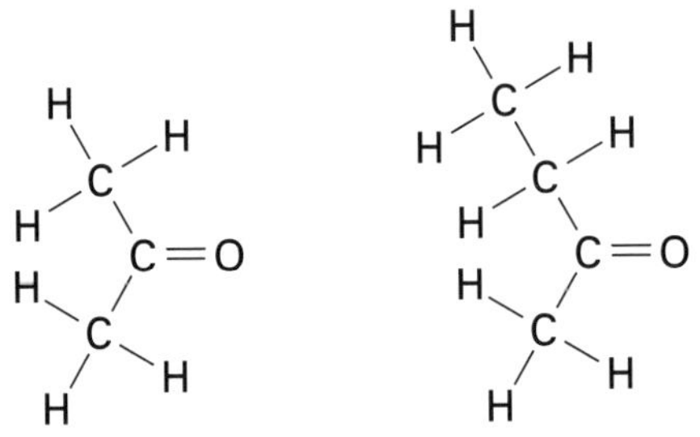

propanone

butanone

The displayed formulae of propanone and butanone. Propanone is also called acetone. It is a solvent in nail polish remover.

Ketones

Ketones are named after the number of carbon atoms they contain, including the carbon atom in the carbonyl group, and their names end in *one*. The carbonyl group in ketones is shown as –CO in shortened structural formulae. The structural formula of propanone is CH_3COCH_3 and the structural formula of butanone is $CH_3CH_2COCH_3$.

Position isomers of ketones

Ketones with more than four carbon atoms have position isomers. For example, hexanone has two position isomers. Hexan-2-one is $CH_3(CH_2)_3COCH_3$ and hexan-3-one is $CH_3(CH_2)_2COCH_2CH_3$. Note that you pronounce the end of the name to rhyme with "stone".

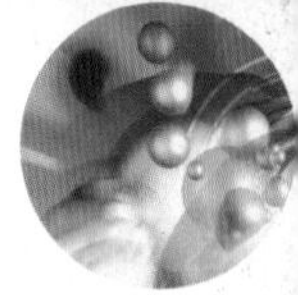

Testing for aldehydes

Tollens' reagent can distinguish ketones from aldehydes.

- A small volume of sodium hydroxide solution is added to some aqueous silver nitrate.
- A precipitate forms, which is changed to aqueous $[Ag(NH_3)_2]^+$ ions by adding some ammonia solution.
- On warming, aldehydes reduce the $[Ag(NH_3)_2]^+$ ions to silver.

The silver coats the inside of the test tube, forming a *silver mirror*. Aldehydes produce a silver mirror with Tollens' reagent but ketones do not.

Fehling's solution is a blue solution containing a complex of Cu^{2+} ions. Aldehydes reduce Fehling's solution when heated with it to form a brick-red precipitate of copper(I) oxide, Cu_2O. Ketones do not react with Fehling's solution, so no change is observed when they are heated with it.

Aldehydes produce a silver mirror when they are warmed with Tollens' reagent.

test substance	Tollens' reagent	Fehling's solution
aldehyde	silver mirror forms	blue solution changes to red precipitate
ketone	no change	no change

A summary of the changes seen with Tollens' reagent and Fehling's solution

Aldehydes produce a brick-red precipitate when they are heated with Fehling's solution.

Carboxylic acids

Carboxylic acids also contain the carbonyl group. This time it is attached to a carbon atom that also has a hydroxyl group. The new functional group is called the **carboxyl** group, from *carb*onyl and hydr*oxyl*. It is shown in shortened structural formulae as –COOH. Carboxylic acids are named after the number of carbon atoms they contain, including the carbon atom in the carboxyl group, and their names end in *oic acid*. The structural formula of methanoic acid is HCOOH and the structural formula of ethanoic acid is CH_3COOH. Vinegar is a dilute solution of ethanoic acid.

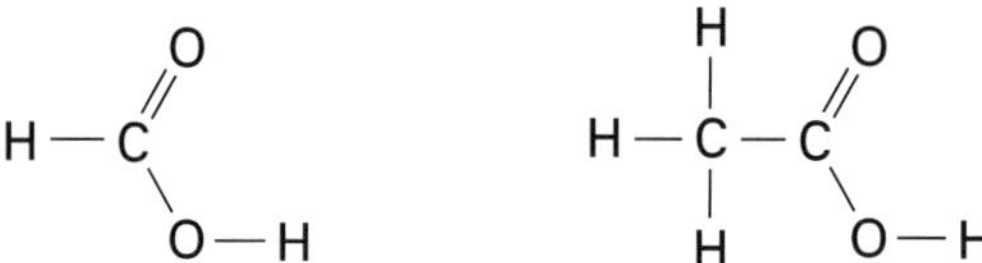

The displayed formulae of methanoic acid and ethanoic acid.

Ant stings contain methanoic acid. The common name for methanoic acid is formic acid, after the Latin word formica, *which means ant. In the past, methanoic acid was extracted from ants and distilled.*

Check your understanding

1. What is the difference in the position of the carbonyl group in aldehydes and ketones?
2. Name these compounds:
   - **a** $CH_3(CH_2)_3CHO$
   - **b** $CH_3COCH_2CH_3$
   - **c** $CH_3(CH_2)_4COOH$
3. Describe a simple laboratory test to distinguish between an aldehyde and a ketone.

14.06 Oxidizing alcohols

OBJECTIVES

already from AS Level, you know

- that alcohols can be classified as primary, secondary, or tertiary

and after this spread you should understand that

- primary alcohols can be oxidized to aldehydes and carboxylic acids using acidified potassium dichromate(VI)
- secondary alcohols can be oxidized to ketones using acidified potassium dichromate(VI)
- tertiary alcohols are not easily oxidized

Acidified potassium dichromate(VI)

Potassium dichromate(VI), $K_2Cr_2O_7$, is acidified by dissolving it in sulfuric acid, forming an orange solution. When it acts as an oxidizing agent, $Cr_2O_7^{2-}$ ions are reduced to chromium(III) ions, forming a blue-green solution:

$Cr_2O_7^{2-}(aq) + 14H^+(aq) + 6e^- \rightarrow 2Cr^{3+}(aq) + 7H_2O(l)$

The sulfuric acid supplies the hydrogen ions needed.

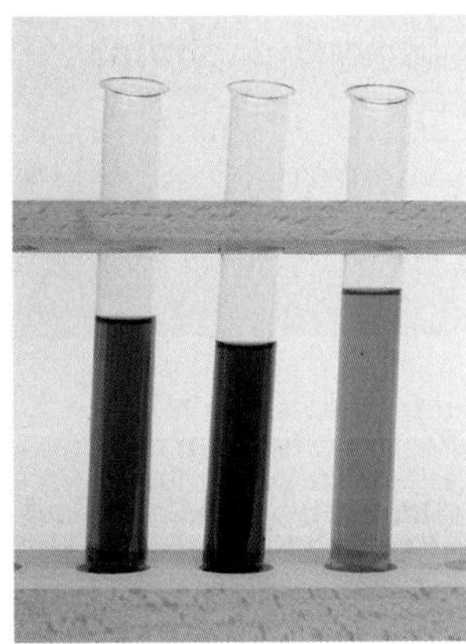

Colour changes on heating alcohols with acidified potassium dichromate(VI): 1° and 2° alcohols reduce orange $Cr_2O_7^{2-}(aq)$ to blue-green $Cr^{3+}(aq)$. Tertiary alcohols (on the right) produce no change.

Alcohols such as ethanol burn in air to produce carbon dioxide and water vapour:

$$CH_3CH_2OH(l) + 3O_2(g) \rightarrow 2CO_2(g) + 3H_2O(g)$$

The alcohol is oxidized in this reaction, but this is not the only way to oxidize alcohols. Remember that:

- Oxidation is the loss of electrons or the gain of oxygen atoms.
- Reduction is the gain of electrons or the loss of oxygen atoms.

Oxidation and reduction can also involve hydrogen atoms:

- Oxidation is the loss of hydrogen atoms.
- Reduction is the gain of hydrogen atoms.

Oxidizing agents remove electrons, supply oxygen atoms, or remove hydrogen atoms. Acidified potassium dichromate(VI) is an oxidizing agent that can oxidize alcohols to aldehydes, carboxylic acids, or ketones. In the following oxidation reactions, the oxidizing agent is shown by the symbol [O]. This is how you will be expected to show it in your examination.

Primary alcohols

Primary alcohols are oxidized at room temperature to form aldehydes. For example, ethanol is oxidized to ethanal:

$$CH_3CH_2OH + [O] \rightarrow CH_3CHO + H_2O$$

Notice that two hydrogen atoms are removed from each ethanol molecule in the reaction, so it is an oxidation reaction. If the acidified potassium dichromate(VI) is in excess, the aldehyde is oxidized further to form a carboxylic acid. In the example above, ethanal would be oxidized to ethanoic acid:

$$CH_3CHO + [O] \rightarrow CH_3COOH$$

Notice that an oxygen atom is added in this reaction, so it is an oxidation reaction. The overall equation for the reaction when ethanol is heated with excess acidified potassium dichromate(VI) is:

$$CH_3CH_2OH + 2[O] \rightarrow CH_3COOH + H_2O$$

You can prevent further oxidation of the aldehyde to the carboxylic acid by distilling the aldehyde as the reaction proceeds. This removes the aldehyde from the oxidizing agent.

Secondary alcohols

Secondary alcohols are oxidized to form ketones. For example, propan-2-ol is oxidized to propanone:

$$CH_3CH(OH)CH_3 + [O] \rightarrow CH_3COCH_3 + H_2O$$

Notice that two hydrogen atoms are removed from each propan-2-ol molecule in the reaction, as in the oxidation of a primary alcohol to an aldehyde.

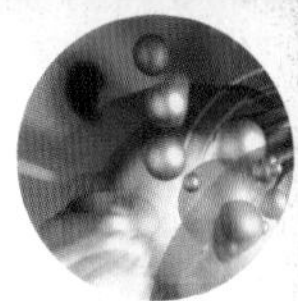

Tertiary alcohols

Tertiary alcohols are not oxidized by acidified potassium dichromate(VI). To understand why, you need to see where the hydrogen atoms come from when primary and secondary alcohols are oxidized. In both cases

- one hydrogen atom is released from the hydroxyl group
- one hydrogen atom is released from its bond with the carbon atom that carries the hydroxyl group

This can happen in primary and secondary alcohols. But in tertiary alcohols there is no hydrogen atom attached to the carbon atom carrying the hydroxyl group. This means that tertiary alcohols resist oxidation by acidified potassium dichromate(VI).

There is no hydrogen atom attached to the carbon atom carrying the hydroxyl group in a tertiary alcohol. Tertiary alcohols resist oxidation by acidified potassium dichromate(VI).

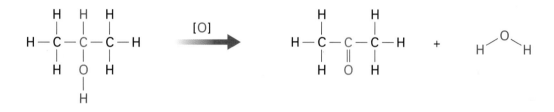

Two hydrogen atoms are removed when a primary alcohol is oxidized by acidified potassium dichromate(VI).

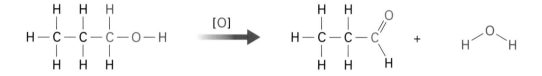

Two hydrogen atoms are removed when a secondary alcohol is oxidized by acidified potassium dichromate(VI).

Distinguishing the three types of alcohol

The products of oxidation can be used to distinguish between primary, secondary, and tertiary alcohols. Given three unknown alcohols, you would attempt to oxidize them using potassium dichromate(VI) acidified with sulfuric acid. You would then separate the oxidation products by distillation and test them with Tollens' reagent or Fehling's solution. The table shows the results expected with the three types of alcohol.

alcohol	colour change during reaction with acidified potassium dichromate(VI)	reaction product with acidified potassium dichromate(VI)	result of testing reaction product with	
			Tollens' reagent	Fehling's solution
primary	orange to green	aldehyde	silver mirror forms	brick-red precipitate
secondary	orange to green	ketone	no change	no change
tertiary	no change	none	no change	no change

Check your understanding

1. What does [O] in this equation represent?

$$CH_3CH_2OH + [O] \rightarrow CH_3CHO + H_2O$$

2. Write equations for the reactions of the following alcohols with acidified potassium dichromate(VI):
   a butan-1-ol
   b butan-2-ol
3. Describe how you could distinguish between primary, secondary, and tertiary alcohols.

14.07 Distillation and organic tests

OBJECTIVES

already from AS Level, you know that

- primary alcohols can be oxidized to aldehydes using acidified potassium dichromate(VI)
- bromine water is used to test for unsaturation
- Tollens' reagent and Fehling's solution are used to test for aldehydes

and after this spread you should know how to

- distil a product from a reaction safely
- carry out some organic tests competently

One of the recommended Practical Skills Assessment tasks is to distil a product from a reaction. This might involve preparing ethanal by oxidizing ethanol or preparing cyclohexene by dehydrating cyclohexanol. Another recommended Practical Skills Assessment task is to carry out some organic tests.

Distilling a product from a reaction

You will be assessed on your ability to

- set up the apparatus correctly
- work safely and competently
- heat carefully and only for as long as necessary
- obtain a satisfactory yield of product

Preparing ethanal

Ethanol can be oxidized to ethanal by heating it with potassium dichromate(VI), acidified with dilute sulfuric acid. The ethanal is prevented from oxidizing further to ethanoic acid if it is distilled as it forms.

Potassium dichromate(VI) is toxic and you should avoid skin contact with it. Take great care around the balance if you are asked to weigh some potassium dichromate(VI) to make your own solution. The bright orange crystals are very obvious if you make a mess. Here are some typical steps you might take to prepare ethanal from ethanol.

- Add about $5\,cm^3$ of $0.1\,mol\,dm^{-3}$ potassium dichromate(VI) solution to a boiling tube.
- Add about $5\,cm^3$ of dilute sulfuric acid, and shake the tube from side to side to mix.
- Add no more than $2\,cm^3$ of ethanol dropwise, shaking the tube between each addition.

You will see a colour change when you add the ethanol.

Distil the ethanal.

- Connect a delivery tube and bung to the boiling tube.
- Apply *gentle* heat to distil $2–3\,cm^3$ of your product into a test tube – do not distil until the boiling tube is empty.

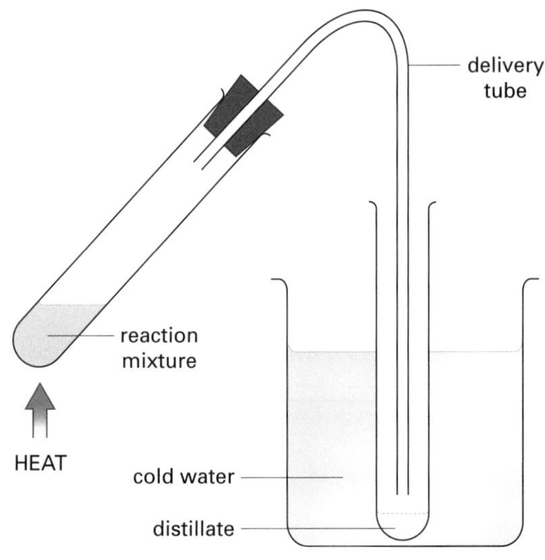

A simple laboratory distillation. Remember to clamp the boiling tube.

Make sure that none of the reaction mixture splashes over. It is very easy for this to happen and you will certainly lose marks if it does. It will be very obvious that you have made a mistake, as the product should be clear and colourless.

Carrying out some organic tests

You will be assessed on your ability to

- work safely and competently
- use sensible volumes of the reagents
- obtain the correct observations for each substance

You know how to use bromine water to test for unsaturation, and Tollens' reagent and Fehling's solution for aldehydes. Since you will be following a set of instructions given to you, it is possible that you might have to carry out tests for alcohols or carboxylic acids, too.

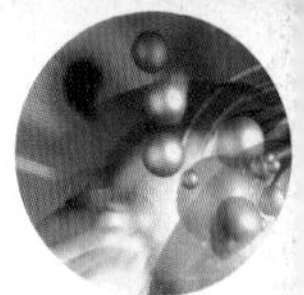

Substance	Test	Observations	Inference
Ethanal	Warmed with fresh Tollen's reagent	Silver mirror forms	Aldehyde present
A			
B			
C			

You may wish to draw a table like this one. It has space to record what you did, your observations, and what you think your observations mean.

Tests for alkenes and aldehydes

Alkenes

Alkenes are unsaturated hydrocarbons. You can test for unsaturation using bromine. But this is a toxic, corrosive, and volatile liquid. Bromine water is safer to use.

- Add about 1 cm^3 of your test liquid to a clean test tube.
- Add a few drops of bromine water and stopper the test tube.
- Shake the test tube from side to side.

observations	inference
brown colour disappears	alkene present
no colour change	alkene not present

Typical observations and inferences for the bromine water test.

Alkenes often give off harmful, irritating, and flammable vapours. You may need to carry out the test in a fume cupboard. Remember to check with your teacher about how to dispose of the contents of the test tube safely. Alkenes are not soluble in water.

Aldehydes

You can test for the presence of aldehydes using Tollens' reagent

- Add a drop of dilute sodium hydroxide to 1 cm^3 of aqueous silver nitrate.
- Add just enough ammonia solution to dissolve the brown precipitate.
- Add a few drops of the test substance and warm in a hot water bath.

observations	inference
silver mirror forms	aldehyde present
no change	aldehyde not present

Typical observations and inferences for a test with Tollens' reagent. Note that the absence of a silver mirror may also mean that the concentration of aldehyde is too low to detect, or your technique is not very good.

Check your understanding

1. **a** Describe two things that would reduce the number of marks obtained when distilling an organic product.
   b Explain how the amount of heat reaching the reaction mixture can be controlled.
2. Describe three precautions you should take to work competently when carrying out tests for alkenes and alcohols.

14.08 Alkenes from alcohols

OBJECTIVES

already from AS Level, you understand

- that alcohols are produced industrially by hydration of alkenes in the presence of an acid catalyst
- concurrent substitution and elimination in haloalkane reactions

and after this spread you should

- know that alkenes can be formed from alcohols by acid-catalysed elimination reactions
- appreciate that this method provides a possible route to polymers without using monomers derived from oil

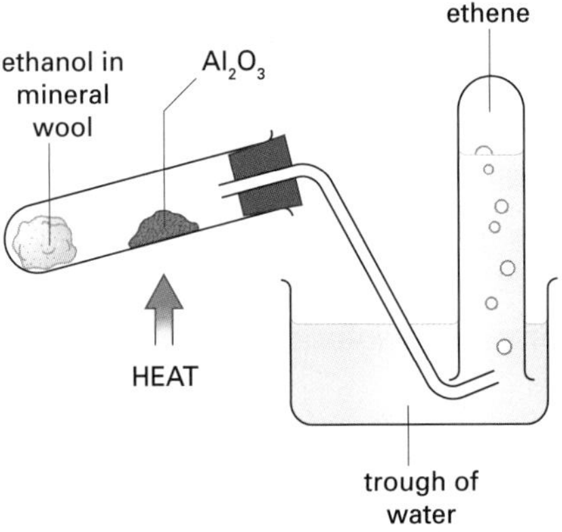

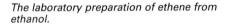

The laboratory preparation of ethene from ethanol.

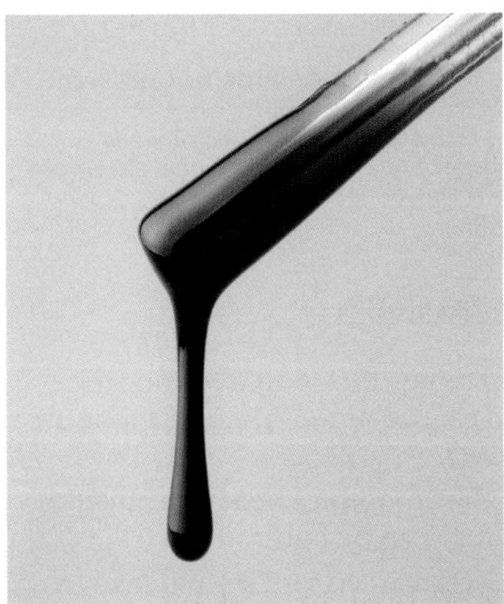

How will we make the polymers we need when oil runs out?

Alcohols can be made from alkenes in two stages using sulfuric acid, or directly using steam. For example, ethanol can be made from ethene:

$$CH_2{=}CH_2 + H_2O \rightarrow CH_3CH_2OH$$

This is an addition reaction because a molecule is added across the double bond in an unsaturated compound to form a saturated compound. The reaction will work in reverse, too. For example, ethene can be made from ethanol:

$$CH_3CH_2OH \rightarrow CH_2{=}CH_2 + H_2O$$

This is an elimination reaction because a small molecule is removed from a saturated compound to form an unsaturated compound. In this case the small molecule is water, so the reaction is also a **dehydration** reaction.

Elimination in alcohols

Alcohols can be dehydrated by

- heating them to about 180°C with concentrated sulfuric acid, which acts as a catalyst for the reaction.
- boiling them and passing the vapour over aluminium oxide heated to about 300°C.

In each case, the ethene can be collected over water.

You do not need to know the mechanism for elimination in alcohols. But it is essentially the reverse of the mechanism for the direct hydration of ethene with steam, with a carbocation intermediate. This means that elimination reactions are most likely to happen in tertiary alcohols, because these produce a tertiary carbocation. Tertiary carbocations are more stable than secondary carbocations, which in turn are more stable than primary carbocations.

The reaction conditions are chosen so that thcy favour the elimination reaction, rather than the reverse process. Elimination in alcohols is an endothermic process, so it is favoured by a high temperature.

Using the dehydration of alcohols

The dehydration of alcohols provides a possible route to polymers without the need to use crude oil. Ethanol is produced by the fermentation of sugars from plant materials. This can be converted to ethene using elimination. The ethene can then be used as the monomer for producing poly(ethene). Since the plant materials are renewable resources, this process would let us make poly(ethene) without the need for crude oil.

Only around 8% of crude oil is used directly or indirectly for manufacturing polymers, so the process may not make a significant contribution to conserving oil reserves. But it may come to our rescue when oil becomes very expensive because of dwindling supplies, or when it runs out completely. Poly(ethene) is not the only polymer in common use. Could other polymers such as poly(propene) be made this way, too?

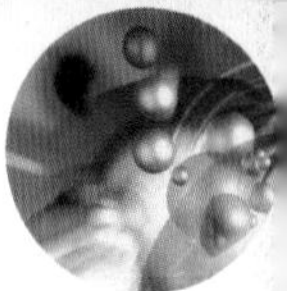

A possible route to poly(propene)

You have learned enough organic chemistry in this Unit to understand how poly(propene) might be manufactured using ethanol from fermentation as a raw material. Here is a possible route.

In this way it is possible to make propene from ethanol. The propene can then be used as the monomer for making poly(propene).

Step 1 Convert ethanol into ethene

Heat ethanol to around 180°C (with sulfuric acid as a catalyst) to produce ethene:

$$CH_3CH_2OH \rightarrow CH_2{=}CH_2 + H_2O$$

This is an elimination reaction.

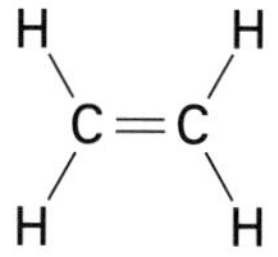

Ethene.

Step 2 Convert ethene into bromoethane

Warm ethene with hydrogen bromide:

$$CH_2{=}CH_2 + HBr \rightarrow CH_3CH_2Br$$

This is an electrophilic addition reaction.

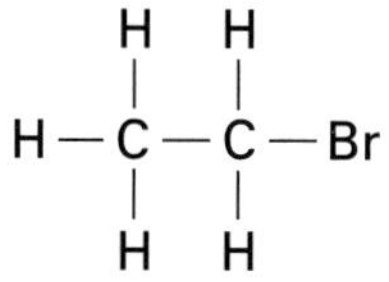

Bromoethane.

Step 3 Convert bromoethane into propanenitrile

React bromoethane with hot ethanolic sodium hydroxide solution:

$$CH_3CH_2Br + CN^- \rightarrow CH_3CH_2CN + Br^-$$

This is a nucleophilic substitution reaction. Note that the carbon chain length increases in this reaction.

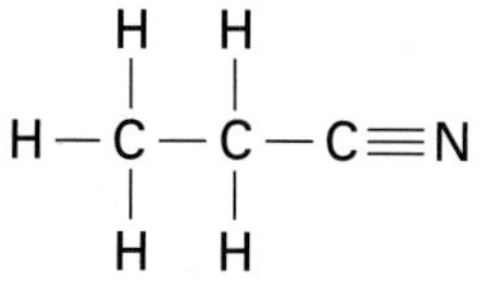

Propanenitrile.

Step 4 Convert propanenitrile into propanoic acid

Reflux propanenitrile with dilute hydrochloric acid:

$$CH_3CH_2CN + 2H_2O + HCl \rightarrow CH_3CH_2COOH + NH_4Cl$$

This is a hydrolysis reaction.

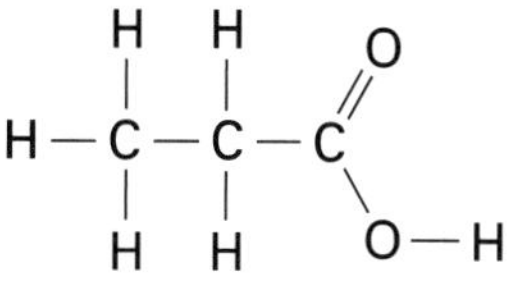

Propanoic acid.

Step 5 Convert propanoic acid into propan-1-ol

You do not need to know this reaction, but it is the reverse of oxidizing a primary alcohol to a carboxylic acid with excess acidified potassium dichromate(VI):

$$CH_3CH_2COOH + 4[H] \rightarrow CH_3CH_2CH_2OH + H_2O$$

The [H] in the equation stands for the reducing agent, just as [O] stands for an oxidizing agent in oxidation reactions.

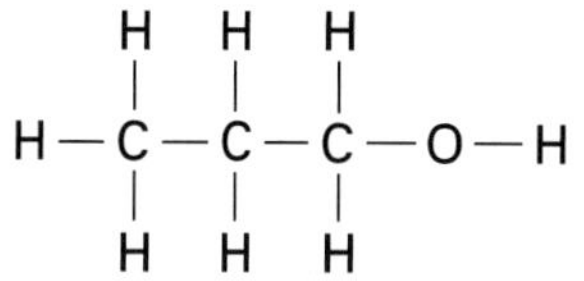

Propan-1-ol.

Step 6 Convert propan-1-ol into propene

This is step 1 but using propan-1-ol instead of ethanol:

$$CH_3CH_2CH_2OH \rightarrow CH_3CH{=}CH_2 + H_2O$$

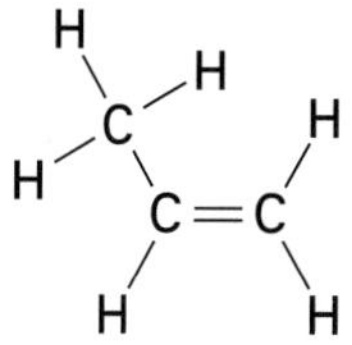

Propene.

This is another elimination reaction.

Check your understanding

1. Suggest why poly(propene) is currently manufactured using propene obtained by cracking crude oil fractions, rather than using propene obtained by processing ethanol from fermentation.
2. The overall reaction for producing propene from ethanol is:

   $$CH_3CH_2OH + HCN + HCl + 4[H] \rightarrow NH_4Cl + CH_3CH{=}CH_2 + H_2O$$

   a Calculate the atom economy for producing propene this way.

   b Suggest why the yield of propene might be low.

15.01 High-resolution mass spectrometry

OBJECTIVES

already from AS Level, you

- can describe how the mass spectrometer works
- know how to work out relative isotopic mass from mass spectrum data
- can use mass spectrum data to work out relative atomic mass and relative molecular mass
- can recall that mass spectrometers in planetary space probes can be used to identify elements

and after this spread you should

- understand that high-resolution mass spectrometry can be used to determine the molecular formula of a compound from the precise mass of the molecular ion

The mass spectrometer and the infrared spectrometer (described on the next spread) provide two very useful instrumental methods of analysis. Instruments have several advantages over traditional methods of chemical analysis, such as titration using a burette. These include

- *Sensitivity*: instruments can analyze tiny amounts of samples. This is useful if a sample is scarce or expensive, or if it is a forensic sample that cannot be replaced.
- *Accuracy*: instruments are very accurate once they have been calibrated correctly. The use of international standard substances means that results are valid from one laboratory to another.
- *Speed*: instruments can analyse samples continuously, 24 hours a day if necessary. They do not get bored and they rarely make mistakes.

Instruments can also be miniaturized and automated for use in space probes. It is not possible to send a chemist to other planets such as Mars yet. But it is possible to send analytical instruments there on space probes.

The Phoenix mission to Mars was launched in 2007. The lander carried a mass spectrometer to help in the search for ice in the soil at the planet's North Pole.

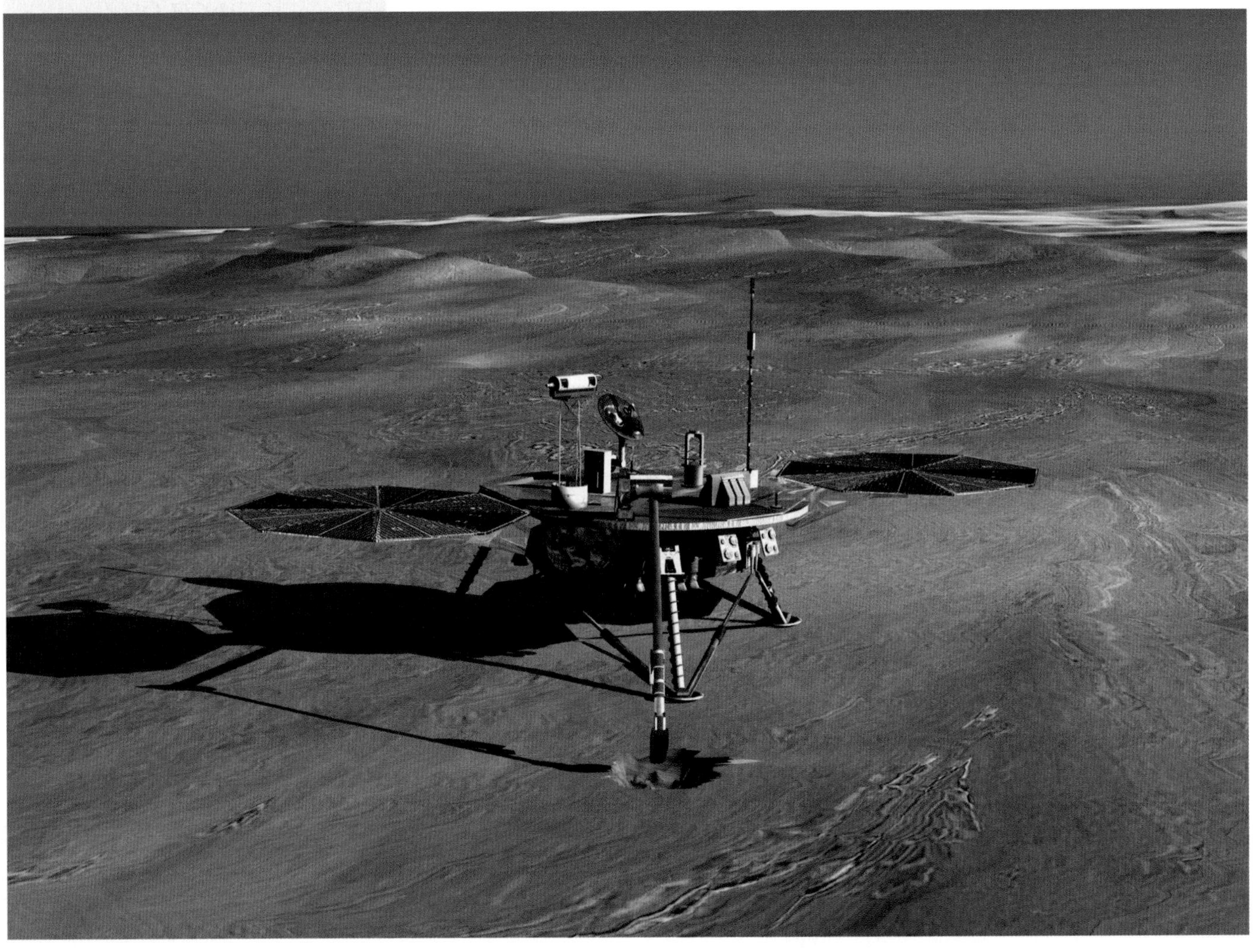

High-resolution mass spectrometry

You discovered in Unit 1 how mass spectrum data from the mass spectrometer can be used to work out relative isotopic mass, relative atomic mass, and relative molecular mass. The **high-resolution mass spectrometer** can measure the mass of a molecular ion to several decimal places. Using very precise relative atomic masses, molecules with very similar relative molecular masses can be distinguished from each other.

		relative atomic mass, A_r	
element	**symbol**	**1 dp**	**4 dp**
hydrogen	H	1.0	1.0079
carbon	C	12.0	12.0107
nitrogen	N	14.0	14.0067
oxygen	O	16.0	15.9994
chlorine	Cl	35.5	35.4532

Some common elements with their one decimal place and four decimal place relative atomic masses. Remember that $A_r(^{12}C) = 12.0000$ by definition.

For example, methanal and ethane have the same relative molecular masses in low-resolution mass spectrometry:

- methanal, $M_r(HCHO) = (2 \times 1.0) + 12.0 + 16.0 = 30.0$
- ethane, $M_r(C_2H_6) = (2 \times 12.0) + (6 \times 1.0) = 30.0$

Their molecular ions would have the same m/z ratio in the mass spectrum. You could not know whether the sample was methanal or ethane. It is a different situation in their high-resolution mass spectra:

- methanal, $M_r(HCHO) = (2 \times 1.0079) + 12.0107 + 15.9994 = 30.0259$
- ethane, $M_r(C_2H_6) = (2 \times 12.0107) + (6 \times 1.0079) = 30.0688$

The difference in the two precise relative molecular masses is very small, but it is enough to tell the two compounds apart in a high-resolution mass spectrum.

Check your understanding

1. The molecular ion peak in a high-resolution mass spectrum has an *m/z* ratio of 58.0822.
   a Explain why the sample was not butane, C_4H_{10}, or propanone, CH_3COCH_3.
   b Explain why the sample could have been ethene-1,2-diamine, $C_2N_2H_6$.
2. A forensic scientist tests a small sample of white powder using a high-resolution mass spectrometer. Explain how she could decide whether the sample was glucose, $C_6H_{12}O_6$, or aspirin, $C_9H_8O_4$. Include calculations in your answer.
3. Some natural samples of organic compounds have slightly different isotopic abundances. Explain how this might affect conclusions drawn from analysing a high-resolution mass spectrum.

Worked example

The molecular ion peak in a high-resolution mass spectrum has an m/z ratio of 60.0551. The sample could be ethanoic acid, CH_3COOH, or urea, NH_2CONH_2. Which compound was in the sample?

$M_r(CH_3COOH) = (2 \times 12.0107) + (2 \times 15.9994) + (4 \times 1.0079) = 60.0518$

$M_r(NH_2CONH_2) = (2 \times 14.0067) + (4 \times 1.0079) + 12.0107 + 15.9994 = 60.0551$

The compound must have been urea.

$H_3C-C(=O)-OH$ ethanoic acid

$H_2N-C(=O)-NH_2$ urea

Ethanoic acid and urea have the same M_r values in low-resolution mass spectra but different M_r values in high-resolution mass spectra.

Precise relative atomic masses

Using high-resolution mass spectrometry to determine the molecular formula of a compound in this way depends on having very precise A_r values. Relative atomic masses are constantly reviewed as technology advances. New figures are published in scientific journals. This allows other scientists to study the work so it can be criticized if necessary, or the work repeated to validate it. IUPAC's Commission on Isotopic Abundances and Atomic Weights meets to consider new data. IUPAC publishes its updated tables of relative atomic masses every two years.

15.02 Infrared spectroscopy

OBJECTIVES

already from AS Level, you

- know that carbon dioxide, methane, and water vapour are greenhouse gases, and that these gases *may* contribute to global warming

and after this spread you should understand

- that certain groups in a molecule absorb infrared radiation at characteristic frequencies
- that fingerprinting allows identification of a molecule by comparison of spectra
- the link between absorption of infrared radiation and global warming

Molecular models often use rigid plastic connectors to represent covalent bonds. In reality, covalent bonds can vibrate in different ways, including stretching and bending. Different bonds in different molecules vibrate at different **frequencies**. These frequencies are in the same range as infrared radiation. A bond vibrates even more when it absorbs infrared radiation of the correct frequency. This explains why increasing amounts of greenhouse gases in atmosphere lead to global warming.

The bonds in greenhouse gases such as carbon dioxide, methane, and water vapour absorb infrared radiation emitted from the Earth's surface. They then re-emit it in a random direction. Some infrared radiation eventually escapes into space after being randomly absorbed and re-emitted, but the rest warms the atmosphere up.

The absorption of infrared radiation by covalent bonds is the basis for an analytical technique called **infrared spectroscopy**, often just called IR spectroscopy. An **infrared absorption spectrum** lets you identify a compound by the bonds it contains.

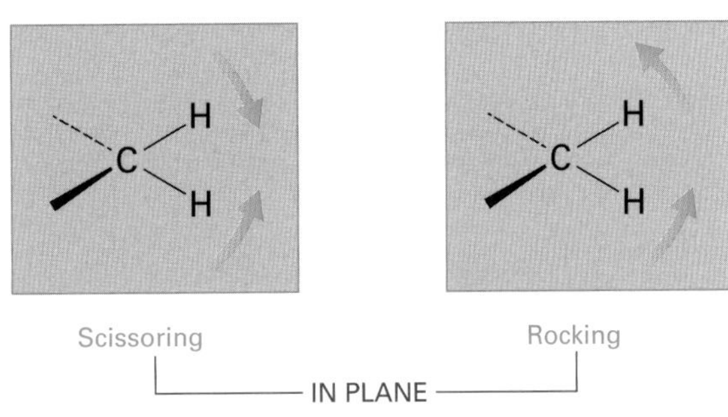

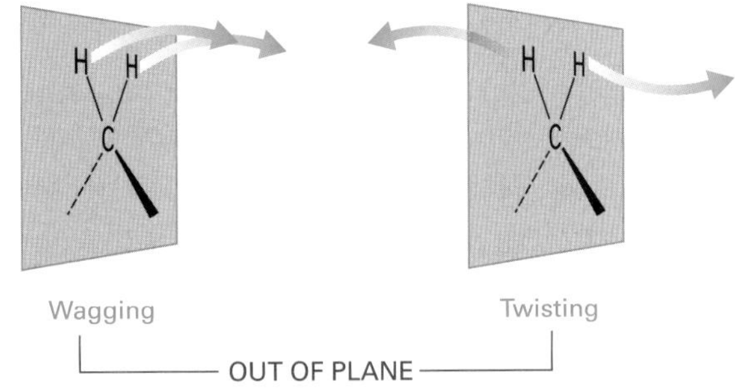

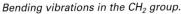

Bending vibrations in the CH_2 group.

The infrared spectrometer

The infrared spectrometer is a device that sends infrared radiation through a sample. It analyses the proportion transmitted through the sample at different frequencies. Certain frequencies are absorbed by the covalent bonds in the sample, so the percentage transmission for those frequencies is low. An IR spectrum contains troughs rather than peaks.

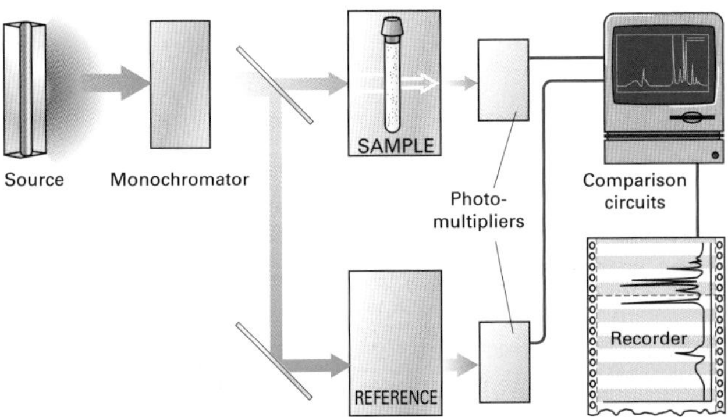

An outline of how the infrared spectrometer works.

The sample is usually dissolved in a solvent, and then sandwiched between two disks made from sodium chloride or potassium bromide. These are ionic compounds, transparent to infrared radiation. The infrared beam is split. It is passed through the sample and a *reference*. The reference consists of the disks and the solvent. The machine compares readings at different frequencies from the sample and the reference. This way, the final spectrum consists of troughs due to the sample only.

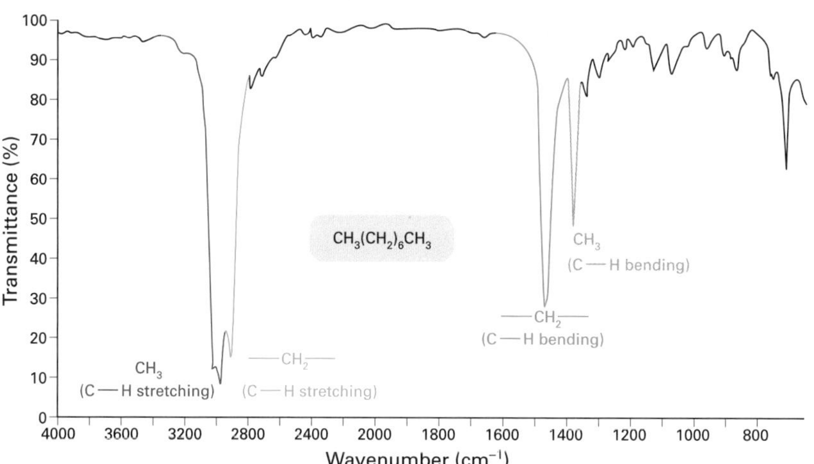

This is an infrared spectrum for octane. The troughs represent absorption of infrared radiation by covalent bonds.

Wavenumber

The horizontal axis on an IR spectrum does not show frequency, as you might expect it to. Instead it shows the **wavenumber** in units of cm^{-1}. The wavenumber is the number of waves that would fit into 1 cm. The higher the frequency, the more waves would fit into 1 cm. So 4000 cm^{-1} on the left of the spectrum represents the highest frequencies. The frequency decreases from left to right.

The fingerprint region

The **fingerprint region** is the part of the IR spectrum from 1500 cm^{-1} to 400 cm^{-1}. It is unique to each compound. An unknown compound is identified by comparing its IR spectrum with a database of IR spectra from known compounds. The unknown compound is identified when its fingerprint region is an exact match to the fingerprint region of a known compound.

Check your understanding

1. Explain why the absorption of infrared radiation can lead to global warming.
2. Explain why infrared spectra have
   a troughs instead of peaks
   b a horizontal axis measured in wavenumber
3. Suggest why samples for infrared spectroscopy are not made up in aqueous solution.
4. a Explain what is meant by the 'fingerprint region'.
   b Explain how you could use the fingerprint region to identify an unknown compound.

15.03 Using infrared spectroscopy

OBJECTIVES

already from AS Level, you understand

- that certain groups in a molecule absorb infrared radiation at characteristic frequencies
- that fingerprinting allows identification of a molecule by comparison of spectra

and after this spread you should

- be able to use spectra to identify particular functional groups and to identify impurities, given data presented in wavenumber form

You can identify particular functional groups using information from an IR spectrum. To do this, you need a table of wavenumbers to work out which bonds are present in the test compound. You may also need some additional information, such as the molecular formula or the results of a simple laboratory test carried out on the compound.

Identifying functional groups

The table shows some bonds and their typical wavenumbers in IR spectra.

bond	wavenumber (cm^{-1})
C–H	2850–3300
C–C	750–1100
C=C	1620–1680
C=O	1680–1750
C–O	1000–1300
O–H in alcohols	3230–3550
O–H in acids	2500–3000

You do not need to remember these numbers but you do need to know how to use them.

You will be given infrared absorption data like these in your examination. Here are some common ways to use information like this.

Alkanes and alkenes

Alkanes and alkenes will both contain C–H bonds. They will probably also contain C–C bonds too, but only alkenes contain C=C bonds. A compound with an absorption band at 1620–1680 cm^{-1} will probably be an alkene.

Aldehydes and ketones

Aldehydes and ketones contain the carbonyl group >C=O. They will have an absorption band at 1680–1750 cm^{-1}. But there are two things to be careful about:

- Carboxylic acids also contain the carbonyl group. You would have to check for the absence of the O–H bond to be sure you had an aldehyde or ketone.
- You cannot distinguish between aldehydes and ketones on the basis of the 1680–1750 cm^{-1} absorption band alone. Tollens' reagent or Fehling's solution would distinguish between them. Aldehydes react with these but ketones do not.

Alcohols

Alcohols contain the hydroxyl group –OH. They will have an absorption band at 3230–3550 cm^{-1} due to the O–H bond. This is very distinctive in IR spectra. The presence of hydrogen bonds causes the trough to be deep and wide.

Worked example

Two species X and Y have the molecular formula C_3H_6O. They have a strong absorption band at 1700 cm^{-1}. Y reacts with Tollens' reagent but X does not. Deduce their structural formulae.

They must contain a C=O bond, so they could be aldehydes or ketones.
Y must be an aldehyde because it reacts with Tollens' reagent, and X must be a ketone because it does not.
X is CH_3COCH_3 and Y is CH_3CH_2CHO.

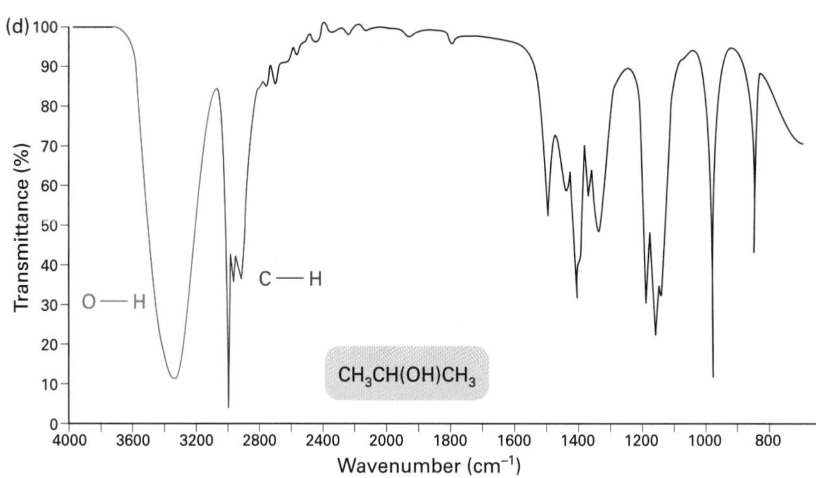

The infrared absorption spectrum of propan-2-ol. Notice the strong trough due the O–H bond.

Carboxylic acids

Carboxylic acids contain the carboxyl group –COOH. This means that they will have two characteristic absorption bands, one at 1680–1750 cm^{-1} due to the C=O bond, and one at 2500–3000 cm^{-1} due to the O–H bond. There are two things to note about this:

- Aldehydes and ketones also have the C=O bond, so they will also have an absorption band at 1680–1750 cm^{-1}. But they will not have one at 2500–3000 cm^{-1} as well, because they do not have O–H bonds.
- Alcohols also have the O–H bond, but the absorption band is at 3230–3550 cm^{-1} and not at 2500–3000 cm^{-1}.

Ethers

Ethoxyethane, $CH_3CH_2OCH_2CH_3$, is an ether. Ethers are compounds that contain the functional group C–O–C. So they contain the C–O bond. Ethers will have an absorption band at 1000–1300 cm^{-1} due to this bond. Note that:

- alcohols and carboxylic acids also have the C–O bond, as in C–O–H. But they will have an absorption band due to their O–H bond that ethers will not have. Alcohols will have an absorption band at 3230–3550 cm^{-1} and carboxylic acids will have one at 2500–3000 cm^{-1}.

Impurities

Impurities can be identified because they will produce absorption bands that should not be there. For example, a sample of ethanol contaminated with ethanoic acid will show an extra absorption band at 2500–3000 cm^{-1} due to the O–H bond in the ethanoic acid.

Check your understanding

1. Two species **A** and **B** have the molecular formula C_4H_8. **A** has an absorption band at 1650 cm^{-1} but **B** does not. Deduce their structural formulae.
2. Two species **C** and **D** have the molecular formula C_2H_6O. Both have an absorption band at 1150 cm^{-1} but **D** also has one at 3300 cm^{-1}. Deduce their structural formulae.
3. A sample of propan-1-ol produced an IR spectrum with two broad absorption bands at 2750 cm^{-1} and 3300 cm^{-1}. Suggest an impurity present in the sample.

OBJECTIVES

already from GCSE, you know that

- elements can be displayed in a periodic table
- elements with similar properties are found near to each other in the periodic table

and after this spread you should be able to

- understand how elements are arranged in the periodic table
- recall the properties of some of the groups in the periodic table

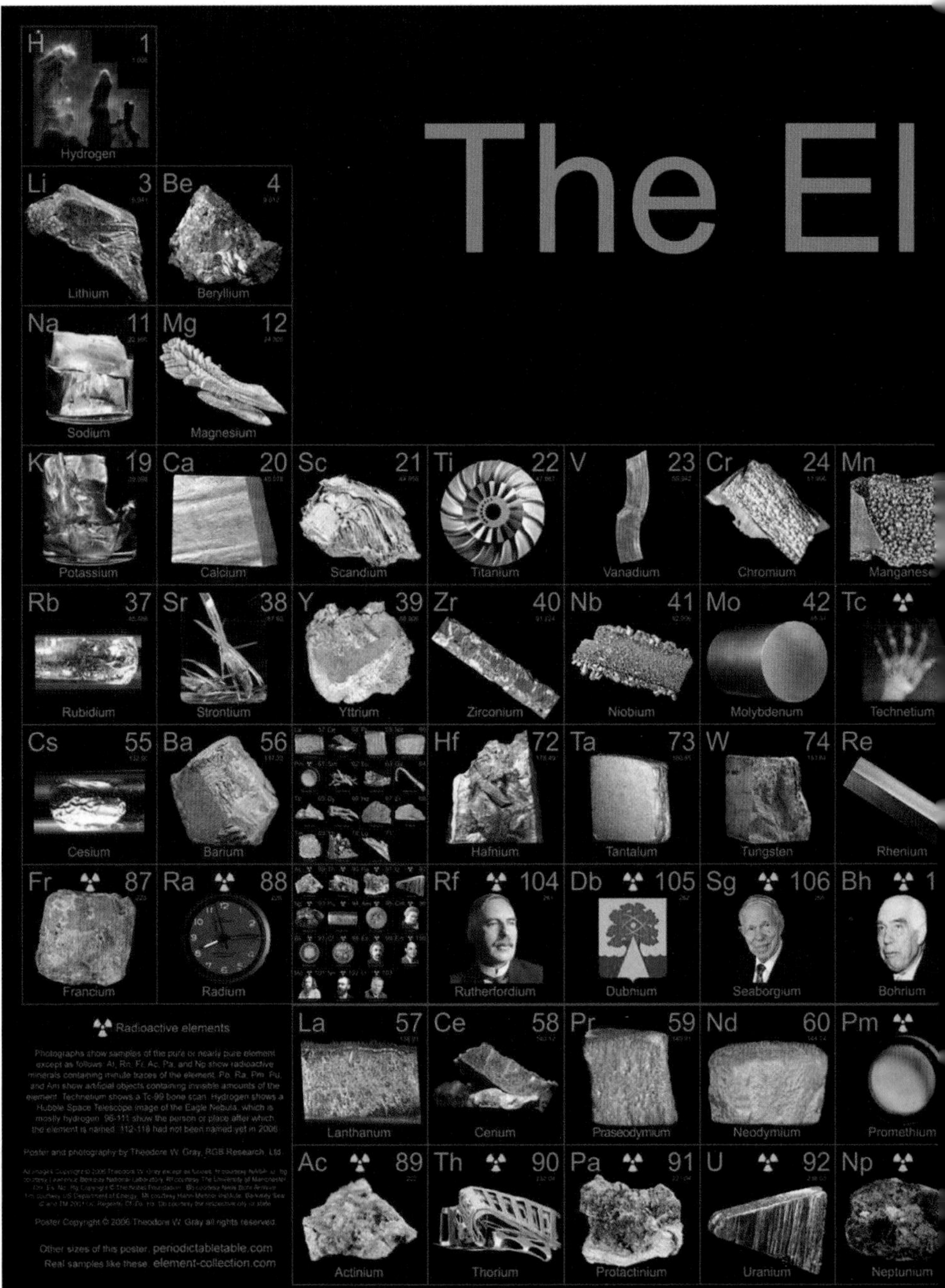

Groups and periods

Elements consist of just one type of atom. There are about a hundred different elements and they are often displayed in a periodic table.

In the modern periodic table the elements are arranged in rows in order of increasing atomic number, so that elements with similar properties occur periodically. This leads to the name the periodic table.

The vertical columns are called *groups* and the horizontal rows are called *periods*. Elements in the same group have similar chemical properties.

Metals

Metallic elements are found on the left hand side of the periodic table. The elements in group 1, including sodium and potassium, are known as the alkali metals. All the alkali metals are very reactive. They react vigorously with water to form a metal hydroxide solution and hydrogen. Reactivity increases down group 1.

Transition metals such as iron and copper are found in the central block of the periodic table. Transition metals are good thermal and electrical conductors and most can be hammered into shape. Mercury is the only metallic element which is liquid at room temperature.

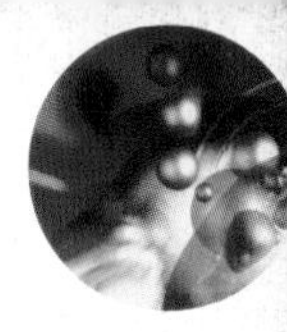

ments

									He 2 Helium
				B 5 Boron	C 6 Carbon	N 7 Nitrogen	O 8 Oxygen	F 9 Fluorine	Ne 10 Neon
				Al 13 Aluminum	Si 14 Silicon	P 15 Phosphorus	S 16 Sulfur	Cl 17 Chlorine	Ar 18 Argon
Co 27 Cobalt	Ni 28 Nickel	Cu 29 Copper	Zn 30 Zinc	Ga 31 Gallium	Ge 32 Germanium	As 33 Arsenic	Se 34 Selenium	Br 35 Bromine	Kr 36 Krypton
Rh 45 Rhodium	Pd 46 Palladium	Ag 47 Silver	Cd 48 Cadmium	In 49 Indium	Sn 50 Tin	Sb 51 Antimony	Te 52 Tellurium	I 53 Iodine	Xe 54 Xenon
Ir 77 Indium	Pt 78 Platinum	Au 79 Gold	Hg 80 Mercury	Tl 81 Thallium	Pb 82 Lead	Bi 83 Bismuth	Po 84 Polonium	At 85 Astatine	Rn 86 Radon
Mt 109 Meitnerium	Ds 110 Darmstadtium	Rg 111 Roentgenium	Uub 112 Ununbium	Uut 113 Ununtrium	Uuq 114 Ununquadium	Uup 115 Ununpentium	Uuh 116 Ununhexium	Uus 117 Ununseptium	Uuo 118 Ununoctium
Eu 63 Europium	Gd 64 Gadolinium	Tb 65 Terbium	Dy 66 Dysprosium	Ho 67 Holmium	Er 68 Erbium	Tm 69 Thulium	Yb 70 Ytterbium	Lu 71 Lutetium	
Am 95 Americium	Cm 96 Curium	Bk 97 Berkelium	Cf 98 Californium	Es 99 Einsteinium	Fm 100 Fermium	Md 101 Mendelevium	No 102 Nobelium	Lr 103 Lawrencium	

Non-metals

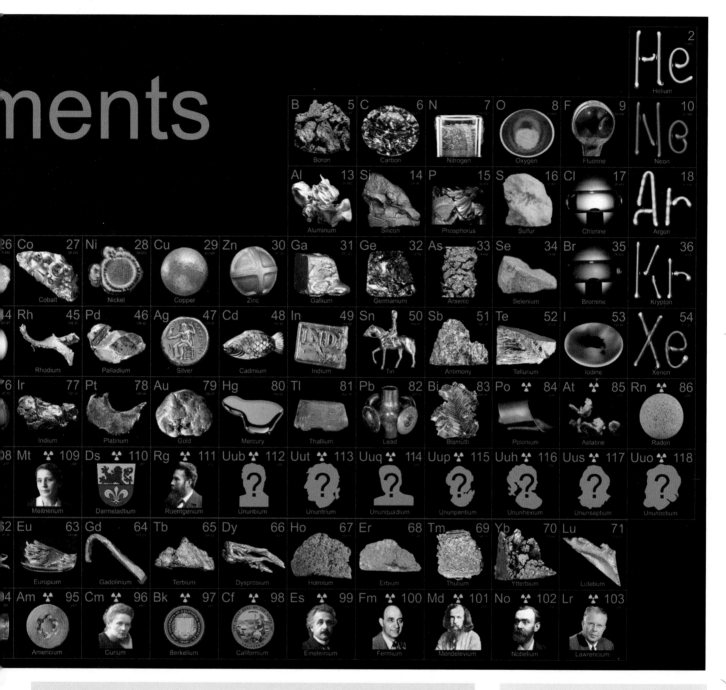

Non-metallic elements are found on the right hand side of the periodic table. The elements in group 7 are known as the halogens. The halogens, which include chlorine and bromine, are typical non-metallic elements. They react with metals to form halides, e.g. sodium reacts with chlorine to form sodium chloride. Bromine is the only non-metallic element which is a liquid at room temperature.

The elements of group 0 are known as the noble gases. They are very unreactive.

Check your understanding

1. How are elements arranged in the periodic table?
2. What are the vertical columns in the periodic table called?
3. Name the two elements which are liquid at room temperature.

S2 Equations

OBJECTIVES

already from GCSE, you know that

- elements are represented by symbols
- that word and symbol equations can be used to represent what happens during chemical reactions

and after this spread you should be able to

- balance symbol equations
- recall some different types of reaction

Word equations

You can use word equations to represent what happens during a chemical reaction. The reaction between sodium and chlorine to form sodium chloride can be represented by the word equation:

sodium + chlorine → sodium chloride

Balanced symbol equations can also be used to represent reactions but they also tell us the proportions in which the substances react together.

The balanced symbol equation for the reaction between sodium and chlorine is:

$$2Na + Cl_2 \rightarrow 2NaCl$$

This equation shows us that two sodium atoms are required to react with each chlorine molecule.

Balancing equations

During chemical reactions atoms are not created or destroyed, they are just rearranged. This means that there must be the same number of each type of atom on both sides of the equation.

Magnesium reacts with oxygen to form magnesium oxide.

magnesium + oxygen → magnesium oxide

You can use the word equation to write the symbols for the reactants and products:

$$Mg + O_2 \rightarrow MgO$$

But the equation doesn't balance; you need to have the same number of each type of atom on both sides of the equation.

There are two oxygen atoms on the reactants side of the equation but only one oxygen atom on the products side so a 2 is placed in front of the MgO:

$$Mg + O_2 \rightarrow 2MgO$$

The oxygen atoms balance but the magnesium atoms don't, so a 2 is placed in front of the magnesium giving:

$$2Mg + O_2 \rightarrow 2MgO$$

so the equation now balances.

State symbols

You can add state symbols to equations to give extra information about the reactants and products. The symbols used are

- (s) for solid
- (l) for liquid
- (g) for gas
- (aq) for aqueous or dissolved in water

Neutralization reactions

The reaction between an acid and a base is called neutralization. Hydrogen ions, H^+, make solutions acidic and hydroxide ions, OH^-, make solutions alkaline.

During neutralization reactions H^+ ions react with OH^- ions to form water molecules:

$$H^+(aq) + OH^-(aq) \rightarrow H_2O(l)$$

Oxidation reactions

During an oxidation reaction a species loses electrons. When magnesium reacts with oxygen to form magnesium oxide, the neutral magnesium atoms lose electrons to form magnesium ions, which have a 2+ charge.

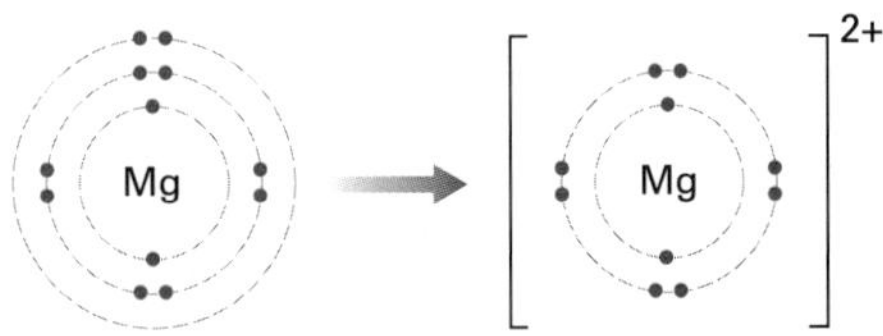

This magnesium atom loses electrons so it is oxidized.

Reduction reaction

During a reduction reaction a species gains electrons.

When magnesium reacts with oxygen to form magnesium oxide, the oxygen atoms gain electrons to form oxide ions, which have a 2– charge.

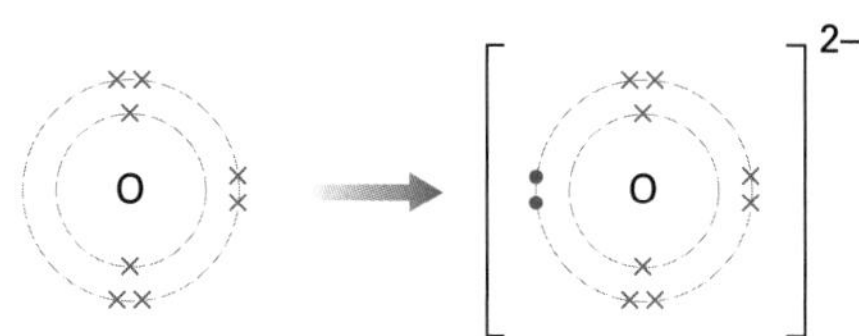

This oxygen atom gains electrons so it is reduced.

The attraction between the positively charged magnesium ions and the negatively charged oxide ions is an example of ionic bonding.

Oxidation and reduction always occur together, so these reactions are sometimes called redox reactions.

Double decomposition reactions

Insoluble salts can be made by reacting together solutions of soluble salts that contain the appropriate ions. The insoluble salt barium sulfate can be made by reacting a solution of barium chloride with a solution of copper(II) sulfate.

Barium chloride + copper(II) sulfate → barium sulfate + copper(II) chloride

$$BaCl_2(aq) + CuSO_4(aq) \rightarrow BaSO_4(s) + CuCl_2(aq)$$

You would see a white precipitate as the barium sulfate is formed. Effectively the two sets of ions have swapped partners and this is called a double decomposition reaction.

Check your understanding

1. Balance these equations
   - a $H_2 + Cl_2 \rightarrow HCl$
   - b $Ca + O_2 \rightarrow CaO$
   - c $K + I_2 \rightarrow KI$
2. Add state symbols to the equation:

$$2Mg + O_2 \rightarrow 2MgO$$

3. Give the name or symbols of the ion that makes solutions acidic?

S3 Formulae

OBJECTIVES

already from GCSE, you know that

- an element is made of one type of atom
- formulae can be used to represent the atoms in a compound
- ionic compounds are formed when metals react with non-metals

after this spread you should know be able to

- understand how to interpret formulae
- be able to work out the formulae of some compounds from their names

Understanding formulae

Chemical formulae are used to show us the atoms in the smallest particle of a substance. You use formulae to represent the atoms in molecules, e.g. carbon dioxide which has the formula CO_2, or giant ionic compounds such as sodium chloride, which has the formula NaCl.

- CO_2 means one atom of carbon to two atoms of oxygen.
- NaCl means one atom of sodium to one atom of chlorine.

Brackets

Some formulae contain brackets. When this happens the small number placed after the bracket indicates what everything inside the brackets must be multiplied by to find the atoms in the smallest particle of the compound. Copper nitrate has the formula $Cu(NO_3)_2$. This means that everything inside the brackets must be multiplied by 2 so in the smallest particle of copper nitrate there is one atom of copper, two atoms of nitrogen, and six atoms of oxygen.

Strontium hydroxide has the formula $Sr(OH)_2$. This means in the smallest particle of strontium hydroxide there is one atom of strontium, two atoms of oxygen, and two atoms of hydrogen.

Naming compounds

Compounds are formed when atoms of two or more elements are chemically combined. When atoms of two elements join together to form a compound the name of the compound is found by combining the names of the two elements. If the compound is formed from a metal and a non-metal, the name of the metal is placed first and the name of the non-metal is changed to end in 'ide'.

Sodium reacts with chlorine to form the compound sodium chloride.

Iron reacts with sulfur to form iron sulfide.

The names of some compounds end in 'ate'. This means that the compound also contains oxygen.

Copper(II) sulfate contains copper, sulfur, and oxygen.

Learn to recognize groups of atoms in ionic formulae and the names used for them. Here are some groups of atoms and how they are named in formulae:

- CO_3 is carbonate
- SO_4 is sulfate
- NO_3 is nitrate
- OH is hydroxide
- NH_4 is ammonium

Writing formulae for ionic compounds

Metals react with non-metals to form ionic compounds. Metal atoms lose electrons to form positive ions, non-metal atoms gain electrons to form negative ions.

positive ions		negative ions	
name	formula	name	formula
hydrogen	H^+	chloride	Cl^-
sodium	Na^+	bromide	Br^-
silver	Ag^+	fluoride	F^-
potassium	K^+	iodide	I^-
lithium	Li^+	hydroxide	OH^-
ammonium	NH_4^+	nitrate	NO_3^-
barium	Ba^{2+}	oxide	O^{2-}
calcium	Ca^{2+}	sulfide	S^{2-}
copper(II)	Cu^{2+}	sulfate	SO_4^{2-}
magnesium	Mg^{2+}	carbonate	CO_3^{2-}
zinc	Zn^{2+}		
lead	Pb^{2+}		
iron(II)	Fe^{2+}		
iron(III)	Fe^{3+}		
aluminium	Al^{3+}		

The formulae of some common ions.

Compounds are neutral so we can use the table of ions to predict the formulae of compounds.

Calcium carbonate consists of calcium ions which have a 2+ charge and carbonate ions which have a 2– charge. The compound must be neutral overall so each Ca^{2+} ion is balanced by one CO_3^{2-} ion giving calcium carbonate the formula $CaCO_3$.

Magnesium chloride consists of Mg^{2+} and Cl^- ions. Each magnesium ion, Mg^{2+} must be balanced by two Cl^- ions giving magnesium chloride the formula $MgCl_2$.

Transition metal ions

Many transition metal atoms can form different ions. Iron commonly forms both Fe^{2+} and Fe^{3+} ions.

When naming transition metal compounds you show the charge on the transition metal ion by writing it in roman numerals in brackets after the name of the metal.

Fe_2O_3 consists of two Fe^{3+} ions and three O^{2-} ions so this compound is named iron(III) oxide.

Check your understanding

1. Identify all the elements, and how many of their atoms are present in the smallest particle of each of these compounds.
   - **a** CO
   - **b** Na_2O
   - **c** $NaNO_3$
2. Name these compounds
   - **a** $MgCO_3$
   - **b** NaOH
   - **c** $CaSO_4$

S4 Structure and bonding

OBJECTIVES

already from GCSE, you know that

- atoms can join together by sharing electrons or giving and taking electrons
- metals react with non-metals to form ionic compounds
- non-metal atoms can join together to form covalent structures
- metals have special properties

and after this spread you should be able to

- describe what ionic bonds are and how they form
- recall some properties of ionic compounds
- describe how covalent structures form
- recall some properties of simple molecules and giant covalent structures
- describe metallic bonding and how it leads to the special properties of metals

Shells and levels

At GCSE you may have answered questions using *shells of electrons*. At AS/A2 you must refer instead to energy levels. So, the outer shell becomes the *highest occupied energy level* and a full shell becomes a *complete energy level.*

Ionic bonding

Ionic bonding involves the transfer of electrons from metal atoms to non-metal atoms. As electrons have a negative charge, the metal atoms become positively charged ions and the non-metal atoms become negatively charged ions. Both sets of ions have a full outer shell of electrons, like a noble gas atom.

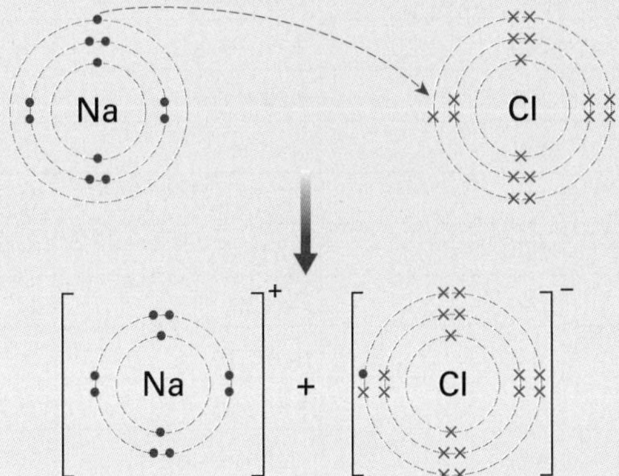

The metal sodium reacts with the non-metal chlorine to form the ionic compound sodium chloride.

Properties of ionic compounds

Ionic compounds have a giant ionic structure. Ionic bonding is the electrostatic attraction between the positively charged metal ions and the negatively charged non-metal ions. Ionic compounds have high melting points and boiling points because a lot of energy must be supplied to overcome the strong forces of attraction between the ions. Ionic compounds do not conduct electricity when solid, but they do conduct electricity when molten or dissolved in water because the ions are able to move.

Covalent bonding

Covalent bonding occurs between non-metal atoms. These atoms join together by sharing pairs of electrons so that all the atoms have a full outer shell, like a noble gas atom.

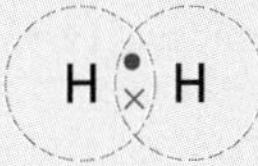

Hydrogen atoms can join together to form a hydrogen molecule.

Properties of simple molecules

Simple molecules are formed when small numbers of non-metal atoms are joined together by covalent bonds. Most simple molecules are gases or liquids at room temperature and those that are solid tend to have quite low melting points. Although there are strong covalent bonds within the molecules, there are only very weak forces of attraction between molecules (intermolecular forces). Only the weak, intermolecular forces must be overcome for the substance to melt or boil, so simple molecules tend to have quite low melting and boiling points.

Simple molecules do not conduct electricity because they do not have an overall electrical charge.

Properties of giant covalent structures

Diamond and graphite are both allotropes of the element carbon. They both have giant covalent structures. In diamond each carbon atom is joined to four other carbon atoms by strong covalent bonds so diamond has a very high melting point and is very hard.

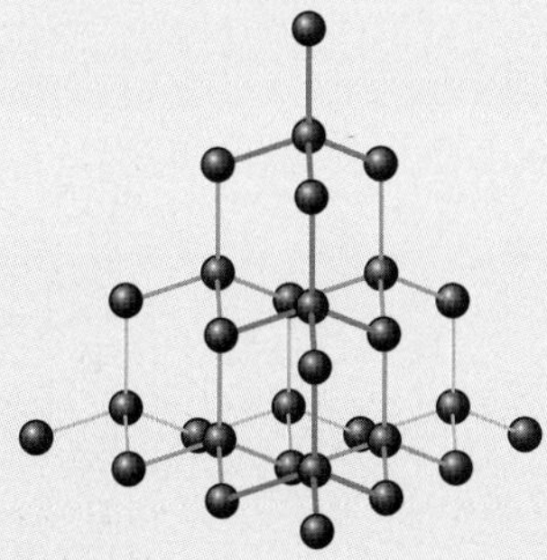

The structure of diamond.

In graphite each carbon atom is bonded to three other carbon atoms in the same layer by strong covalent bonds but the bonding between layers is much weaker. Graphite has a high melting point but the layers can slip over each other quite easily so it is soft and slippery. It can conduct heat and electricity because it contains free electrons within its layers.

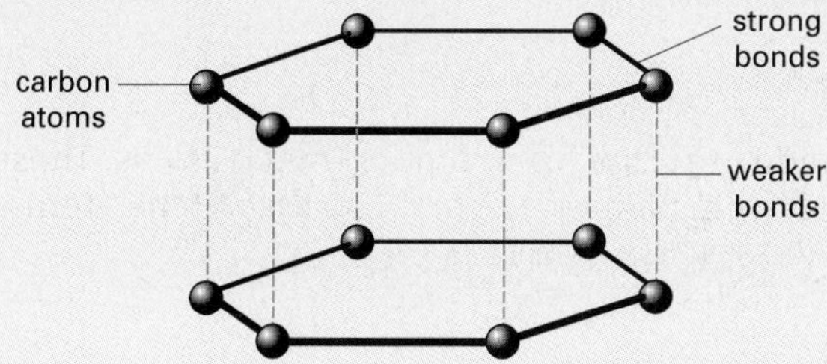

The structure of graphite.

Check your understanding

1. What is an ionic bond?
2. What is a covalent bond?
3. Oxygen is a simple molecule. Why does it have quite a low boiling point?

Metallic bonding

Metals have a giant structure. In metal atoms the outermost electrons are delocalized. This leads to positive metal ions surrounded by negative delocalized electrons.

Metallic bonding is the attraction between these positive metal ions and the negative electrons. Metals are good thermal and electrical conductors because the delocalized electrons are free to move through the whole structure. Metals can be hammered into shape because the layers of atoms can slide over each other.

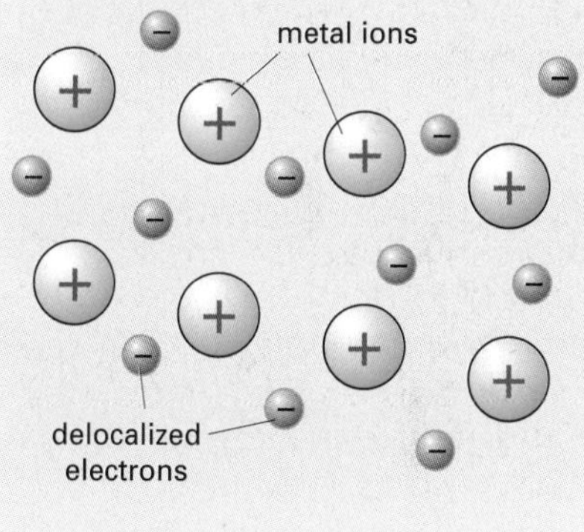

Metallic bonding is the attraction between the positive metal ions and negative delocalized electrons.

S5 Organic chemistry

OBJECTIVES

already from GCSE, you know

- what a hydrocarbon is
- the names of some hydrocarbon molecules
- that some hydrocarbon molecules make useful fuels

and after this spread you should be able to

- recognize some alkane and alkene molecules from their full structural displayed formula
- recall the general formula of alkanes and alkenes
- recall some uses of alkanes and alkenes

The importance of carbon

Organic chemistry is the study of the compounds formed by carbon.

A carbon atom has four electrons in its outer shell. This means that it can form four covalent bonds with other atoms.

Alkanes

The alkanes are a family of organic compounds. All alkane molecules have a similar structure. Alkanes are saturated hydrocarbons because they only contain carbon and hydrogen atoms, and they do not have any carbon–carbon double bonds. The carbon–carbon bonds and the carbon–hydrogen bonds are both very strong, so alkanes are quite unreactive.

Alkanes are represented by the general formula C_nH_{2n+2}. Owing to their similar structures when different alkane molecules do react, they react in a similar way.

The simplest alkane molecule is called methane. It has the formula CH_4.

```
    H
    |
H — C — H
    |
    H
```

Each line represents a covalent bond.

Ethane, propane, and butane are also alkanes.

```
    H   H
    |   |
H — C — C — H
    |   |
    H   H
```

Ethane has the formula C_2H_6.

```
    H   H   H
    |   |   |
H — C — C — C — H
    |   |   |
    H   H   H
```

Propane has the formula C_3H_8.

```
    H   H   H   H
    |   |   |   |
H — C — C — C — C — H
    |   |   |   |
    H   H   H   H
```

Butane has the formula C_4H_{10}.

Uses of alkanes

Many alkane molecules are good fuels. When they are burnt in a good supply of oxygen they produce carbon dioxide and water vapour and release a lot of energy. Methane is used as a fuel in Bunsen burners.

methane + oxygen → carbon dioxide + water

$$CH_4 + 2O_2 \rightarrow CO_2 + 2H_2O$$

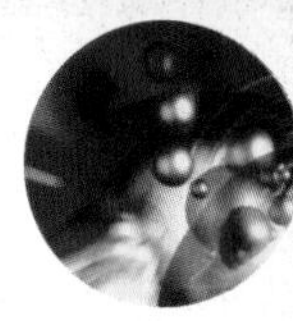

Alkenes

The alkenes are another hydrocarbon family. They all contain carbon–carbon double bonds so they are known as unsaturated hydrocarbons. Alkene molecules are represented by the general formula C_nH_{2n}. Owing to the similarities in their structures all alkene molecules react in a similar way.

```
H       H
 \     /
  C = C
 /     \
H       H
```

Ethene has the formula C_2H_4.

```
    H   H   H
    |   |   |
H — C — C = C — H
    |
    H
```

Propene has the formula C_3H_6.

Uses of alkenes

Alkene molecules are much more reactive than alkane molecules and they have a wide range of uses.

$$n\left(\begin{array}{c} H \quad H \\ C=C \\ H \quad H \end{array}\right) \longrightarrow \left(\begin{array}{c} H \quad H \\ | \quad | \\ -C-C- \\ | \quad | \\ H \quad H \end{array}\right)_n$$

The ethene molecules are called monomers.

Some alkene molecules can be used to make very large molecules called polymers. In this way many ethene molecules can be joined together to make a polymer called poly(ethene) or polythene.

The alkene ethene can also be reacted with steam to form ethanol. The reaction requires a high temperature and a catalyst.

$$\text{ethene} + \text{steam} \rightarrow \text{ethanol}$$

$$C_2H_4 + H_2O \rightarrow C_2H_5OH$$

Check your understanding

1. What is the general formula for
   a alkanes
   b alkenes?
2. How are many alkane molecules used?
3. Name the polymer made from a large number of ethene molecules
4. What is the symbol equation for the reaction between ethene and steam?

S6 Calculations

OBJECTIVES

already from GCSE, you know

- that different atoms have different masses
- because atoms are very small we use their relative atomic masses
- that we can look up the relative atomic mass of an element on a periodic table

and after this spread you should be able to find

- the relative formula mass of a substance
- the number of moles of a substance
- the percentage mass of an element in a compound

Relative atomic mass

The *Relative atomic mass* is used to compare the masses of different atoms. As most naturally occurring elements consist of a mixture of different isotopes, the relative atomic mass of an element is the average mass of an element compared with 1/12 of the mass of a carbon-12 atom. Relative atomic masses do not have units.

Relative formula mass

The formula of a compound shows us the type and number of atoms present in the smallest particle of a substance. The relative formula mass of a substance is worked out by adding together the relative atomic masses of all the atoms present in the formula of the substance.

The relative formula mass of carbon dioxide, CO_2, is worked out by adding together the relative atomic mass of one carbon atom and two oxygen atoms

$$= (1 \times 12.0) + (2 \times 16.0) = 44.0$$

The relative formula mass of calcium carbonate, $CaCO_3$

$$= (1 \times 40.1) + (1 \times 12.0) + (3 \times 16.0) = 100.1$$

The relative formula mass of copper nitrate, $Cu(NO_3)_2$

$$= (1 \times 63.5) + (2 \times 14.0) + (6 \times 16.0) = 187.5$$

You can use a periodic table to find the relative atomic mass of an element. (This Periodic table is photocopiable in the purchaser's institute.)

Times of discovery
- before 1800
- 1800–1849
- 1849-1899
- 1900-1949
- 1949-1999

Period / Group	1	2											3	4	5	6	7	0
1	1.0 H Hydrogen 1																	4.0 He Helium 2
2	6.9 Li Lithium 3	9.0 Be Beryllium 4											10.8 B Boron 5	12.0 C Carbon 6	14.0 N Nitrogen 7	16.0 O Oxygen 8	19.0 F Fluorine 9	20.2 Ne Neon 10
3	23.0 Na Sodium 11	24.3 Mg Magnesium 12											27.0 Al Aluminium 13	28.1 Si Silicon 14	31.0 P Phosphorus 15	32.1 S Sulfur 16	35.5 Cl Chlorine 17	39.9 Ar Argon 18
4	39.1 K Potassium 19	40.1 Ca Calcium 20	45.0 Sc Scandium 21	47.9 Ti Titanium 22	50.9 V Vanadium 23	52.0 Cr Chromium 24	54.9 Mn Manganese 25	55.8 Fe Iron 26	58.9 Co Cobalt 27	58.7 Ni Nickel 28	63.5 Cu Copper 29	65.4 Zn Zinc 30	69.7 Ga Gallium 31	72.6 Ge Germanium 32	74.9 As Arsenic 33	79.0 Se Selenium 34	79.9 Br Bromine 35	83.8 Kr Krypton 36
5	85.5 Rb Rubidium 37	87.6 Sr Strontium 38	88.9 Y Yttrium 39	91.2 Zr Zirconium 40	92.9 Nb Niobium 41	95.9 Mo Molybdenum 42	98.9 Tc Technetium 43	101.1 Ru Ruthenium 44	102.9 Rh Rhodium 45	106.4 Pd Palladium 46	107.9 Ag Silver 47	112.4 Cd Cadmium 48	114.8 In Indium 49	118.7 Sn Tin 50	121.8 Sb Antimony 51	127.6 Te Tellurium 52	126.9 I Iodine 53	131.3 Xe Xenon 54
6	132.9 Cs Caesium 55	137.3 Ba Barium 56	138.9 La Lanthanum 57	178.5 Hf Hafnium 72	180.9 Ta Tantalum 73	183.9 W Tungsten 74	186.2 Re Rhenium 75	190.2 Os Osmium 76	192.2 Ir Iridium 77	195.1 Pt Platinum 78	197.0 Au Gold 79	200.6 Hg Mercury 80	204.4 Tl Thallium 81	207.2 Pb Lead 82	209.0 Bi Bismuth 83	210.0 Po Polonium 84	210.0 At Astatine 85	222.0 Rn Radon 86
7	(223) Fr Francium 87	(226) Ra Radium 88	(227) Ac Actinium 89															

140.1 Ce Cerium 58	140.9 Pr Praseodymium 59	144.2 Nd Neodymium 60	144.9 Pm Promethium 61	150.4 Sm Samarium 62	152.0 Eu Europium 63	157.3 Gd Gadolinium 64	158.9 Tb Terbium 65	162.5 Dy Dysprosium 66	164.9 Ho Holmium 67	167.3 Er Erbium 68	168.9 Tm Thulium 69	173.0 Yb Ytterbium 70	175.0 Lu Lutetium 71
232.0 Th Thorium 90	231.0 Pa Protactinium 91	238.0 U Uranium 92	237.0 Np Neptunium 93	239.1 Pu Plutonium 94	243.1 Am Americium 95	247.1 Cm Curium 96	247.1 Bk Berkelium 97	252.1 Cf Californium 98	(252) Es Einsteinium 99	(257) Fm Fermium 100	(258) Md Mendelevium 101	(259) No Nobelium 102	(260) Lr Lawrencium 103

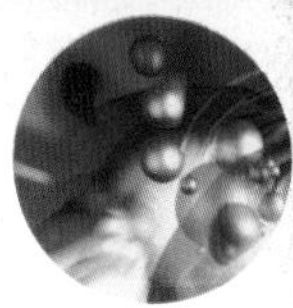

Moles

One mole of any substance contains the same number of particles. The relative formula mass of a substance, in grams, is known as the mass of one mole of that substance. This means;

- one mole of carbon dioxide, CO_2, has a mass of 44.0 g
- one mole of calcium carbonate, $CaCO_3$, has a mass of 100.1 g
- one mole of copper nitrate, $Cu(NO_3)_2$, has a mass of 187.5 g.

We cannot measure the number of moles in a substance directly but we can calculate the number of moles present by dividing the mass of the substance by its relative formula mass.

$$\text{number of moles} = \frac{\text{mass of substance}}{\text{relative formula mass of substance}}$$

The number of moles in 11.0 g of carbon dioxide, CO_2

= 11.0/44.0 = 0.25 moles

The number of moles in 125 g of calcium carbonate, $CaCO_3$

= 125.0/100.1 = 1.25 moles

The number of moles in 375 g of copper nitrate, $Cu(NO_3)_2$

= 375/187.5 = 2.0 moles

GCSE vs AS/A2

GCSE = moles

AS/A2 = mol

A_r numbers

GCSE = whole numbers (except copper, chlorine)

AS/A2 = one decimal place

Percentage composition

We also use the relative formula mass of a substance when want to find out the percentage by mass of an element in a compound.

$$\begin{array}{l}\text{percentage of}\\\text{an element in}\\\text{a compound}\end{array} = \frac{\begin{array}{c}\text{relative atomic}\\\text{mass of}\\\text{the element}\end{array} \times \begin{array}{c}\text{number of atoms}\\\text{of the element}\\\text{in the formula}\end{array}}{\begin{array}{c}\text{relative formula mass of}\\\text{the compound}\end{array}} \times 100$$

The percentage of nitrogen by mass in the compound ammonium nitrate, NH_4NO_3

$$= \frac{14.0 \times 2}{80.0} \times 100 = 35\%$$

Check your understanding

1. What is the relative atomic mass of
   a carbon **b** chlorine **c** argon?
2. Calculate the relative formula mass of
   a carbon monoxide, CO **b** ethene, C_2H_4 **c** butane, C_4H_{10}
3. Calculate the number of moles present in
   a 2.8 g of ethene, C_2H_4 **b** 9.0 g of water, H_2O
   c 2.3 g of ethanol, C_2H_5OH
4. Calculate the percentage by mass of carbon in
   a carbon monoxide, CO **b** carbon dioxide, CO_2

Glossary

absolute temperature: The temperature measured in kelvin, K.

absolute zero: The lowest possible temperature theoretically achievable, –273.15 °C or 0 K.

acceleration: The stage in mass spectrometry where positive ions are speeded up by an electric field.

acid rain: Rain that has been acidified by pollutants such as sulfur dioxide.

actual yield: The mass of product obtained in a reaction.

addition polymer: Polymer formed when many small unsaturated molecules join together.

addition reaction: The adding together of two or more molecules to form one larger molecule.

aldehyde: An organic compound containing the –CHO functional group.

alkaline earth metal: An element from group 2 of the periodic table.

alkane: A saturated hydrocarbon with the general molecular formula C_nH_{2n+2}.

allotrope: Allotropes are different forms of the same element that exist in the same physical state. Diamond and graphite are allotropes of carbon.

alloy: A mixture of two or more metals, or a mixture of a metal and a non-metal.

alpha particle: A particle consisting of two protons and two neutrons, ejected from a nucleus.

amine: An organic compound that contains a nitrogen atom joined to one or more carbon atoms, and two or less hydrogen atoms.

amount of substance: The number of particles present, symbol *n*. It is measured in mole, mol.

anion: A negatively charged ion, attracted to the anode during electrolysis.

antacid: A base taken as a medicine to neutralize excess stomach acid.

aromatic: Containing one or more benzene rings.

atom: The smallest particle of an element that has the properties of that element.

atom economy: The proportion of reactants that are converted into useful products rather than waste products.

atomic number: The number of protons in the nucleus of an atom, symbol *Z*.

atomic radius: The distance from the centre of the nucleus to the outer electrons of an atom.

Aufbau principle: The building up process that describes the filling of atomic orbitals in order of increasing energy.

Avogadro constant: The number of particles in one mole of a substance, 6.022×10^{23}.

Avogadro's principle: The idea that equal volumes of gases contain the same number of molecules, under the same conditions of temperature and pressure.

axial atom: Atom positioned at the top or bottom of a trigonal bipyramidal molecule.

barometer: An instrument for measuring atmospheric pressure.

base peak: The tallest peak in a mass spectrum.

batch process: An industrial process that is started and stopped at intervals. Production of ethanol by fermentation is a batch process.

bent line: The shape of a molecule that contains two bonding pairs and two lone pairs of electrons.

biofuel: A fuel made from the products of living things.

bond angle: The angle between two adjacent bonds on the same atom.

bond: Attractive force between two atoms, ions or molecules.

Boyle's law: The volume of a fixed mass of gas (at a constant temperature) is inversely proportional to its pressure.

burette: Laboratory apparatus used to add precise volumes of liquid during a titration.

carbocation: Ion with a positively charged carbon atom.

carbonyl: An organic compound containing the >C=O functional group, found in aldehydes and ketones.

carboxyl: An organic compound containing the –COOH functional group.

carboxylic acid: Organic acids with the general formula RCOOH. Their names end in -oic acid.

cation: A positively charged ion, attracted to the cathode during electrolysis.

CFC: Abbreviation for chlorofluorocarbon, a hydrocarbon in which some or all the hydrogen atoms are replaced by chlorine and fluorine atoms.

chain isomerism: A type of structural isomerism in which compounds have identical molecular formulae but their carbon atoms are joined together in different arrangements. Chain isomers involve branched and unbranched carbon chains.

charge density: The charge:size ratio of an ion. Small ions with a high charges have large charge densities.

Charles's law: The volume of a fixed mass of gas (at a constant pressure) is proportional to its absolute temperature.

chlorofluorocarbon: See CFC.

coke: Solid produced by heating coal in the absence of air, almost pure carbon.

complete combustion: Burning a fuel in excess oxygen. Carbon dioxide and water vapour are produced from the complete combustion of hydrocarbons.

concordant results: Titres that are in agreement, usually within 0.10 cm^3 of each other.

continuous process: An industrial process in which products are made all the time without any break. Production of iron in the blast furnace is a continuous process.

co-ordinate bond: A covalent bond in which the shared pair of electrons is provided by only one of the bonded atoms. In the bond X→Y, X provides both electrons.

covalent bond: A shared pair of electrons.

cracking: A process used by the petroleum industry to produce shorter alkanes and alkenes from longer alkanes.

cyclic: Hydrocarbons in which there are closed rings of carbon atoms are described as cyclic.

d block: The central section of the periodic table between groups 2 and 3, containing the transition metals.

dative covalent bond: Another name for a co-ordinate bond.

decomposition reaction: A reaction where one substance is broken down into two or more different substances.

deflection: The stage in mass spectrometry where positive ions are moved from their original path by a magnetic field.

dehydration reaction: A reaction where the elements hydrogen and oxygen are removed from a reactant in the ratio of 2:1, effectively the removal of water.

delocalized: Electrons that are free to move between all atoms in a structure are delocalized. Delocalized electrons are found in metals and graphite.

detection: The stage in mass spectrometry where positive ions reach a detector and produce an electrical signal.

diastereoisomers: Stereoisomers that are not mirror-images of each other, sometimes called geometrical isomers.

diatomic: A molecule containing just two atoms.

dimer: A molecule consisting of two monomer molecules joined together.

dipole: Opposite charges separated by a short distance in a molecule or ion.

displace: To replace an atom or ion in a compound in a chemical reaction. For example, chlorine displaces iodine in sodium iodide.

displayed formula: A chemical formula showing all the atoms in a compound and their bonds.

disproportionation: The simultaneous oxidation and reduction of a species.

dot and cross diagram: A diagram showing all of the bonding electrons in a molecule. The electrons in one atom are shown as dots and the electrons in the other atom are shown as crosses.

double covalent bond: A bond in which two atoms are joined by two shared pairs of electrons.

ductile: Easily pulled into a thin wire.

electrolysis: The decomposition of a compound into simpler substances using an electric current.

electron gun: The source of high-energy electrons used to ionize the sample in mass spectrometry.

electron-deficient: An atom with a vacant orbital.

electronegativity: The power of an atom to withdraw electron density from a covalent bond.

electron: Sub-atomic particle with a negative electric charge.

electrophile: A species that can accept a pair of electrons.

electrophilic addition: A reaction in which an electrophile is attracted to a region of high electron density, such as a carbon-carbon double bond, and adds on to an atom or group.

electrostatic: Involving opposite charges.

element: A substance containing atoms which all have the same atomic number.

elimination reaction: A reaction in which a small, simple molecule is removed from a compound, forming a double covalent bond.

empirical formula: A formula that gives the simplest whole number ratio of atoms of each element in a compound.

endothermic: A reaction in which heat energy is absorbed from the surroundings.

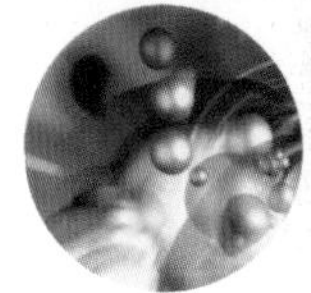

endpoint: Where the indicator just changes colour in a titration.

energy level: A certain fixed amount of energy that electrons in an atom can have, also called a shell.

enzyme: A biological catalyst.

equatorial atom: Atom positioned around the middle of a trigonal bipyramidal molecule.

equivalence point: In an acid-base titration, the point where equal numbers of moles of hydrogen ions and hydroxide ions have reacted.

excess: More than the amount of reactant needed in a reaction.

feedstock: Raw material used in a manufacturing process.

fermentation: The process for making ethanol from sugar using yeast.

fingerprint region: The part of an infrared spectrum that is unique to a particular compound.

first ionization energy: The energy needed to remove one mole of electrons from one mole of gaseous atoms, forming one mole of ions with a single positive charge.

flue gas desulfurization: Removing sulfur compounds from waste gases produced by combustion.

flue gas: Waste gases from the combustion of a fuel, for example in a power station.

fraction: A part of a mixture collected at a particular temperature range by fractional distillation. A crude oil fraction contains hydrocarbons with a similar chain length.

fractional distillation: A method of separating mixtures of liquids or gases according to their boiling temperatures.

free radical: A species that contains an unpaired electron, produced by the homolytic fission of a covalent bond.

frequency: The number of waves per second, measured in hertz, Hz.

full equation: A chemical equation showing the formulae and correct amounts of all the reactants and products in a reaction.

functional group: An atom, or group of atoms, in a molecule which determines its chemical properties.

gas constant: The constant used in the ideal gas equation. It has the symbol R and is approximately 8.31 $J\ K^{-1}\ mol^{-1}$.

gas syringe: A glass syringe used for collecting gases and measuring their volumes.

giant covalent: A structure in which very many atoms are joined by covalent bonds to form a regular structure. Diamond and graphite have giant covalent structures.

global warming: Increasing worldwide average temperatures.

greenhouse effect: Absorption of thermal energy by certain gases in the atmosphere, keeping the planet warmer than it would otherwise be.

greenhouse gas: An atmospheric gas that traps infrared radiation that would otherwise be radiated from the Earth's surface into space.

halide ion: A negatively charged ion formed when a halogen atom gains an electron.

high resolution mass spectrometer: Device capable of measuring relative atomic masses and relative molecular masses to a high degree of precision.

homologous series: A series of compounds with the same general formula and functional group. Each member differs from the next by the presence of one more $-CH_2$ group.

homolytic fission: Breaking of a covalent bond so that each atom takes one electron from the shared pair, becoming a radical.

Hund's rule: Only when all the orbitals in a particular sub-level contain an electron do electrons begin to occupy the orbitals in pairs.

hydration: The addition of water across a double bond.

hydrocarbon: A compound containing hydrogen and carbon atoms only.

hydrogen bond: An intermolecular force between a lone pair of electrons on an N, O or F atom in one molecule, and an H atom joined to an N, O, or F atom in another molecule.

hydrolysis: The splitting of a compound by reaction with water.

hydroxyl group: An –OH group, the functional group found in alcohols.

ideal gas equation: The equation that describes the relationship between pressure, volume, amount of substance, and absolute temperature of a gas: $pV = nRT$.

incomplete combustion: Burning of a fuel in a restricted amount of oxygen.

indicator: A substance that changes colour according to the pH of a solution.

induced dipole: An uneven distribution of charge in a molecule or atom, caused by a charge in an adjacent particle.

infrared radiation: Electromagnetic radiation with a lower frequency than visible light, felt as heat.

infrared absorption spectrum: Spectrum produced when infrared radiation of various frequencies is absorbed by covalent bonds in a molecule.

infrared spectroscopy: Method used to analyze compounds by their absorption of infrared radiation.

initiation: The first stage in a free radical reaction.

intermediate: Unstable species produced during a reaction before the final product is made.

intermolecular force: Weak attractive force between molecules.

ionic bond: Electrostatic force of attraction between oppositely charged ions.

ionic compound: A compound made up of oppositely charged ions.

ionic equation: Chemical equation showing the separate ions in a chemical reaction and any essential reactant or product.

ionic radius: The distance from the centre of the nucleus to the outer electrons of an ion.

ionization: Producing an electrically charged particle by adding or removing electrons from an atom or molecule.

ionized: An atom or molecule that has gained or lost electrons is said to be ionized.

ion: A charged particle formed when an atom or molecule gains or loses one or more electrons.

isoelectronic: Having the same electron configuration as another species.

isomer: Compounds with the same molecular formula but different structural formulae.

isotope: Atoms with the same number of protons but different numbers of neutrons in their nuclei.

kelvin: The unit of absolute temperature, symbol K.

ketone: An organic compound containing the carbonyl functional group >C=O, and with the general formula R_1COR_2.

lattice: A regular arrangement of atoms, ions or molecules in a structure.

limiting reactant: A reactant that is completely used up before the other reactants are converted into products.

linear molecule: A molecule with all of its atoms in a straight line.

lone pair: A pair of electrons in the highest occupied energy level that are not used in bonding.

macromolecular: See giant covalent.

main chain: The longest chain of carbon atoms in an organic compound.

malleable: Can be bent or hammered into shape without breaking.

mass number: The total number of protons and neutrons in the nucleus of an atom, symbol *A*.

mass spectrometer: An instrument used to determine the relative atomic mass of an element or relative molecular mass of a compound. The structure of a complex molecule can be worked out by analysis of a mass spectrum.

mass spectrum: The output from a mass spectrometer, plotting relative abundance against mass to charge ratio, *m/z*.

mass to charge ratio: The mass of an ion divided by its charge, symbol *m/z*.

metallic bond: The electrostatic force of attraction between metal ions and the delocalized electrons in a metallic lattice.

metalloid: Element with properties that are intermediate between the properties of a metal and the properties of a non-metal.

mol: The symbol for amount of substance measured in moles.

molar volume: The volume occupied by one mole of gas at a specified temperature and pressure.

molarity: The concentration of a solution measured in moles of solvent per cubic decimetre of solution, $mol\ dm^{-3}$.

mole: The amount of substance that contains as many particles as there are atoms in exactly 12 g of ^{12}C.

molecular crystal: Covalent molecules held together in a regular arrangement by intermolecular forces.

molecular formula: A formula that gives the actual number of atoms of each element in a molecule.

molecular ion: In mass spectrometry, the ion that produces a peak in the mass spectrum at the highest *m/z* value.

molecular sieve: Porous materials, such as zeolites, that let some molecules pass through but not others.

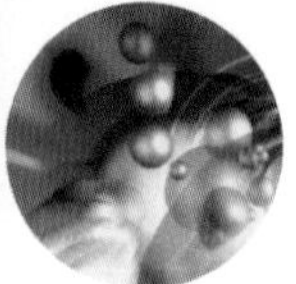

molecule: A particle containing two or more atoms joined together.

monomer: A small molecule that can join together to make a polymer.

monoprotic: An acid containing one replaceable hydrogen ion, such as HCl and HNO_3.

neutron: A neutral sub-atomic particle found in the nucleus of an atom.

nitrile: An organic compound containing the –C≡N functional group.

non-polar: Having no dipole. A molecule with polar bonds may be non-polar if its shape is such that the dipoles cancel each other out.

non-renewable resource: Resource that cannot be replaced once it has all been used up. Fossil fuels and metal ores are non-renewable resources.

nucleophile: A species with a lone pair of electrons that is available to form a co-ordinate bond.

nucleophilic substitution: A chemical reaction in which one nucleophile replaces another in a molecule.

nucleus: The central part of the atom, made from protons and neutrons.

octahedral: The shape of a molecule containing six bonding pairs of electrons.

orbital: The volume of space in an atom where one or two electrons are most likely to be found.

ore: A mineral from which metals can be extracted and purified.

ozone: An allotrope of oxygen with the formula O_3.

ozone layer: The part of the atmosphere with the greatest concentration of ozone. Found in the stratosphere, it absorbs harmful ultraviolet radiation from the Sun.

p block: The part of the periodic table containing groups 3 to 0.

Pauli exclusion principle: The idea that an orbital cannot hold more than two electrons.

Pauling electronegativity scale: A scale showing the ability of elements to withdraw electron density from a covalent bond. The larger the number, the more electronegative the element.

percentage composition: The percentage mass of a compound due to a particular element.

percentage difference: The difference between an experimental result and an expected result, shown as a percentage of the expected result.

percentage yield: The actual yield of a product shown as a percentage of the expected yield.

permanent dipole–dipole forces: Attractive forces that exist between polar molecules.

plum pudding model: A disproved model of the atom in which electrons move in sea of positive charge.

polar bond: A covalent bond between atoms with different electronegativities.

polarized: Having opposite charges, separated by a small distance.

polyatomic ion: An ion containing more than one atom.

polymer: A large molecule made up of many repeating units or monomers.

position isomerism: A type of isomerism where the functional group can be joined at different places on the carbon skeleton.

precipitate: An insoluble solid formed when two solutions are mixed.

primary haloalkane: An haloalkane with one carbon atom, or where the carbon atom carrying the halogen atom is directly attached to just one other carbon atom.

product: A substance made in a chemical reaction.

propagation: The stage in a free radical mechanism where a particular radical is used in one reaction then produced again in a subsequent reaction.

proton: A positively charged sub-atomic particle found in the nucleus of an atom.

radioactive: A substance that produces radiation is said to be radioactive.

raw material: Substance in its natural state intended for use in a chemical process.

reactant: A substance used in a chemical reaction.

reaction mechanism: A step by step description of how a reaction happens.

relative abundance: The proportion of a particular species in a sample.

relative atomic mass: The mean mass of an atom of an element compared to one-twelfth the mass of a ^{12}C atom, symbol A_r.

relative charge: The charge of a sub-atomic particle compared to the charge on a proton, taken as +1.

relative formula mass: The mean mass of a unit of a compound compared to one-twelfth the mass of a ^{12}C atom, symbol M_r.

relative isotopic mass: The mass of an atom of a particular isotope compared to one-twelfth the mass of a ^{12}C atom.

relative mass: The mass of a sub-atomic particle compared to the mass of a proton, taken as 1.

relative molecular mass: The mean mass of a molecule compared to one-twelfth the mass of a ^{12}C atom, symbol M_r.

repeating unit: The short sequence of atoms that is repeated many times in a polymer.

roast: Heating strongly in a stream of air, usually applied to metal ores.

s block: The part of the periodic table containing groups 1 and 2.

saturated: A compound containing only single covalent bonds between carbon atoms.

semiconductor: A substance that is an electrical insulator at room temperature, but a conductor when warmed or when other elements are added to it.

shield: In an atom with more than one occupied energy level, a decrease in the force of attraction between an electron and the nucleus because of electrons in lower energy levels.

shortened structural formula: Abbreviated structural formula in which the arrangement of atoms and group is shown without drawing bonds. For example, hexane would be $CH_3(CH_2)_4CH_3$.

side chain: A shorter chain of carbon atoms attached to a longer main chain, a branch.

simple molecule: A molecule containing just a few atoms, such as O_2 and NH_3.

skeletal formula: A type of displayed formula in which the symbols for carbon atoms are left out.

solute: The substance that will dissolve in a solvent.

solvent: The substance in which a solute will dissolve.

sparingly soluble: Almost insoluble but a very small amount will dissolve.

species: An atom, molecule or ion.

spectator ion: An ion that appears on both sides of an equation but does not take part in the reaction.

square pyramidal: The shape of a molecule that contains five bonding pairs and one lone pair of electrons.

standard solution: A solution whose exact molarity is known.

standard temperature and pressure: 273 K and 100 kPa.

state symbol: Symbols used in chemical equations to show the state of the substance: solid (s), liquid (l), gas (g), aqueous or dissolved in water (aq).

states of matter: Solid, liquid, and gas.

stereoisomer: Molecules with the same structural formula but a different arrangement of their bonds in space.

stratosphere: Upper part of the atmosphere, approximately 17 km to 50 km high, containing the ozone layer.

structural formula: A formula showing the atoms present and the bonds between them.

structural isomerism: When two or more compounds have the same molecular formulae, but different structures.

sub-atomic particle: A particle found within an atom, for example the proton, neutron, or electron.

sub-level: Part of an energy level in an atom, containing pairs of electrons: s sub-levels contain up to one electron pair, p sub-levels contain up to three electron pairs, and d sub-levels contain up to five electron pairs.

sublime: To pass directly from the solid state to the gas state.

substitution: The replacement of one atom or group of atoms in a molecule by another atom or group of atoms.

sustainable development: Living in such a way that we meet our needs without damaging the ability of future generations to meet their own needs.

temporary dipole: The asymmetrical distribution of the electron pair in a covalent bond.

termination: The final stage in a free radical substitution reaction.

tetrahedral: The shape of a molecule with 4 bonding pairs of electrons.

theoretical yield: The maximum mass of product possible, calculated using the mass of reactants and the balanced equation.

thermal cracking: The thermal decomposition of hydrocarbons to produce shorter alkanes and alkenes.

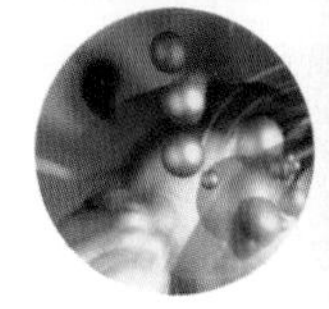

titrant: The solution added from the burette in a titration.

titration: Method used to find the concentration of a sample using a reactant of known concentration.

titre: The volume of titrant added to reach the end-point in a titration.

Tollens' reagent: A reagent used to distinguish between aldehydes and ketones. A silver mirror is formed when aldehydes are warmed with it.

transfer pipette: A piece of glassware used to add an accurate volume of liquid.

trigonal bipyramidal: The shape of a molecule with 5 bonding pairs of electrons.

trigonal planar: The shape of a molecule with 3 bonding pairs of electrons.

trigonal pyramidal: The shape of a molecule with 3 bonding pairs and 1 lone pair of electrons.

triple covalent bond: A bond in which two atoms are joined by three shared pairs of electrons.

unburned hydrocarbons: Pollutants in the exhaust from car engines, due to incomplete combustion of the fuel.

unsaturated: Containing at least one carbon-carbon double bond.

vacant orbital: An orbital that can accept a pair of electrons.

valence shell: The energy level in an atom that is involved in forming bonds.

van der Waals' forces: Temporary, induced dipole-dipole attractions between covalent molecules.

volatile: A liquid that easily vaporizes is said to be very volatile.

volumetric flask: An item of glassware used to make up a standard solution. Also called a standard flask or graduated flask.

VSEPR: Valence Shell Electron Pair Repulsion theory. The theory used to predict the shape of a covalent molecule using the idea of repulsion by pairs of electrons.

wavenumber: A measure of frequency used in infrared spectroscopy. It has units of cm^{-1}.

yield: The mass of product formed in a reaction.

zeolite: Compounds of aluminium, silicon, and oxygen with microscopic pores. Zeolites are used as catalysts and molecular sieves in the petrochemical industry.

Answers: Check your understanding

Answers to questions with calculations

1.01 Atoms p16:

2 Atomic number 18, mass number 40

1.05 Detecting isotopes p24:

1 **a** 7 **b** 14 **c** 28 **d** 14

4 24

1.06 More from the mass spectrometer p26:

1 207.2

2 16

2.01 Masses of atoms and molecules p38:

1 **a** 32.0 **b** 159.8 **c** 36.5 **d** 18.0 **e** 16.0 **f** 100.0

2 **a** 40.0 **b** 63.0 **c** 98.1 **d** 74.1 **e** 96.0 **f** 213.0

2.02 Amount of substance p40:

1 **a** 24.3 g **b** 50.5 g **c** 9.0 g **d** 200.2 g **e** 180.0 g

2.03 More moles p42:

1 **a** 1 mol **b** 0.5 mol

2 **a** 0.25 mol **b** 0.5 mol

3 **a** 0.25 mol **b** 2.0 mol

2.04 The ideal gas equation p44:

1 **a** 0.5 dm^3 **b** 2.0 dm^3

2 **a** 2.0 dm^3 **b** 0.5 dm^3

4 **a** 273 K **b** 423 K **c** 1808 K **d** 233 K

2.05 Using the ideal gas equation p46:

1 **a** 2500 Pa **b** 100 Pa

2 **a** 1.25 m^3 **b** 0.5 m^3 **c** 2.5×10^{-5} m^3 (0.000 025 m^3

3 0.0243 m^3 (24.3 dm^3)

4 9 972 000 Pa (9.97×10^6 Pa

5 2.41 K

2.06 Experiments to find M_r p48:

1 **a** 0.174 g **b** 3.01×10^{-3} mol **c** 57.8

2 **a** 86 **b** 1.4%

4 16.0

5 40.0

2.10 Reacting masses p56:

1 **a** 11.2 g **b** 3.57 tonnes

2 **a** 1.74 g **b** 2.17 g

2.11 Percentage yield p58:

1 **a** 9.2 g **b** 13%

2 **b** 0.50 g **c** 98%

2.12 Atom economy p60:

1 20%

2 **b** 51.1%

3 **a** 1.31% **b** 11.1%

2.14 Reacting volumes p62:

1 **a** 100 cm^3 **b** 200 cm^3 **c** 100 cm^3

2 76.4 g

2.15 Concentrations of solutions p66:

1 **a** 0.50 mol dm^{-3} **b** 8.0 mol dm^{-3} **c** 2.0 mol dm^{-3}

2 **a** 1.25 mol **b** 0.0125 mol

3 2.65 g

2.16 Acid-base titrations p68:

1 0.22 mol dm^{-3}

2 **b** 0.30 mol dm^{-3} **c** 36.5

5.01 Enthalpy changes and standard conditions:

3 **a** 890 kJ **b** 4450 kJ **c** 1668.75 kJ **d** 92.7 kJ

4 –95 kJ mol^{-1}

5.02 Calculating enthalpy of reaction p124:

1 for 350 cm^3: 138985 J; for 1700 cm^3 of water: 675070 J

2 37620 J; moles burnt = 0.03; 1153680 J; enthalpy of combustion = -1153.68 kJ mol^{-1}

3 moles of $MgBr_2$ = 0.038; heat energy released = 7160 J; 4.28 °C

5.03 Energy cycle diagrams p126:

1 $\Delta H_r^\ominus$ = -176.5 kJ

2 $\Delta H_r^\ominus$ = -1368 kJ

5.04 Calculating enthalpy changes p128:

1 $\Delta H_r^\ominus$ = -2816 kJ

2 –95 kJ mol^{-1}

3 –106 kJ mol^{-1}

5.05 Bond enthalpies p130:

1 **a** -1181 kJ **b** -2703 kJ **c** -2163 kJ

2 464.5 kJ mol^{-1}

14.08 Alkenes from alcohols p236:

2 **a** 37%

S6 Calculations p254:

1 **a** 12 **b** 35.5 **c** 40

2 **a** 28 **b** 28 **c** 58

3 **a** 0.1 **b** 0.5 **c** 0.05

4 **a** 42.9% **b** 27.3%

Index

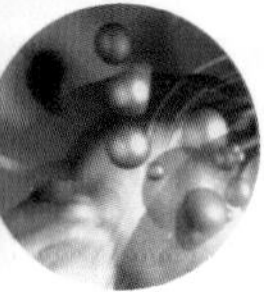

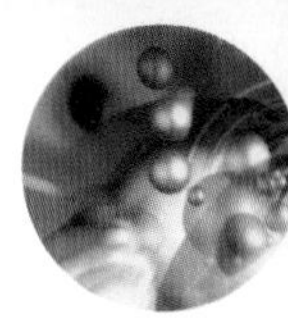

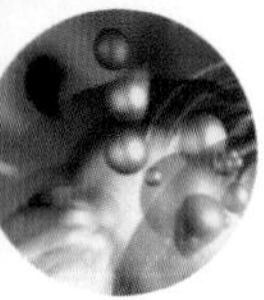

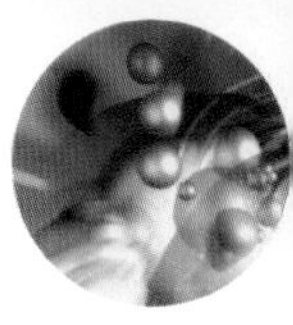

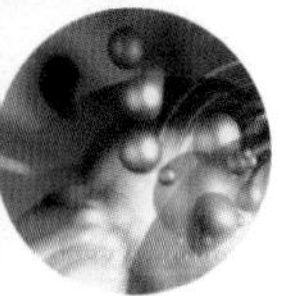

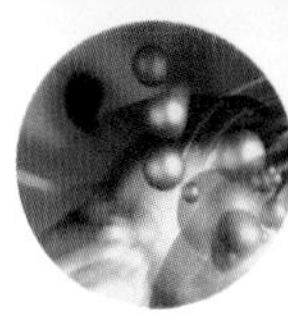

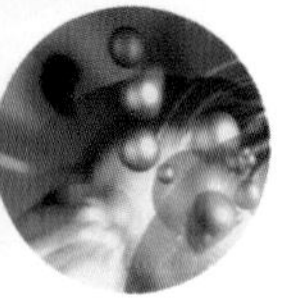

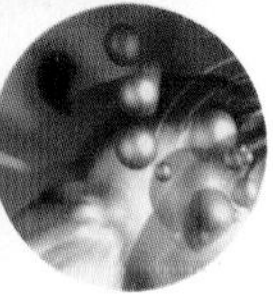